The Anatomy *of* Blackness

The Anatomy *of* Blackness

Science & Slavery in an Age of Enlightenment

ANDREW S. CURRAN

The Johns Hopkins University Press
Baltimore

Printed in the United States of America on acid-free paper

Johns Hopkins Paperback edition, 2013
2 4 6 8 9 7 5 3 1

The Johns Hopkins University Press
2715 North Charles Street
Baltimore, Maryland 21218-4363
www.press.jhu.edu

The Library of Congress has cataloged the hardcover edition of this book as follows:
Curran, Andrew S.
The anatomy of blackness : science and slavery in an age of Enlightenment /
Andrew S. Curran.
p. cm.
Includes bibliographical references and index.
ISBN-13: 978-1-4214-0150-8 (hardcover : acid-free paper)
ISBN-10: 1-4214-0150-9 (hardcover : acid-free paper)
1. French literature—18th century—History. 2. French literature—19th century—History. 3. Blacks—Race identity. 4. Travel writing—History and criticism.
5. Blacks in literature. 6. Africa—In literature. I. Title.
PQ265.C87 2011
840.9'352996—dc22 2010050254

A catalog record for this book is available from the British Library.

ISBN-13: 978-1-4214-0965-8
ISBN-10: 1-4214-0965-8

Special discounts are available for bulk purchases of this book. For more information, please contact Special Sales at 410-516-6936 or specialsales@press.jhu.edu.

The Johns Hopkins University Press uses environmentally friendly
book materials, including recycled text paper that is composed of at least
30 percent post-consumer waste, whenever possible.

To the memory of my father,
Thomas C. Curran

Contents

Preface

From its inception, I was convinced that this book would require an explanatory preface. After all, it confronts the reader with the same set of eighteenth- and nineteenth-century beliefs about black Africans that the Academy has been deconstructing for the past forty years. The people who ultimately read this book will often find the unpleasantness of the past in these pages, and without the same mediation that university professors generally provide their students as a matter of course.

The Anatomy of Blackness has three overlapping narratives. The first relates how eighteenth-century naturalists and philosophes drew from travel literature to discuss the perceived problem of human blackness within the nascent human sciences. The second describes how a number of now-forgotten anatomists revolutionized the era's understanding of the black African by emphasizing both the supposed liabilities of this group and the corresponding "advantages" of whiteness. The third charts the shift of the slavery debate itself, from the moral, mercantile, and theological realms toward that of the black body itself.

Not unexpectedly, such an approach reveals more about Europeans (and their secondary construction of themselves) than it does about real Africans. Readers should bear in mind that this is not a book about black African agency, or about how Africans grappled with the realities of European aggression and mercantile exploitation in Africa. Nor is it a book about how men and women of African descent undertook their own revolution at the end of the eighteenth century, appropriating and deploying a series of republican ideals that certainly had not been imagined with them in mind. In a word, this study focuses on the *textualization* of the black African.

During the course of this project, several people asked me if I thought my method might inadvertently "ventriloquize or replicate" some of the era's structures of oppression; one person suggested that I should use more sensitive racial or ethnic categories when writing about black Africans. Such "enlightened" terminology, of course, serves a number of worthwhile functions in contemporary society. Not only do these expressions explicitly acknowledge the right of ethnic or racial groups to choose their own names; they also allow these same constituencies to construct their identities as they see fit. But progressive terminology is, alas, poorly suited to a study that seeks to recover the full texture of what French eighteenth-century thinkers referred to as the *nègre*. And this brings me to the wounding one-syllable word that I have just typed: *nègre*. In his now well known book *Nigger: The Strange Career of a Troublesome Word*, Randall Kennedy explains that his primary intention in examining this derogatory term was to put "a tracer on [it]" in order to "report on its use, and assess the controversies to which it gives rise."[1] While Kennedy's combination of history and individual testimony differs from my own approach, his desire to force his reader to think through the content and contexts of racialized terms certainly echoes my intent as well.

Like Kennedy, I have avoided both euphemism and paraphrasing when citing eighteenth-century writers. I have, for example, specifically steered clear of translating *nègre* into "Negro," because to do so would be to inflect what was a French and therefore Catholic-inspired construct with its British-Protestant equivalent. Likewise, in keeping with eighteenth-century typography and practice, I generally do not capitalize *nègre*, which would confer a false and misleading dignity on a concept that often had very little at the time.[2] In fact, I capitalize *nègre* only when this is unequivocally done in the text being discussed. This is the case, for example, in Abbé Henri Grégroire's *De la littérature des Nègres* (1808), where capitalization presumably serves the author's larger politics of humanization. This also seems to obtain when the word *nègre* was cited alongside other, more often capitalized, human categories, such as *Chinois*.

Similarly, in discussing eighteenth-century thinkers, I have also employed the specific language that they themselves used to refer to the so-called *nègre*. When I cite the monogenist Pierre-Louis de Maupertuis's discussion of the black African, for example, I use the term "variety." When I quote from Voltaire, I cite the more trenchant markers that he used, "race" or "species." Although perhaps initially a little confusing, this ultimately allows the reader to absorb the meanings of these troublesome categories (variety, race, species) within the specific eighteenth-century contexts in which they appeared. I would hope that this approach is much more effective than simply stating that some (but not all) biblically minded thinkers initially referred to the *nègre* as a "people"; that many Buffon-influenced naturalists tended to use the botanical term "variety"; that a number of thinkers employed the zoological term "race" in order to emphasize the anatomical or conceptual separation of human categories; and that the most extreme polygenists often claimed that the African was *a different species*.

This clinical treatment of the European representation of black Africans may raise objections among certain scholars. Some may argue that what I have dubbed the *anatomizing* of Africanist discourse emphasizes and explains European ideas at the expense of the damage done to Africans themselves. Others might add that this method tends to exculpate the Enlightenment era from the legacy of slavery and racism, from what Emmanuel Eze rightly identifies as the "highly ambiguous relationship of Enlightenment philosophic and scientific reason to racial diversity in the eighteenth century."[3]

My intention in writing this book actually has nothing to do with either of these objectives. Indeed, by freeing Enlightenment-era thinkers from what is often portrayed as a monolithic thought system, I hope to recover the eighteenth-century individual's ability not only to be a passive participant in Africanist discourse, but to absorb, react, and contribute to the overall representation of the African. In my opinion, this method has several advantages over more rigid genealogies or single-legacy histories. By examining the representation of Africans on the level of individual thinkers or groups of thinkers, one can more effectively chart the ambiguous relationships among Enlightenment universalism, the equally strong constraints of the era's proto-ethnography, and the economic imperatives of slavery. In contrast, by envisioning the writings of the philosophes—not to mention the most offensive pro-slavery thinkers—as nothing more than a linear and overdetermined set of ideas, we not only underestimate the intellectual autonomy of these writers or philosophers but mitigate the responsibility of the so-called Enlightenment-era mind. While it may seem paradoxical, moving away from full-scale indictments of the era and focusing on the "exquisite sophistication of eighteenth-century writing" may actually be the best way of illuminating the obscure aspects of an era whose primary metaphor was one of light.[4]

Acknowledgments

During the past four years, I have been lucky enough to discuss this book with a wide range of superb scholars, among them Charlemagne Amegan, Wilda Anderson, Stephen Angle, Srinivas Aravamudan, Sophie Audidière, Robert Bernasconi, Dan Brewer, Michel Delon, David Diop, Madeleine Dobie, Dr. Alan Douglass, M.D., Julia Douthwaite, James Delbourgo, Alex Dupuy, Rick Elphick, Demetrius Eudell, Brian Fay, Lynn Festa, Alden Gordon, Ruth Hill, Marianne Hobson, Jean-Marc Kehrès, Joan B. Landes, Natasha Lee, Christopher L. Miller, Jill Morawski, Ourida Moustefai, Dr. James Norris, M.D., John C. O'Neal, Julie Perkins, Michael Roth, Elena Russo, Pierre Saint-Amand, Joanna Stalnaker, Jim Steintrager, Tony Strugnell, Mary Terrall, Ann Thomson, Kate Tunstall, Flore Villemin, Roxann Wheeler, Jim Windolf, and Kari Weil. Conversing about this project with such irreplaceable friends and talented colleagues has been immensely useful and pleasurable.

Among those people who helped me during the final stages of this project, I must single out Josiah Blackmore, Catherine Gallouët, Paul Halliday, Jason Harris, Thierry Hoquet, Peter Mark, and Pernille Røge, for reading significant portions of the typescript. Each of these careful readers (a specialist of early-modern Portugal, a French eighteenth-century specialist, a legal historian, a scholar of Christian thought, a philosopher, an Africanist, and a historian of colonial trade practices) pushed me in new directions. Special thanks also go to Peter Dreyer, whose exemplary copyediting extended well beyond the norm. I also want to recognize Patrick Graille who, as was the case for my first book, engaged deeply with the whole project by phone, e-mail, and in person over the past four years. This manuscript also benefitted from the tenacious and intelligent readership of Gedney Barclay, Delia Casa, Emma Drew, Katie Horowitz, Julia-Jonas-Day, and Nathan Marvin. Among those close friends and family who played an important role in this process, I would like to recognize Carolyn Curran, Jennifer Curran, Anne and Walter Mayo, Tom Cushing, and Zhaoxun and Yulan Zhang.

Much of what is found in this book would not have been possible without generous support from various foundations. Initial research was made possible by a National Endowment for the Humanities grant. Later stages of the book, undertaken both in France and in the United States, were facilitated by fellowships and travel money from the Mellon Foundation, the Thomas and Catherine McMahon fund, the Société Valmont-Bomare, and the New York Academy of Medicine. My home institution, Wesleyan University, also provided substantial institutional support for this project.

I have also received precious help from numerous librarians at the Bibliothèque nationale de France, the New York Public Library, and the respective libraries at Brown, Cornell, Yale, Dartmouth, and New York Universities during the past four years. In addition to recognizing these institutions and their generous staff members, I would also like to single out Suzy Taraba and Valerie Gillispie at Wesleyan University's Special Collections and Archives for helping me out on numerous occasions. I also want to thank Arlene Shaner at the New York Academy of Medicine's rare book room for her attentive and knowledgeable help.

My colleagues in Wesleyan University's Department of Romance Languages have also been terrific interlocutors during the past four years. In particular, I would like to thank Michael Armstrong-Roche, Robert Conn, Fernando Degiovanni, Antonio Gonzalez, Christine Lalande, Typhaine Leservot, Ellen Nerenberg, Julie Solomon, and Daniela Viale, for taking an active interest in my project.

I shall always be indebted to Trevor C. Lipscombe and Matt McAdam at the Johns Hopkins University Press for supporting this book so strongly. Along the same lines, I would also like to express my gratitude to the Voltaire Foundation at the University of Oxford for permission to reprint portions of two articles that appeared in *SVEC*. I am equally indebted to the editors of *History and Theory* for the same courtesy.

The Anatomy *of* Blackness

Introduction

Tissue Samples in the Land of Conjecture

Those who have wanted to disinherit *les Nègres* have used anatomy to their advantage, and the difference of color gave rise to their first observations.

ABBÉ HENRI GRÉGOIRE, *DE LA LITTÉRATURE DES NÈGRES* (1808)

In 1618, the influential Parisian anatomist Jean Riolan the Younger became the first person to seek out the precise source of the blackness within African skin. Borrowing his general method from Vesalius himself, Riolan blistered the skin of a black African man with a chemical agent. He then removed the seared specimen and painstakingly examined its various strata.[1] Some sixty years after Riolan conducted his experiment, Marcello Malpighi used a similar method to identify the actual layer of skin where dark pigmentation is found: the *rete mucosum,* or Malpighian layer. During the same decade, the famous Dutch anatomist Antoni van Leeuwenhoek also examined a sample of black skin under a microscope, asserting that blackness came from what he identified as dark scales.[2] And, as the eighteenth century began, the French anatomist Alexis Littré added his own contribution to this series of experiment-based speculations regarding African physiology. Recounting his dissection of the sexual organ of a black African male in the 1702 *Histoire de l'Académie royale des sciences,* Littré asserted a causal relationship between air and the appearance of blackness.[3]

While climate theory and scripture had long provided the overarching frameworks for the explanation of blackness, pointed enquiry into the specific nature of African pigmentation began with early-modern anatomy. During the Enlightenment era, much of this investigation into the supposed structures of blackness increasingly took place alongside more sweeping conjecture regarding the precise nature and significance of the African variety. Although there had been earlier and isolated speculation regarding the origin of the *nègre*, by the 1730s an increasing number of naturalists, anatomists, and religious writers began debating this question much more intensely and from a variety of perspectives. In 1733, in the *Journal de Trévoux,* or *Mémoires pour servir à l'histoire des sciences et des beaux-arts*, the Jesuit priest Auguste Malfert advanced a potentially heretical theory of blackness that seemingly asserted that humans with different morphologies and pigmentations had different origins.[4] Three years later, a certain Monsieur de J*** also published a widely read theory in the same religious journal positing the deleterious effects of climate on the African (and the African's color).[5] Interest in this subject reached a high point in 1739 when the Académie royale des sciences de Bordeaux focused Europe's attention on human blackness by offering a prize for the best essay addressing the following question: "What is the physical cause of *nègres'* color, of the quality of their hair, and of the degeneration of the one and of the other?"[6]

The Bordeaux prize inspired a wide range of European thinkers to speculate on the anatomical and conceptual status of blackness. Among the many biblical, anatomical, and environmentalist explanations for this phenomenon that were submitted to the Académie royale (I shall return to these essays in chapter 2), the most influential was provided by Pierre Barrère, a Perpignan physician. In contrast to many of the other essayists, Barrère eschewed biblical exegesis and vague environmental generalizations in lieu of a pointed analysis of the African body based on dissection studies that he had conducted on slaves while in Cayenne; this authoritative approach allowed Barrère to transcend the limitations of skin analysis and write about blackness as an overall physiological phenomenon. Rejecting earlier static theories of pigmentation that limited themselves to the existence of particular cutaneous structures (e.g., Malpighi's *rete mucosum*), Barrère theorized that the *nègre* had an "abundant" darkened bile, which coursed through his body, staining the blood and the epidermis. His essay, entitled the *Dissertation sur la cause physique de la couleur des nègres*, ultimately appeared in print in 1741; a number of naturalists, most

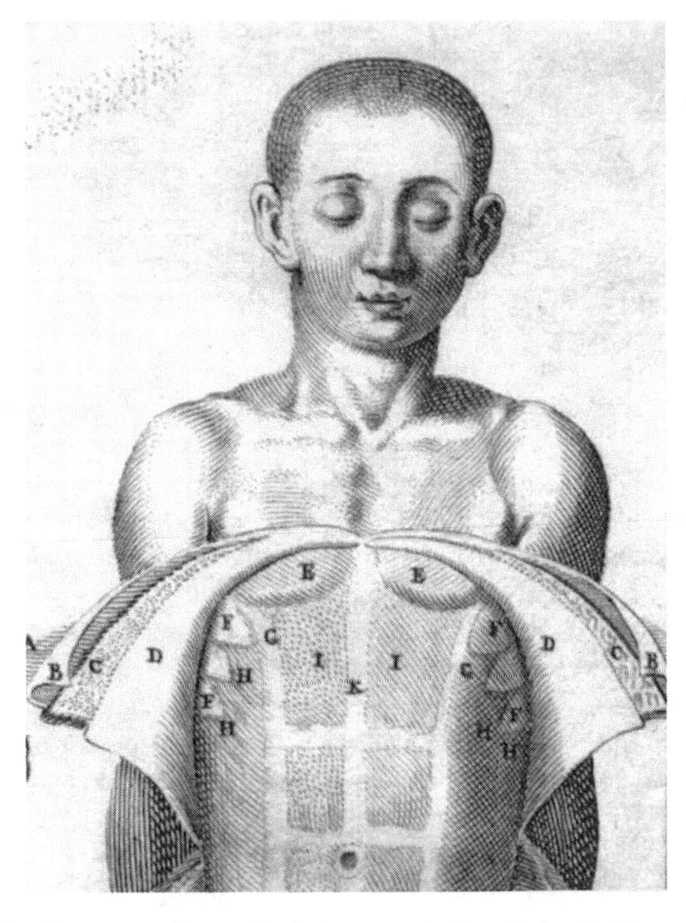

Human skin layers according to the Paris anatomist Jean Riolan the Younger. Riolan affirmed that black pigmentation was found in the *cuticula*, or thin outer layer of skin, marked *A*. Closely adhering to this is the *cutis*, or "real skin" (*B*), which contains the pores, through which disease supposedly entered. *C* is the *pinguedo*, or fat layer. Finally, *D* is the *panniculus carnosus*, or fleshy subcutaneous tissue. Riolan believed that the inner "whiteness" of black Africans, which was exactly like that of whites, pointed to an environmental explanation for the different human varieties. From Jean Riolan the Younger, *Encheiridium anatomicum et pathologicum* (1649). Courtesy New York Academy of Medicine.

notably Georges-Louis Leclerc, comte de Buffon, readily acknowledged this fascinating discovery of dark bile and blood.

In addition to propagating spurious conceptions regarding the specificity of African bodies, Barrère's monograph also prompted a new generation of European anatomists to seek out, measure, and track human blackness. Much of this research turned to the brain after the 1750s. At the Berlin Royal Academy of Sciences, the hugely influential (but now forgotten) Johann Friedrich Meckel asserted that African brains had a comparatively darker or bluish hue. Meckel's findings were quickly accepted in both France and Germany, and in 1765, his work was further advanced by Claude-Nicolas Le Cat, a Rouen surgeon. In addition to confirming Meckel's "discovery" of dark brain tissues, Le Cat also posited the existence of an elemental fluid that he dubbed *oethiops* (ethiops), which supposedly originated in the African's darkened brain and flowed through the nerves and into the skin.[7] Not unexpectedly, the implications of "black brains" and the *circulation* of a "brain-generated blackness" extended well beyond the realm of color. Many anatomists and naturalists simply assumed that these new anatomical discoveries justified the long-standing prejudice that the *nègre* had a comparatively limited cognitive potential. Most notoriously, the Göttingen savant Samuel Thomas Soemmering proclaimed triumphantly in 1784 that science had found a demonstrable link between the African's brutal stupidity and the supposedly coarse "strings" linking his brain to the rest of his body.[8]

Anatomy, especially as it was practiced by influential thinkers like Soemmering, produced the most authoritative statements regarding the particularities of the black African body during the eighteenth century. Although the discipline had previously been considered inferior to the core medical subjects of physiology, pathology, and therapeutics, by the mid-eighteenth century, anatomists could claim to have explained some of the most important corporeal structures of the human body, be it white or black.[9] Looking back in 1795 on what the field's practitioners had accomplished, the anatomist Aubin-Louis Millin asserted proudly that anatomy had "advanced almost to the highest degree of perfection [in this century, because] the most renowned anatomists and physiologists felt that it was time to direct their research toward explaining the movements of animals and the hidden mechanisms behind their sensations."[10]

Rife with mechanistic metaphors and persuasive rhetorical arguments that claimed to explain the placement and form of and relationship among hu-

Frontispiece to Daniel Tauvry's *Nouvelle anatomie raisonnée* (New reasoned anatomy) (1690, 1720). The goddess of wisdom, perhaps Minerva Medica, presides over the rise of the new anatomy, pointing skyward to progress and new knowledge, while trampling underfoot old works of anatomy, perhaps Galen. Courtesy New York Academy of Medicine.

man body parts, eighteenth-century anatomy was indeed compelling, and not only for scientific readers. Anatomists were seen as able to pronounce on what was most appropriate to humankind's "way of life" and the fulfillment of its needs.[11] To paraphrase Thomas Laqueur, the springs and pulleys of anatomy were increasingly understood to reflect human destiny.[12] This development was unfortunate for the so-called *nègre*; anatomists had arrogated to themselves the right, not only to identify the black African's corporeal "liabilities," but to explain why this particular category of human was fundamentally inferior to the highest expression of humankind, the European.[13]

Defining *le Nègre*

As its title suggests, much of this book focuses on the anatomical concepts that played a key role in the categorization of black humans during the early-modern era. In addition to recovering a sphere of science within which the notion of a black human "variety" slowly crystallized into that of the black *race*, this approach also examines the way in which the anatomy of the black body became a site of heated debate in the era's thought; what is more, it also highlights the definition of human whiteness that emerged concurrently.

While the history of French and European anatomy underpins much of this study, its scope extends well past the era's life sciences. Indeed, one of the primary intentions of my work is to recapture the wider (and forgotten) interplay between the era's anatomy and the more theoretical conceptualizations of black Africans taking place in fields that we would now label sociology, ethnography, anthropology, psychology, and political science. This disciplinary intersection—between anatomical "discoveries" and the budding human sciences—contributed significantly to the era's discussion of the black African; in addition to intriguing naturalists and proto-anthropologists, the science of the black African was the catalyst for debates among slavers, missionaries, and colonists, as well as pro- and anti-slavery writers, philosophes, religious thinkers, and, ultimately, Africans themselves.

Not unexpectedly, the overall image of the black African during the Enlightenment era is more a shifting mosaic than a fixed portrait. Besides the fact that there were numerous inconsistencies and divergences within Africanist discourse, many of the era's foundational assumptions regarding the black African shifted from decade to decade. Even more vexing, many ideas associated with the *nègre* during the eighteenth century had already moved across eras and, sometimes, genres.

Whether in the field of anatomy, natural history, or theology, tracking a specific genealogy within Africanist thought is a daunting task. Consider how African "ethnography" often morphed after its initial appearance. In Duarte Lopes's 1591 *Relação do reino do Congo e das terras circunvizinhas* (*A Report of the Kingdom of Congo and the Surrounding Countries*), for example, the overall image of the black African is relatively positive, and even quite laudatory at times. This "ethnography" took on a much more deterministic flavor 150 years later, however, in Abbé Prévost's *Histoire générale des voyages* (1746–70), when it was bundled with writings from other eras that described the same African peoples. Along similar lines, when the Dutch compiler Olfert Dapper adapted first-person travelogues by other writers for his *Naukeurige Beschrijvingen der Afrikaensche gewesten* (Exact descriptions of the African lands) (1668), he aimed at an open-minded breakdown of black African ethnicities that avoided sensationalism. Several decades later, his unattributed writings were often taken out of context and completely revised into prejudicial truisms for both pocket-sized "geographies" and giant compendiums such as Diderot's and d'Alembert's *Encyclopédie*.

There are even more curious instances of how "facts" regarding black Africans changed over time. Here again, Abbé Prévost provides a telling example. In 1744, this avid reader of African travelogues and future translator of John Green's *A New General Collection of Voyages and Travels* (1745–47), published a fictional account of the adventures of an English ship captain, Robert Lade, in which he not only recounted African anecdotes evoking the pleasures and pitfalls of cross-racial romance, but sketched ethnographic portraits of the inhabitants of West and South Africa. Several years later, while writing his *Histoire naturelle*, Buffon mistook Prévost's novel *Voyages du capitaine Robert Lade* for a "travelogue" and repeatedly cited it in his assessment of African mores. While perhaps a unique case, this transmission of knowledge from Prévost to Buffon (and from novel to natural history) demonstrates the ability of African proto-ethnography to move across permeable borders.[14]

Although modern scholars who discover such revealing examples of the *imagining* of black Africa and Africans tend to believe that they are the first to have done so, a significant number of Enlightenment-era skeptics also acknowledged the shortcomings of the representation of black Africa. In one of the best-known examples, Jonathan Swift mocked cartographers who drew on their imaginations to fill in unexplored lands on their maps:

> Geographers, in *Afric* maps,
> With savage pictures fill their gaps,
> And o'er unhabitable downs
> Place elephants, for want of towns.[15]

In the introduction to his 1735 translation of Jerónimo Lobo's Itinerário, under the title *A Voyage to Abyssinia,* the young Samuel Johnson also scoffed at his era's exaggerated evocations of Africa's geography and peoples.[16] Among eighteenth-century writers, however, the most vociferous condemnation of the reliability of African travelogues came from Jean-Jacques Rousseau in his *Discours sur l'origine et les fondements de l'inegalite parmi les hommes* (1755). From the philosophe's perspective, the Africa described by travel writers—and popularizers who quoted them—was a hodgepodge of contradictions and half-truths produced by inferior minds.

Throughout the eighteenth century, the problem of veracity loomed large over the question of black Africa and Africans. And yet, it is imperative to recall that even the most skeptical eighteenth-century thinkers believed that there was a great deal of truth in the prevailing set of negative ideas about the *nègre*. This is quite understandable when one considers the extent to which withering views of Africans were diffused on all levels of society. Nicolas Lenglet du Fresnoy's popular *Géographie des enfans* (first ed., 1736) provides an unsettling example of how a censorious image of Africa and Africans was inculcated among the youngest members of (literate) French society. In this book of teacher-student dialogues, which were designed to be role-played, children were supposed to repeat the salient details about far-off lands, among them southern Africa:

> [Child's Question]: What do you mean by *la Cafrerie?*
>
> [Adult's Answer]: *La Cafrerie* or the country of the *Caffres* is a place on the sea, inhabited by the most barbarous and stupid people of Africa. The most significant places are the Dutch Cape of Good Hope and Sofala, [the latter of] which belongs to the Portuguese.[17]

One finds a similar, dogmatic tone in a 1787 issue of the *Bibliothèque universelle des dames*. Designed for the era's budding female readership, this particular volume of the *Bibliothèque* described significant advances in the era's anatomy and natural history, among them a new and astonishingly race-based definition of the *nègre* and what might be called body-based *négritude*:

> One could say that a *nègre* is [a *nègre*] in all the parts of his body, except for the teeth. All the organs carry the mark of this color to a lesser or greater degree; the medullary substance of the brain is blackish and this color dominates more or less in diverse parts of this organ, the semen, the blood, etc.[18]

Other well-worn, well-diffused, and generally accepted notions about black Africans can be found in the era's recycled dictionary definitions, whether in Louis Moréri's *Grand dictionnaire historique* (1759), Diderot's and d'Alembert's *Encyclopédie*, or the successor to the latter, Charles-Joseph Panckoucke's *Encyclopédie méthodique*.[19]

This is also the case for the often-consulted Jesuit *Dictionnaire de Trévoux*. Initially published in 1704 and reedited and updated on numerous occasions throughout the eighteenth century, this dictionary made several predictable and recurring points regarding the *nègre*. In the first section found under this headword, this particular group of black humans is associated with "La Nigritie" or the *Pays des nègres*.[20] Suggesting that this *peuple d'Afrique* must have spread across the continent, the *Dictionnaire* then asserts, without transition or proof, that Africans are effectively responsible for their own enslaved condition and subsequent misery by recounting the "fact" that black men readily sell their own wives and children to European traders.[21] Having reduced all *nègres* to these general truths, the author of this article then qualifies the notion of the *nègre*, at first enumerating the different ethnicities and colors found among the *nègres*, as well as specifying that different subgroups are either Muslim, pagan, or supposedly without any religion.[22] The next entry (added in 1732) underscores the foundational tautology of the term *nègre*: it was understood as perfectly synonymous with *esclave*. A version of this concordance between "black African" and "slave" also shows up in Pierre-Charles Berthelin's 1762 abridged version of the *Dictionnaire de Trévoux* as well, which notes that the word *nègre* is commonly used in a simile-based phrase to designate someone ill-treated or overworked, as in "on l'a traité comme un *nègre*" (he was treated like a *nègre*).[23] While this dictionary definition of *nègre* might seem to suggest that compassion should be extended to the suffering and subjugated black African slave, it should be emphasized that the person who elicits sympathy in this particular locution is not the *nègre* himself, but the unfortunate non-black being treated *like* a *nègre*.

At first glance, the *Dictionnaire de Trévoux*'s treatment of the term *nègre* is seemingly quite straightforward, but much is obscured in such dictionary definitions. Take, for example, the curious gendering of what was ostensibly a gender-neutral category. While the concept being defined in this dictionary—*le nègre*—theoretically conjured up all African men, women (*les négresses*), and children, *Trévoux*'s definition also reflects an overdetermined gendering that deemphasizes both African women and children and prioritizes the *black male slave*.[24] The *nègre*, in short, was primarily imagined as a man. While it is certainly true that the era's West and Central African travelogues provided detailed portraits of West African women—indeed, they gave to understand that the *négresse*'s indefatigable toil in the fields encouraged the *nègre*'s sloth while her hypersexuality encouraged his lasciviousness—outside of the African context, the term *nègre* generally evoked a subjected male human whose very essence was associated with slavery and the mechanical functioning of the colonial enterprise.[25] This was, after all, the most fundamental (if unrecognized) ethnographic trait of the black African: enslavement.

If eighteenth-century dictionary definitions tended to reduce black Africans to an enslaved male, they also glossed over another important fact: the word *nègre* generally functioned both as a general and singular concept in the era's thought. When the term was accompanied by a definite article—as in "*le nègre* is a vicious species of man"—the word was used to conjure up all members of this group. This is *le nègre* that is generally discussed in natural history treatises, geographies, and books on anatomy. In contrast to this all-encompassing and necessarily reductive view, many writers of travel literature, novels, and even slavers produced portraits of individual black Africans that call into question the depictions of a "universal" black African. These images of intelligent, staid, or clever *nègres* function in two ways. For abolitionists such as the American Quaker Anthony Benezet, this ethnography raised the possibility that the overall image of the black African was not only reductive but erroneous. For the vast majority of readers, however, such images were simply exceptions that proved the rule.

The New Africanist Discourse after 1740

As a topic of conversation, the *nègre* would ultimately fascinate a significant number of eighteenth-century writers, philosophes, and naturalists. Few issues generated as many curious anecdotes, proto-ethnographical assertions, and, at times, heated discussions regarding the undeniable horror of slavery. And yet, it would be misleading to begin a discussion of the overall discourse on the black African without first acknowledging that the subject of the *nègre* initially remained irrelevant to most people in France and Europe—including the figureheads of the so-called High Enlightenment—until the 1750s. Denis Diderot illustrates this development. To come up with material for the imposing headword "Afrique"—"one of the four principal parts of the Earth"—in the first volume of the *Encyclopédie* (1751), Diderot pilfered two hundred uninspired and unoriginal words from Jacques Savary des Brûlons's 1723 *Dictionnaire universel de commerce.* Although Diderot ultimately provided several intriguing articles on subjects including African cannibalism and religious practices in the early volumes of the *Encyclopédie,* he was unmistakably much more concerned with exploring theories of knowledge and refuting constructive metaphysics in 1750 than he was with engaging with issues such as Africanness or chattel slavery. Compared to his miniscule article "Afrique," for example, Diderot contributed three meticulously crafted folio-length pages to the *supplément* to the *Encyclopédie* article on the human soul ("Ame").

Diderot's treatment (or non-treatment) of black Africans in the article "Afrique" illustrates the dual position of the *nègre* around mid-century: simultaneously marginalized and yet critical to France's economic prosperity. This seeming lack of concern slowly gave way, however, as a series of new, more comprehensive or speculative assessments of blackness and black Africans began appearing, many of them in the years subsequent to the Académie royale des sciences de Bordeaux debate on blackness. In 1745, for example, Pierre-Louis de Maupertuis published his anonymous *Vénus physique,* in which he explained the black variety of man as something of a fluke of nature.[26] Two years later, in 1747, Abbé Prévost began publishing what would amount to four-and-a-half *in-quarto* volumes of annotated African travelogues organized by region, providing Enlightenment-era thinkers with a virtual encyclopedia of African ethnography. And, in 1749, volume 3 of Buffon's *Histoire naturelle,* containing his "Variétés dans l'espèce humaine," ushered in what would become a new era in the interpretation of the African.

Not only did Buffon's "Variétés" provide an authoritative explanation of the *nègre* that was totally free of metaphysics, it also supplied an influential natural history definition of the *nègre* that cleared up some of the ambiguity produced by the lexicographers and geographers who had been recycling definitions of the word for decades. Surveying the different types of dark-skinned peoples found on the African continent, Buffon affirmed that the blacker an African was, the more he corresponded to the category of the *nègre*: "true *nègres*" were "les plus noirs de tous les Noirs" (the blackest of all the Blacks).[27] This measurable blackness criterion— true *négritude*—functioned in multiple fashions in the *Histoire naturelle*. In addition to providing a counterpoint to new, latent definitions of whiteness, the amount of *blackness* became the decisive factor in the classification of humans living on the African continent. In fact, according to Buffon, the *nègre*'s darkness was so distinct that he was to be distinguished conceptually from the other large group of black humans found in Africa, the Khoisan peoples. Explaining that these aboriginal hunters and pastoral peoples were lighter-skinned than the *nègre*, Buffon affirmed that these *Caffres* or *Hottentots* were another *variété* within the *race of blacks*.[28] They were not, as he put it, true *nègres*.[29]

In a sense, Buffon's belief that there was a real and measurable *nègre* within the larger category of the "black race" simply confirmed more vague breakdowns that had existed in travel literature for centuries. And yet, in an increasingly anthropological era that was becoming less restricted by biblical accounts of humankind's disparate peoples, Buffon's authoritative "processing" of longstanding ethnographical beliefs into more concrete pigmentation-based categories invited further investigation into the essence and origins of blackness. Like other complex puzzles of the early-modern era, the pressing reality of this dark ethnicity had to be reconciled with the presumed regularities, not only of nature, but of whiteness. Whether armchair naturalist, Caribbean planter, or philosophe, those authors who took the time to write about black Africans in light of the *Histoire naturelle* implicitly or explicitly wondered to what extent the *nègre* represented a "limit case" within the overall understanding of humankind.[30] Contiguous to this natural history question was, of course, a related problem that would become increasingly important in subsequent decades: to what extent did the new material understanding of blackness overlap with the political construction of human bondage?

Diderot himself would come to ask many of these same questions. As the scope of his interests increased, he eventually had occasion to evoke the so-called *nègre* in a number of works. In 1765, Diderot asserted in the *Encyclopédie* article "Humaine espèce" that the *nègre* was a sensitive yet not terribly intelligent "type" of man who suffered under slavery.[31] Two years later, in the *Salon de 1767,* he brought up the African in a discussion regarding the relativity of taste, asserting: "I think that *les nègres* are less attractive to themselves, than whites are to *les nègres,* or *nègres* are to whites."[32] In 1773, after meeting Petrus Camper in Holland, Diderot also let it be known that he was fascinated with the Dutchman's system of facial angles, which "demonstrated" an esthetic (but not biological) hierarchy that descended from the "visage of Gods" to any given national physiognomy, and finally to the "head of man, of the *nègre,* to that of the monkey."[33] Most famously, of course, Diderot provided a series of anti-colonial assessments of the African and African chattel slavery for the third edition of Abbé G.-T. Raynal's *Histoire philosophique et politique des établissements et du commerce des Européens dans les deux Indes* (1780). Some thirty years after penning the nondescript article "Afrique" for the *Encyclopédie,* the philosophe, who was now sixty-seven years of age, became not only one of the strongest anti-slavery voices of his generation but also an apologist for a group of people that many of his contemporaries asserted to be "born for abjection and dependence, for work and punishment."[34]

Such a multiplicity of postures may lead us to wonder what the African represented for a thinker such as Diderot, or, more generally, what the African meant within Enlightenment thought itself. For the first scholars to think through this quandary during the 1960s, the answer was often assumed to be part of a larger chronology: to understand the conflicting images of Africa in the thought system of a philosophe like Diderot, one had only to situate his writings within a periodized historical trajectory, moving from an era of indifference regarding slavery (until around 1750) to an era of guilt (until about 1770), and, finally, to a time of activism.[35] In other words, Diderot's views of the African, whether degrading or positive, have often been seen as the product of larger shifts within a monolithic Enlightenment-era *mentalité.*

To a large degree, this misleading teleology is the extension of the classical narrative of "Enlightenment," a progress-driven narrative that follows the intellectual and political developments of the eighteenth century through to their seemingly inevitable conclusion: a time of revolution, republic, and, in the case of the African, the (temporary) emancipation of 1794. Diderot's interventions on the black African clearly belie this interpretative framework. Although his portrait of the *nègre* did evolve over time, his overall relationship to the question of blackness remained more tied to context than to chronology. When Diderot treated the African from the point of view of natural history, he echoed the *diagnostic* understanding of blackness that was becoming increasingly rampant during his era. When constructing his defense of this oppressed people, he put forward his era's sentimentalized version of classical liberalism. In both of these cases, Diderot's so-called convictions regarding the black African were perhaps less real beliefs than they were the reflection of specific intent, conventions of genre, and competing Enlightenment-era epistemologies.

Diderot is far from the only Enlightenment-era thinker whose beliefs regarding the African can be characterized as syncretic. Voltaire's varied writings on the African are even more so. By far the most race-oriented thinker of his generation, Voltaire repeatedly asserted in his natural-history musings that Africans' morphology and supposedly limited powers of reason had conceptual significance. As he put it in the 1756 *Essai sur les moeurs et l'esprit des nations*: "their round eyes, their flat noses, their invariably fat lips, the wool on their head, even the extent of their intelligence reflects prodigious divergences between them and other species of men."[36] Such a view seems at odds with Voltaire's much better known moral indictments of chattel slavery, the most prominent being that voiced by the *nègre de Surinam* in *Candide* (1759). Here, in contrast to the essentializing Voltaire, we find the more celebrated and universalist Voltaire: the voice of reason and critic of intolerance who ventriloquizes the suffering African slave: "When we work in the sugar refinery, and when the millstone catches our fingers, they cut off our hands; when we want to flee, they cut off our legs: I found myself in both situations. It is at this cost that you eat sugar in Europe."[37]

This latter outburst, one of the most famous moments in *Candide*, functioned as a moral litmus test for Voltaire's era. To witness the pathetic scene that Voltaire conjures up was seemingly to be forced to choose between two sets of values: those of common sense, universalism, and empathy and those of the planter, namely, greed, cruelty, and a lack of feeling. And yet, this is perhaps a false binary. After all, this celebrated accusation of slavery camouflages the fact that, during his entire career, Voltaire never understood the link between his sneering representations of Africans and the justifications of human bondage espoused by pro-slavery thinkers. As Lynn Festa has perceptively written about the sentimentalized portraits of the African that appeared during the eighteenth century, be they in *Robinson Crusoe* or *Candide*, such moments were often less about moral condemnation than they were about providing the era's readers with an opportunity to disengage themselves from the undeniable horrors of the colonial enterprise by sympathizing with the suffering *nègre*.[38]

The Contexts of Representation

The textualized African produced by eighteenth-century writers is a protean construct engineered by both anonymous popularizers and intellectual Olympians like Buffon. While it cannot be emphasized enough that the brutal imperatives of slavery tended to homogenize any and all knowledge relative to the *nègre*, it should nonetheless be kept in mind that the overall representation of black Africans was the product of different contexts, psychologies, and geographies. The first two variables are dealt with in depth in the body of this book. As for the geographical context, it is perhaps useful to review (briefly) some of the spaces within which the African was evoked, as well as the telling white-black demographics that shaped how Europeans saw and described real Africans.

Actual contact between Europeans and Africans (and thus first-hand representation of the so-called *nègre*) took place in three major arenas: in Africa, in slave-based colonies, and in the homelands of colonizing powers such as France.[39] As I make clear in chapter 1, the vast majority of French-African interactions obviously took place in French colonies, particularly in the Caribbean.[40] While African slaves ultimately came to significantly outnumber whites on these islands, eventually by more than ten to one on Saint-Domingue (Haiti) by the late 1780s, the number of European colonists in the French Antilles was nonetheless quite substantial. By 1789, there were, by most estimates, more than 60,000 white colonists (and well over 600,000 slaves) living on the three major French islands, Guadeloupe, Martinique, and Saint-Domingue. Not unexpectedly, the portraits of the *nègre* coming from this context reflect, not only the constant fear of revolt, but also the highly regimented racial and behavioral codes that gave shape to plantation life.[41]

The French "encounter" with and representation of black Africans in West Africa also mirrored structural and demographic realities. While the numerous illustrations of European trading forts (*comptoirs*) found in works such as the mid-century *Histoire générale des voyages* may give the impression that tens of thousands of Europeans were making their lives in Africa by the first half

The castle of St. George de Mina (Elmina), founded by the Portuguese in the late fifteenth century on Africa's Gold Coast (now Ghana), taken from Barbot and Dapper, and nearby Fort Conradsbourg on Mont St. Jago, from which the Dutch captured Elmina in 1637, in Antoine François Prévost, *Histoire générale des voyages* (1746–70), vol. 3. Author's collection.

of the eighteenth century, the reality was actually markedly different. In general, the fortified trading structures where Africans and Europeans exchanged slaves, gum, sugar, liquor, ivory, wax, and sometimes gold had limited (white) populations, ranging from fifteen men in small outposts to perhaps two hundred in major forts.[42] This was even true of French strongholds in Senegal. During his 1786 trip to Saint-Louis, Sylvain Meinrad Xavier de Golbéry reported 2,400 free blacks and people of mixed descent, 2,400 slaves, and 660 whites, only some 60 of whom were permanent residents.[43] Gorée's population distribution was similar, with 70 or 80 Europeans living among 600 free blacks and 1,000 slaves.[44] Not surprisingly, in both cases these white minorities mixed more freely with the local population than was the case in the Caribbean; they also had little interest in establishing segregated societies of the type often found in Caribbean colonies, and were even less able to do so than their counterparts in the latter. What these white populations did do, however, was participate in an ongoing market economy for slaves that fundamentally altered the internal politics of the lands in which they were living. This, too, as we shall see, was reflected in the characterizations of African life by slavers, traders, and missionaries.

As for French-African contact in France itself, estimates put the number of blacks residing in France in 1750 at approximately 4,000–5,000.[45] This was a varied population that included both servants and slaves who had returned (often temporarily) with their masters from the colonies, as well as the freed or abandoned Africans who lived permanently among the poor populations of cities like Nantes and Paris. While evidence of this population can perhaps be best seen in the era's painting—the mode of representing the *négrillon* servant comes to mind—this black population actually generated very little sustained interest among writers and philosophes. Where black Africans engendered more debate was among the colonists who had come back to France with their *nègres*. This particular constituency rightfully feared that the blacks that they were taking to France might be emancipated since, according to the *Code noir*, slaves were supposed to remain in the colonies and, more important, any enslaved African who set foot on French soil was supposed to benefit from the so-called Freedom principle.[46] As Sue Peabody's and Pierre Henri Boulle's important studies on this subject have demonstrated, planters and colonists worked hard to find a legal solution to this problem. In 1716, their lobbying led to an important edict that effectively legalized the presence of Africans on French territory. While this early *déclaration* established specific conditions for importing black Africans into France (e.g., for religious or vocational training), later statutes regarding France's blacks were much more broad in scope. Responding to the notion that black Africans somehow constituted a threat to France, the 1738 *déclaration du roi* sought, for example, to encourage the departure of *nègres* through a series of controls including forbidding interracial marriage. The duc de Choiseul's directive of 1763 was even more forceful in this respect. Citing the possibility of miscegenation and an increasingly larger black population in France, Louis XV's minister of the marine sought to deport all black Africans to the colonies.[47] This fear of a "deluge" of blacks was, as Sue Peabody has demonstrated, more paranoia than anything else. Although there were far more blacks in France than there were Frenchmen living in West Africa during the same era, this imported population represented only a minuscule fraction (.0002) of the 25 million people living in France in 1750.[48]

African-French demographics reflect one of the fundamental realities of the discourse on Africans during this era: excluding the geographically isolated white inhabitants of the Caribbean—who lived as a privileged minority within much larger black populations—the vast majority of the (primarily rural) French population probably never saw or met an African. Indeed, to the extent that people living in France were familiar with the so-called *nègre*, most inhabitants of the mother country surely derived their "information" from word of mouth and/or the era's written and visual representations of Africans. This group included many of the thinkers who would usher in a major transformation of the notion of the *nègre* at mid-century.

Representing Africanist Discourse

To help readers understand the textual representation of the *nègre* in eighteenth-century thought, this book replicates the reading practices of an imagined eighteenth-century reader. Thus, the first chapter of this study does not begin straightaway with a discussion of the era's debates on racial categories or slavery; rather, it starts where most Enlightenment-era people presumably did: with travelers' accounts and compilations.

The first real African travelogues to which eighteenth-century thinkers had general access date from the Renaissance, the era when Portuguese caravels first reported their spectacular *descobrimentos* along the Senegalese littoral.[49] As the Portuguese progressively explored the West and Central African coast in the fifteenth century, this small country undertook the first sustained European engagement with black African peoples.[50] The information derived from travels to this heretofore terra incognita changed Europeans' relationship to the rest of the world. In addition to disproving medieval ideas about the limits of the habitable Earth, the written and visual accounts of black Africans that were produced as a result of Portuguese exploration—particularly after the advent of the printing press—became epistemological and ethnographic benchmarks against which Europeans could increasingly measure their own identity.

Chapter 1, "Paper Trails: Writing the African, 1450–1750," chronicles the evolution of the textualized African that Europeans created in early-modern travel literature. In addition to familiarizing the reader with the wide range of authors (e.g., Leo Africanus, Lopes and Pigafetta, and Dapper) who were well known to Enlightenment-era naturalists and philosophers, chapter 1 explores the importance of generic and geographic conventions on the evolving pan-European construction of the black African. In many first-person accounts of Africans produced during the sixteenth and seventeenth centuries, for example, authors provided episodic (and belittling) views of the continent's inhabitants. In contrast, by the early seventeenth century, a new group of European writers including the Englishman John Pory was able to synthesize a number of texts into compilations whose breadth allowed for more schematic ethnographic categorizations of African peoples. The inconsistent quality of this information notwithstanding, the publication of these more complete Africanist works allowed Europeans to rethink the continent in terms of an expansive and pessimistic human geography that stretched from Cape Bojador to Abyssinia.

The second half of chapter 1 examines the interplay between African travel writing and the increasingly authoritative Caribbean "ethnography" flowing back from the colonial world after the mid-seventeenth century. Through readings of authors including the Caribbeanist-turned-Africanist Jean-Baptiste Labat, chapter 1 demonstrates how the overall portrait of the African provided to Enlightenment-era readers was the product of differing contexts and evaluative criteria. If properly Africa-oriented texts dedicated most of their pages to African morphology, occupations, institutions, tools, pastimes, religions, sexual mores, and trading practices, Caribbean texts such as the Jesuit historian Pierre-François-Xavier de Charlevoix's 1730–31 *Histoire de l'isle Espagnole ou de Saint-Domingue* had their preoccupations as well, particularly the utility, ingenuity, docility, pliability, and brute strength of their African labor force.

To summarize, this first portion of the book demonstrates how seventeenth- and eighteenth-century writers repackaged elements from Africanist and Caribbeanist assessments of the African into a new view of the so-called *nègre,* one with a distinctly Atlantic orientation that combined information on African mores and culture with the suitability of different ethnicities to various tasks.

Chapters 2 and 3 engage in a more pointed examination of the progressively material and scientific view of the African. In particular, these two chapters explore the link between speculative micro-physiology and the era's more conceptual representations of the African. In a sense, these two chapters examine the way in which the era's naturalists and anatomists sought to reconcile blackness with existing paradigms; they also demonstrate how existing thought structures were adapted to reflect the perceived realities of blackness.

Not surprisingly, Buffon is a key figure in both of these chapters. As well-known Buffon scholars, including Thierry Hoquet, Claude Blanckaert, Robert Wokler, and Jacques Roger, have all asserted, Buffon's *Histoire naturelle* had a significant effect on the overall representation and understanding of the so-called *nègre.*[51] Indeed, the naturalist reconciled what had seemingly been conflicting tendencies within the overall presentation of the African in 1749, namely: (1) a vague belief in the essential unity of the human species and (2) an increasingly long list of data regarding the specificity of the African variety. Buffon's major innovation, in short, was setting forth a theory of degeneration that posited a shared human lineage that readily acknowledged and explained the morphological and moral differences of the *nègre.*

Buffon's belief in degeneration—which posited an explicit *cousinage* between a prototype race of whites and the darker varieties of humans—contrasted with several other contemporary schools of thought, all of which tended to see human varieties as more fixed or separate entities. The first was the Scripture-based (polygenist) theory of separate human origins, which in France was initially articulated by the "pre-Adamist" Isaac de la Peyrère in 1655. The second was the more taxonomical breakdown of humankind into "races" or "species" by François Bernier in his 1684 "Nouvelle division de la terre."[52] Of these two seventeenth-century theories, it was the more trenchant classification of humankind put forward by Bernier that Buffon implicitly refuted. While Buffon did not mention Bernier by name when theorizing a human monogenesis, he was clearly contradicting the breakdown of the human species into "four or five" races, each of which supposedly had a biological integrity. Bernier's categorization of humankind was only the first of Buffon's targets, however. In addition to refuting this schematic separation of the human race, Buffon (and his anti-essentialist and fluid portrait of humankind) implicitly countered a more recent development in the era's thought: Carl Linnaeus's division of the genus *Homo* into four distinct categories (including *Africanus niger*) in his 1735 *Systema naturae*.[53] While it should be noted that Linnaeus was not a polygenist, his *Systema naturae* explicitly claimed the right to project an implacable logic on human phenotypes.[54] This was something that Buffon could not accept. Although his understanding of the black African was as deterministic as Linnaeus's, Buffon nonetheless rejected the absolute categories that underpinned the Swede's understanding of humankind.

Situating Buffon's relationship to both polygenesis and the birth of race classification is a difficult task. To the extent that scholars have tackled this problem, they have generally and understandably sought to evaluate it against two major preoccupations. The most pressing question, as the scholarly divide between Louis Sala-Molins and Jean Ehrard has demonstrated, is: "Was Buffon himself racist?"[55] The second and overlapping query is more genealogical in nature: what was the precise link (or lack thereof) between Buffon's thought and later thinkers who more unambiguously perpetuated the "racializing" of other peoples?[56]

While much of chapters 2 and 3 certainly engage with the relation that existed between Buffon's thought and increasingly clear-cut and zoological categories of the human, my treatment of Buffon initially eschews such debates in lieu of a more thorough contextualization of the genesis of Buffon's ideas on the black African circa 1749. It is with this in mind that chapter 2 ("Sameness and Science, 1730–1750") begins with an examination of the essays on the subject of blackness submitted to the Bordeaux Académie royale des sciences in 1741. In addition to providing a survey of the competing theories on this question before mid-century, this segment of the book reveals the unstated, yet fundamental, problem that most naturalists, including Buffon, sought to address when they took up the topic of blackness. This was not the Kantian query "What is a race?" Rather, the question that perplexed natural historians including Buffon was "What happened to humankind's essential sameness?"

Buffon, of course, provided the most compelling answer to this question in the *Histoire naturelle*. But this chapter also demonstrates that, well before Buffon, Pierre de Maupertuis had seemingly solved the riddle of essential human sameness by examining what was the ultimate example of racial crossover: the

African albino. Strictly speaking, of course, the pathological condition that came to be known as albinism had not yet been defined in the eighteenth century; nor did French thinkers refer to "whites" born to "black" parents as albinos. Revealingly, the term that both Maupertuis and Buffon used to refer this human curiosity was *nègre blanc,* literally a "white negro." Although eighteenth-century thinkers had long been intrigued by reports of these category-defying humans, the full implications of *nègreblancisme* only became clear after Maupertuis published his 1745 *Vénus physique,* the first text that used albinism as a means of thinking through the question of race.

While it may be hard to imagine now, for Maupertuis and those who followed in his footsteps, the albino provided a type of "empirical" proof for the era's vague belief in an essential sameness or shared human origin. Maintaining that the birth of an albino was a fluke occurrence, Maupertuis explained that a *nègre blanc* was a rare case of a member of a black variety of human producing a member of a separate conceptual group, in this case, a white human. In addition to this important "fact"—which implied that the white and black varieties were biologically linked—the albino became the critical construct in a new and increasingly white-centered chronology of the human species. How this theory worked was actually quite curious. Maupertuis interpreted the fact that "blacks" sometimes produced "whites" to mean that whites were the original variety to which blacks sometimes reverted. A less European understanding of albinism might have asserted the contrary, of course. Indeed, the fact that blacks occasionally produced whites could have more logically indicated that blacks were the original race, a race that had, in the past, generated enough of these white "accidents" to form their own race.

Despite its shortcomings, Maupertuis's stunning example of quasi-logic echoed in the life sciences for decades. If, as d'Alembert wrote in the *Encyclopédie,* "[c]hronology . . . places men in time," and "[geography] distributes them across our Earth," the concept of the albino allowed Maupertuis to combine human biology, geography, and chronology into a dynamic narrative that explained both the origin and the source of humankind's differences.[57] Seamlessly accepted into Buffon's overall reconceptualization of the human species in his *Histoire naturelle* several years later, Maupertuis's use of the albino became the hidden foundation of a white-centered *science de l'homme* for the next three decades. What neither Maupertuis nor Buffon understood about the albino, however, was that this strange being functioned dialectically in their thought. On the one hand, the accidentalist explanation for the existence of *nègres blancs* allowed both thinkers to manufacture a more "scientific" theory of a white prototype variety. On the other hand, the fundamentally degenerative genealogy of humankind that the albino suggested opened the door to increasingly pessimistic and diagnostic chronologies of the African. This is one of the most overlooked aspects of the history of the eighteenth-century life sciences: in putting forward this new scientific paradigm of an essential and original human sameness, Maupertuis and then Buffon implicitly invited a new generation of thinkers to identify the specific corporeal changes that now separated the African from the European. In other words, it was against the critical backdrop of an original sameness (and not essential difference) that a belief in the deeply divergent nature of the *nègre* would ultimately emerge.[58]

Much of chapter 3, "The Problem of Difference: Philosophes and the Processing of African Ethnography, 1750–1775," examines the refinement of new theories of the African in light of degeneration theory. This too is a significant departure from the existing historiography. Instead of concentrating on the history of human classificatory schemes, as is often the case in contemporary scholarship, this chapter focuses on those anatomically and physiologically oriented thinkers who saw monogenesis as a call to speculate on just what had gone wrong in the African's body over deep time. Put simply, this chapter recovers anatomists' beguiling contribution to (1) the ongoing reconceptualization of blackness and (2) the attendant redefinition of the notions of variety, species, and ultimately race that anatomical "discoveries" engendered. Rather than trying to identify a precise date when race came into existence (or historicizing this concept in a linear fashion from figurehead to figurehead), this chapter examines the riddle of human difference as it related to biological processes and questions of human categories.[59]

This treatment of the sameness-difference tension in the French life sciences also allows for a fruitful reinterpretation of many of the era's foundational "Enlightenment" thinkers. The goal of this section is not to recount or evaluate the relationship of Enlightenment to the concept of race as a whole, however; rather, it is to examine how the era's philosophes reacted to and often incorporated physiologically oriented understandings of the African into the new human sciences. This portion of the book resituates what have long been "problems" for eighteenth-century specialists, for example, Voltaire's unapologetic castigation of the African, the *Encyclopédie*'s curious ambiguity regarding the *nègre*, and the new "definition" of blackness emerging in natural history dictionaries. Most significantly, perhaps, this section also provides an analysis of Montesquieu's landmark *De l'esprit des lois* (1748) that interprets his famous "denunciation" of slavery in light of his natural history beliefs regarding the African. Much like Buffon's views on the African, Montesquieu's thought has divided Enlightenment specialists. If it is true that Montesquieu was theoretically, morally, and philosophically opposed to the enslavement of Africans, it is also quite clear that, like other thinkers of his era, he gave credence to the unsettling biopolitical implications of natural history. Indeed, while Montesquieu was the first to identify and to debunk the racial stereotypes that justified the overall justification of the slave trade, his understanding of the ethno-physiological reality of the African also seems to reserve a particular geographic space for African chattel slavery. In a century known for its relativism, this is perhaps one of the most disturbingly relativistic moments of the high Enlightenment.

Whereas the first three chapters of this book treat the question of slavery on numerous occasions, the final chapter, "The Natural History of Slavery, 1770–1802," provides an integrated reading of the question of natural history and human bondage. To the extent that scholars including Winthrop D. Jordan, Christopher L. Miller, and Adam Lively have examined the link between anatomy and the discourse on slavery, they have generally done so by documenting how colonial thinkers drew on proto-raciology in order to rationalize human bondage. While this chapter certainly draws on this important scholarship, it also examines the use of natural history by anti-slavery thinkers as well. This more comprehensive approach to the question of anatomy and slavery allows for a subtler understanding of the ambiguously politicized body of the African and the status of natural history itself.

Among the post-1750 texts that are taken up in this section, the multifarious and multi-author *Encyclopédie* provides a telltale example of the peculiar relationship existing between natural history and slavery during this era. Some articles in this huge project contain racialized mercantile assessments of the African that came from the Caribbean; others reproduce demeaning portraits of black Africans from the era's travelogues; and yet, most famously, two articles—among the most cited in abolitionist historiography—call for an end to the slave trade. These divergent treatments of the *nègre* point both to the ad hoc construction of the *Encyclopédie* itself and to the importance of the overarching rubric under which the subject of the black African was being treated. When liberal thinkers examined the *nègre* from the perspective of natural history, for example, they tended to remain within this framework, suggesting that the black African was a reprehensible and degenerate "variety" or "race." Elsewhere and unlinked by *renvoi* (cross-reference), these same philosophes interpreted the black African's plight from the point of view of natural law and often produced a plaintive and sentimental portrait.

If this initial analysis of the *Encyclopédie* serves to introduce many of the contradictions existing among Enlightenment-era thinkers, the remainder of the chapter examines how late-century anti-slavery thinkers began to reconcile or address these incongruities by practicing a new type of ideologically driven natural history. The most illustrative text to grapple with the question of anatomy and slavery is Abbé Raynal's so-called *Histoire des deux Indes* (in its multiple editions). Insofar as this subject has even been treated, the relationship between Raynal's understanding of African physiology and "his" views on slavery has puzzled critics. After all, in the first (1770) edition, the author, who would soon be known as the era's greatest *négrophile*, argued that the African—given his supposedly dark sperm and inferior intelligence—was a different species of human.

Not surprisingly, much of chapter 4's analysis of the *Histoire* hinges on an examination of the third and best-known edition of Raynal's text (1780). In this final iteration of the work, Diderot (as ghostwriter) completely rejected the cutting-edge proto-raciology of the earlier editions, which had seemingly contradicted Raynal's famous anti-slavery paragraphs. This was a significant moment in anti-slavery discourse. By following in the footsteps of the more positive reevaluation of the black African first undertaken by the Quaker Anthony Benezet in the 1760s and by Abbé Pierre-Joseph-André Roubaud ten years later, Diderot provided one of the first entirely coherent anti-slavery arguments in French thought. In addition to discrediting the era's racially essentialist anatomy and reasserting a belief in the shared unity of the human species, Diderot attributed the "shortcomings" of the African "type" to the institution of slavery and the white planters themselves.

Diderot's refutation of a century's worth of negative Caribbean-born stereotypes (and the purported scientific insights into the *nègre*'s physiology that served to justify them) changed the rules of the debate on slavery. Before the publication of the third edition of the *Histoire*, pro-slavery and anti-slavery thinkers had generally accepted a shared set of "facts" regarding the natural history of the African. While elements of these data varied significantly, the African was, for the most part, considered on both sides of the slavery debate as an inferior "variety" or "race" whose pathological physiology was the result of a climate-induced degeneration from an original white prototype. By disallowing the properly essentialist and racist elements of this explanation of the African, Diderot—arguably the most anatomically oriented materialist of the philosophes—not only turned his back on some of his era's most original anatomical "discoveries"; he recognized the nefarious biopolitics of his era's natural history. Like Benezet and Roubaud before him, Diderot understood that to argue against slavery effectively, one had to debunk the negative natural history of the African on which slavery increasingly relied.

The final section of chapter 4 takes up the fate of *négrophile* discourse as it was practiced during and after the era of the Société des amis des noirs (1788–1802). In particular, this segment of the book examines the evolution of *négrophile* politics during the revolutionary era, a time when slave uprisings in the French Caribbean led pro-slavery writers to castigate writers like Raynal as irresponsible ideologues who had incited rebellion on French islands. Ultimately, this section of the book underscores the fact that, as far as rhetorical strategies went, the efficacy of the new, more favorable natural history of the African depended heavily on historical context. If praising the African's potential to become a full-fledged member of the human community was a widespread and successful tactic in the mid-1780s, this particular argument was more difficult to put forward after 1791 when, among other things, Saint-Domingue had erupted into open revolt.

Anatomizing the History of Blackness

Coming to grips with the shifting construction of blackness during the eighteenth century involves a series of perils and methodological pitfalls. To produce a historical narrative on this subject—or any subject, as Hayden White famously suggested—is not only to select, but also to reflect a particular worldview. While I have attempted to produce a neutral historical account, it is nonetheless undeniable that my own work reflects an orientation with "distinct ideological and specifically political implications."[60]

To a large degree, this book can be seen as part of a wider meditation on African representations that began after Christopher Miller published his seminal *Blank Darkness* in 1985. As I have maintained elsewhere, Miller drew from the theoretical foundations of both Michel Foucault and Edward Saïd in order to examine Africanist documents from a novel perspective.[61] Rather than using these sources to establish a chronology of human events in Africa, articulate a figurehead genealogy, or acquire specific knowledge of the African past, Miller studied Africanist texts with two goals in mind. First, he identified a period-crossing discursive structuring of Africa that functioned as a form of grammar, as a highly codified nexus of beliefs that gave shape to the overall representation of Africa. Second, and perhaps more important, he mapped the application of this "knowledge" onto a larger colonial project of representation and exploitation. Miller has built on and distilled this method to great effect in *The French Atlantic Triangle* (2008), a study in which he has demonstrated, for example, how Voltaire's varied opinions on slavery can be seen in terms of three overlapping discourses: a noble savage Peruvian one (*Alzire*), a "French Atlantic Triangle" view (*Candide*), and an orientalized form of human bondage (*Le blanc et le noir*).[62]

Miller's influential contributions to the question of representation have implicitly invited scholars to examine the European discourse on the African with the same intensity that Europeans themselves examined other cultures during the early-modern era.[63] This has been the impetus for a number of studies including Y.V. Mudimbe's *The Invention of Africa*; Linda Merians's and François-Xavier Fauvelle-Aymar's respective books on the representation of the so-called Hottentot, *Envisioning the Worst* and *L'invention du Hottentot*; Roxann Wheeler's admirable analysis of skin color and its role in defining race in eighteenth-century England, *The Complexion of Race*; Adam Lively's charting of racial discourse in both eighteenth-century pseudo-science and modernist culture, *Masks*; and the collection of essays edited by Catherine Gallouët and others entitled *L'Afrique du siècle des Lumières* (2009). To the extent that I can contribute to the work of those scholars who have written on the black African after *Blank Darkness*, I am hoping not only to enhance our understanding of European mentalities, but also to challenge what is becoming an increasingly powerful and unquestioned belief in an anxious and monolithic Enlightenment-era consciousness that supposedly acted as both factory and repository for the representation of black Africans.[64]

Readers of this book should be forewarned. This study does not posit the existence of a pan-national European "mind" engaged in a conscious attempt at removing colonized peoples from history. Rather, it offers an interdisciplinary examination of how Enlightenment-era thinkers living in France (and elsewhere) processed "ethnography" in the context of their own changing preoccupations. While identifying the French and European understanding or representation of Africans admittedly does not shed light on some of the primary questions explored by historians of slavery—how many black Africans perished in *razzias*, or how they contended with conditions on Caribbean and American plantations—such a study does provide insight into an era that factored the torturous labor and murder of black Africans into the costs of production of various crops and commodities. Put bluntly, this book seeks to underscore the relationship between representation and the brutal lives to which Africans were consigned in European colonies.

This brings me back to the use of "anatomy" as it appears in this book's title. In the most literal sense, this term obviously refers to the authoritative scientific discipline that played a key role in diffusing corporeal explanations for what had previously been vague, intuited preconceptions regarding the black African. But the title *The Anatomy of Blackness* is also intended to evoke a second meaning that the term *anatomie* had in eighteenth-century thought. As Jean-François Féraud pointed out in his 1787 *Dictionnaire critique de la langue française*, eighteenth-century thinkers employed the concept metaphorically to describe an incisive analysis of a given body of thought: "one can [also] say *undertake the anatomy of a discourse.*"[65] Implicit in this figurative use of the term is the telos of anatomy itself: to understand a complex *whole*, one is obliged to extract, dissect, and consider the viscera that compose it. Such a method was the point of departure for the study that follows.

monument conjures up a story beginning with the much-earlier Portuguese ventures into "black" Africa during the first half of the fifteenth century.[2]

Considered solely from the point of view of maritime history, the Portuguese *descobrimentos* were undeniably remarkable achievements. Prince Henry himself never ventured to West Africa—on the monument he holds a miniature caravel, not a sextant—but his vision, oversight, and financing of the early voyages inaugurated a new era in the relationship between Europeans and the continent's black inhabitants. Under Henry's tenure, Gil Eanes crossed the daunting Cape Bojador in 1434, Dinis Dias met the inhabitants of what he called Cabo Verde (modern Cap-Vert in Senegal) in 1444, and Pedro de Sintra landed in Sierra Leone in 1460. During this time, Portugal also began importing what would become a significant African slave population.

Exploration initially slowed after Henry died in 1460, but began again in earnest after King João II conceived of moving beyond the southernmost point of Africa and on to spice-rich India.[3] During the next twenty years, the most famous era of Portuguese exploration of Africa, Diogo Cão reached Angola (1484), Bartholomeu Dias passed the Cape of Good Hope (1487–88), and, finally, under João II's successor, Manuel I (pictured on the monument), Vasco da Gama established the European-Asian sea route by sailing to Calcutta by way of Mozambique (1497–99).[4] Less than a century after Henry's father, João I, had ordered his navy to attack and capture the Moroccan city of Ceuta (1415), Portuguese sailors and explorers circumnavigated what would be called, among other things, the "Land of the Negroes."

Needless to say, the Portuguese conception of Africa changed considerably during this early era. In addition to the fact that Portuguese explorers had more contact with the interior parts of Africa than is generally known—they reached Timbuktu and Mali by the 1480s—information gained during the many "Guinea" landfalls was carefully noted by the official scribes and cartographers who ventured forth on the small country's caravels.[5] By 1500, geographers privy to the ethnographic and cartographic information collected at the Casa da India e da Guiné in Lisbon were able to piece together the rough physical contours of this continent for the first time in history.[6] Although Portuguese authorities kept their knowledge of Africa a closely guarded secret, the European textualization of both Africa and black Africans had begun.

This chapter examines the evolution of the textualized African from the fifteenth through the eighteenth centuries. While the primary focus of this book as a whole is clearly the later (Enlightenment) period, identifying the

CHAPTER ONE

Paper Trails

Writing the African, 1450–1750

I saw Africa, but I have never set foot there.

JEAN-BAPTISTE LABAT, *NOUVELLE RELATION DE L'AFRIQUE OCCIDENTALE* (1728)

Across from the Jerónimos Monastery on the north bank of the Tagus River at Lisbon stands the Padrão dos Descobrimentos, an imposing marble monument erected in 1960 to commemorate the five-hundredth anniversary of the death of the *infante* Henrique o Navegador, known to us as "Henry the Navigator." This shiplike memorial gives pride of place to Prince Henry, who commands at the prow.[1] Standing or kneeling behind him are a number of historical and allegorical figures: King Manuel I, the epic poet Camões, the explorers Vasco da Gama, Magellan, Cabral, and Henry's mother, Philippa of Lancaster. Less prominently, there are also the cartographers and missionaries who contributed to the Portuguese *descobrimentos* in their particular fashions. This celebratory rendering of history, which conflates discoveries in Africa and Brazil with the first circumnavigation of the globe, is aptly named a marker, or *padrão*. Like the inscribed stone crosses, or *padrões*, that Portuguese explorers sank into the ground along the African littoral, this testament to the fundamental realignment of the world stakes a distinctly Portuguese claim in Western historiography. In stark contrast to a view of the "Age of Exploration" that has traditionally concentrated on Columbus and the Americas, this giant

The Padrão dos Descobrimentos monument in Lisbon. At the prow, the figure of Prince Henry the Navigator holds a model caravel, the ship that made Portuguese exploration of the world possible. Photo by Bill Curran.

high points of Africanist discourse during earlier eras informs our understanding of subsequent representations of Africa in several fashions. In the first place, writings by travelers to Africa from the fifteenth, sixteenth, and seventeenth centuries—which were often cited in the same breath with much later portraits of the continent and its peoples—continued to play a critical role in the overall understanding of Africa during the eighteenth century. Although contemporary scholars working on the question of representation generally tend to limit the scope of their study to their own period, early-modern naturalists and philosophes who read about or wrote on black Africans during the Enlightenment era did not. By way of example, Samuel von Pufendorf, Edward Gibbon, and the abbé Edme-François Mallet engaged significantly with the writings of the sixteenth-century writer Leo Africanus. Likewise, Jean-Jacques Rousseau, Abbé Prévost, and the authors of the *Universal History* commented on the Congolese ethnography put forward by the Renaissance sailor Andrew Battel. As important as these cross-era links may be for our overall understanding of the representation of black Africans, however, studying the texture of early travelogues on Africa has another distinct advantage: it provides insight into a time before the presentation of the *nègre* was dominated by the overwhelmingly powerful modes of interpretation that blossomed along with the Caribbean plantation.

The Early Africanists: The Episodic and the Epic

To the extent that a real Africanist discourse existed in fifteenth-century Portugal, much of it was a carefully guarded secret enforced by the *política de sigilo*, the so-called policy of secrecy. As the sailors and scribes who returned from West African landfalls pulled into Portuguese harbors, they were obliged to leave all written traces of their voyages—be they maps, logs, or ethnological musings—with port authorities working for the Portuguese crown. Not unexpectedly, the number of people who had access to "useful" or "real" knowledge relative to Africa remained modest during the first decades of the Portuguese expansion.[7]

Nonetheless, Portuguese elites heard a filtered account of the triumphant news of the *descobrimentos* in 1453 when Gomes Eanes de Zurara's *Crónica do descobrimento e conquista da Guiné (Chronicle of the Discovery and Conquest of Guinea)* began circulating.[8] Commissioned by King Afonso V, the *Chronicle*'s episodic recapitulation of the exploration of the African littoral is a product of a Europe not quite beyond its feudal mind-set. At many points, its tone and overall structure recall a medieval *chanson de geste* or epic poem.[9] This is perhaps best exemplified in Zurara's portrayal of the protagonist of this Homeric narrative, Prince Henry. Described as "naturally constrained" by God and his birth sign to "undertake great actions," Henry is portrayed as having the intelligence and the drive to engage in a series of worthy endeavors including trade, information gathering, evangelization, and the quest for the fabled African Christian king "Prester John," who, the Portuguese believed, might participate in the eternal war against the unbelieving Moors.

Characterized by a mixture of xenophobia and an Augustinian sense of destiny, Zurara's text also reveals a disconcerting nonchalance regarding the brutal treatment of black Africans by Portuguese explorers. Whether it be the kidnapping of coastal peoples during *razzias* or the selling off of these humans in Portuguese slave markets, all such events are recounted matter-of-factly, as accepted practice.[10] As is the case with other late-medieval and Renaissance manuscripts chronicling such intercultural encounters—most notably Jean de Bethancourt's account of how Canary Islanders (Guanches) were captured, enslaved, and sent to the Cadiz slave market at the very beginning of the fifteenth century—Zurara's narrative speaks to an era in which the violence of feudal Europe was consciously and self-assuredly exported to the rest of the world, more often than not with the benediction of the reigning popes.[11] In fact, during the same basic era the *Chronicle* appeared, Pope Nicholas V issued papal bulls including the 1452 *Dum Diversas* and 1455 *Romanus Pontifex*, conferring upon Portugal the sacred right to subjugate infidels, meaning non-Christians.

With the benefit of historical hindsight, Zurara's version of the Portuguese outthrust can certainly be read as the initial chapter in the long and tortuous book of Atlantic history. Written only a few years before African labor was actually employed in plantation settings by the Portuguese—on Madeira, São Tomé, and the Cape Verde Islands—Zurara's text reads, at times, as a blueprint for the centuries of regrettable treatment that Europeans would inflict on sub-Saharan Africa and its inhabitants. And yet, while texts such as Zurara's *Chronicle* can be seamlessly woven into a narrative charting the development and consequences of the Atlantic slave trade, other Renaissance accounts of Africa point to a more fragmented and ambiguous history of early contacts and their representations, a history that is worth bearing in mind as we examine later eras.

During the fifteenth century, the writings of actual explorers—as opposed to an official chronicler like Zurara—provide a somewhat more intricate idea of how Europeans may have envisioned initial encounters between Africans and themselves. Many such travelogues belie the notion that views of Africa can be reduced to a series of unifying themes and characteristics, an early master narrative of European domination, as it were. While it is certainly true that fifteenth-century African travelogues often reveal the viciousness of European crews who sought their fortune in Africa, such texts do not reflect the ideological, religious, and military coherence of what would later become the Portuguese empire in Africa. Eustache de La Fosse's account of his 1479 voyage to Sierra Leone provides a clear example of this phenomenon. Although his narrative accurately reflects the greed and brutality of the typical Renaissance sailor—which was often directed at fellow Europeans—the French explorer also spoke of the Africans with whom he traded as "good enough people [who] genuinely trust us."[12] Nowhere to be found in these chronicles are the prejudices of antiquity or the dehumanizing tone that characterizes Zurara's chronicle. Unlike more cohesive and calculating accounts, La Fosse's view of Africa and Africans is quite open-minded, albeit inconsistent and episodic.

The inaccuracy of reducing the Renaissance representation of Africa to a unified discourse becomes even clearer when we examine one of the primary Africanist sources from this era, Alvise da Ca' da Mosto's account of his voyages to Senegal and Gambia in the mid 1450s, published in 1507. A Venetian noble, sailor, navigator, and merchant who had been instructed by Prince Henry to replace the *razzia* with *resgate*—the raid with trade—Ca' da Mosto is generally recognized by contemporary scholars as remarkably forthright, especially in his observations of the Wolof. In line with that other famous Venetian traveler Marco Polo, Ca' da Mosto frequently provides surprisingly broad-minded meditations on encounters with Africans that foreshadow Michel Adanson's writings on the same general area in the 1750s.[13]

On his first trip to Senegal (1455), Ca' da Mosto spent twenty-eight days with a Senegalese monarch, Budomel, presumably to trade Spanish horses for the latter's most alluring commodity, slaves.[14] During this extended encounter, the Venetian traveler admitted to being as curious about these people, inasmuch as he was eager to make this a profitable venture. Accordingly, he carefully discussed a variety of topics: the procedure for electing the king, the source of royal revenues, tributes paid to the monarch, the market for slaves, and the influence that Muslim "priests" had at court.[15] As a rule, Ca' da Mosto remained relatively tolerant regarding various aspects of African life. Although he demonstrated disdain for the material poverty of African royalty, the errors of Islam, and the appalling savagery of African wars, he also revealed himself to be very fair-minded with respect to certain African customs or conventions that would, in the eyes of the missionaries who later journeyed to Africa, provoke strong censure. He dispassionately explained polygamy, for example, as an intelligent way of managing smaller economic systems.[16] Ca' da Mosto also acknowledged Senegalese magic as not only authentic but effective.[17] Yet the most remarkable moments in Ca' da Mosto's account came when he evoked himself as a foreign entity in the eyes of the Africans who came into contact with him. In the first instance, the Venetian related the thrill of interethnic experience at a local Senegalese market:

> They marveled no less at my clothing than at my white skin. My clothes were after the Spanish fashion a doublet of black damask, with a short cloak of grey wool over it. They examined the woolen cloth which was new to them and the doublet with much amazement. Some touched my hands and limbs, and rubbed me with their spittle to discover whether my whiteness was dye or flesh. Finding that it was flesh they were astounded.[18]

Ca' da Mosto also attempted to convey the indigenes' perspective in his account of two brutal skirmishes that he and his crew had with the Serer in Gambia. According to his own version of the events, Ca' da Mosto was eager to trade with this group, and, as he had done with the Senegalese, he sent out an African interpreter, who was, however, immediately slain by several warriors waiting on shore.[19] Sailing farther south with hopes of finding a less warlike people, his caravels came under attack by what he estimated to be 150 Serers shooting poisonous arrows from pirogues. Ca' da Mosto and his men retaliated with cannon and crossbows, which left scores of African bodies in the water and in the pirogues, and he then had the remaining warriors pursued in order to find out why they had attacked his ships in the first place:

> [The surviving Serers] responded that they had had knowledge of our dealings with the Negroes of Senegal, [dealings] that could only be bad if [the Senegalese] were our friends, because they were sure that we Christians ate human flesh and that we were buying Negroes in order to devour them; accordingly they did not want to make our acquaintance at any price and had hoped, rather, to kill us.[20]

Reading such an account some 550 years after it occurred, one cannot help but interpret this violent clash between well-armed Europeans and indigenous peoples as having a racial subtext. And yet, viewed within its own epistemological framework, Ca' da Mosto's account makes it clear that this was not primarily a conflict between races or between belief systems.[21] And herein lies one of the primary realities of early contacts that we must bear in mind as we begin to follow the paper trail of representation during this era. Though early explorers like Ca' da Mosto often conflated their whiteness with Christianity, blackness did not yet unequivocally signal idolatry, intrinsic savagery, or a degenerate race. In the era well before racialized conjectural (biological) histories of humankind or the pseudo-science of craniology, Ca' da Mosto's observations did not derive from a series of pre-formed conclusions on the status of the black African. Instead, as his own text demonstrates, he endeavored to understand (within his own limits) multiple African points of view, even one like the Serers', which accused Europeans like himself of being anthropophagic barbarians worthy of death.

Like all early accounts of the West Coast of Africa, Ca' da Mosto's mid-fifteenth-century narrative is fragmentary and intervallic. Broader attempts at representing or understanding Africa began to appear in subsequent decades, after the continent had been circumnavigated. By the mid sixteenth century, a growing interest in the era's voyages to unknown lands prompted European publishers to edit compilations or multi-volume works organized by geographical region. In Lisbon, João de Barros, who was treasurer of the Casa da India, Casa da Mina, and Casa de Ceuta, as well as a brilliant historian, wrote the *Décadas da Asia* (1552), an account of the Portuguese conquests and discoveries "ranging from Guinea to China and from the fourteenth century to 1538."[22] In England, Richard Hakluyt published his voluminous *Principal Navigations, Voyages, Traffiques and Discoveries of the English Nation*, the second edition (1599) of which had substantial sections dedicated to sub-Saharan Africa.[23] The most important sixteenth-century compilation on Africa was, however, Giovanni Battista Ramusio's *Navigationi e viaggi*, published in Venice in 1563.[24] In addition to publishing Ca' da Mosto's accounts of his voyages to Senegal, Ramusio also included another watershed text: the encyclopedic *Descrittione dell'Africa* by Leo Africanus (Al-Hasan ibn Muhammad al-Wazzan al-Fasi). The influence of this latter work would last well into the eighteenth century.

The exceptional circumstances of Leo Africanus's life made him the perfect "African" translator for his European audience. Born in Granada a few years after it fell to the Reyes Católicos, Ferdinand and Isabella, in 1492, he and his family moved shortly thereafter from what had been the last Muslim kingdom on the Iberian Peninsula to Fez in Morocco.[25] Although historians have discovered very little about his youth, it is known that the future Leo Africanus studied theology and law in Fez and served the Moroccan throne as an ambassador for some ten years, during which time his embassies led him to Tunis, Egypt, Armenia, Constantinople, and the Niger Delta. Captured by Spanish pirates while returning from a voyage to Egypt, he was taken to Rome in 1520, where he was introduced to Pope Leo X.[26] Shortly thereafter he converted to Christianity and took the name Leo. The discovery of the manuscript on which Ramusio based his text has led scholars to believe that during the next five years, Leo Africanus dictated his voluminous notes in Arabic to an

Arabic-speaking Maronite monk, who then translated them into Italian.[27] It was this manuscript that Ramusio polished, edited, and then published some twenty-five years later in his *Viaggi*.

Leo's pioneering treatment of Africa instantly made a hundred years of African travelogues seem outmoded. Unlike other texts on the subject, which generally recount the events related to one or several landfalls, Leo's *Descrittione* provided his readers with thorough, region-specific assessments of what he knew of the African continent. His account of the various societies of North Africa are particularly comprehensive. The section on Egypt, for example, contains detailed information on geography, ethnography, the biblical and political history of the land, climate, major cities, mores, and the military. While such detail was invaluable, Leo's most critical innovation was the fact that he buttressed his region-specific treatments of Africa with an approach derivative of Renaissance cosmologies.[28] Indeed, Leo's text overlaid the personal account of his travels with what would become a long-standing division of Africa into four parts: Barbary, Numidia, Libya, and the Nijar, or "Land of the Negroes."

Leo's description of this "Land"—which he separated into fifteen kingdoms—varies widely in quality. The first chapter on "Guinea," which is not based on personal experience, lacks both geographical and ethnographic detail. Leo's writings on the inner regions of the Niger Delta, however, contain concise assessments of both the "pagan" and Muslim cultures in what is present-day Mali. Not surprisingly, Leo's analysis of pagan peoples, such as the inhabitants of Zanfara, reveal the prejudices of a Muslim versed in theology and law: "the inhabitants are tall in stature and extremely Black, their visages are broad, and their dispositions most savage and brutish."[29] And yet Leo's travelogue also acknowledged the erudition and sophistication of a number of African ethnicities, chief among them the inhabitants of Melli (Mali). This was, according to Leo, a land where a rich people with "plenty of wares" attended "a great store of temples" filled with "priests and professors."[30] In Leo's hierarchical view of the peoples of the Niger Delta, the inhabitants of Melli "excell[ed] all other[s] in wit, civility, and industry."[31] Leo was almost as complimentary regarding Timbuktu, describing it as a place of both fabulous wealth and great learning, where books were prized above all other merchandise, and where a "great store of doctors, judges, priests, and other learned men . . . [were] bountifully maintained at the king's cost and charges."[32] The glory days of this empire had come to an end by 1591, but Leo's account of the great riches of Timbuktu nonetheless lived on well past the fall of the Songhai.

Attracted by the intriguing prospect of this sophisticated black civilization where, Leo had reported, the king's scepter weighed "1,300 pounds," the Paris Société de géographie offered a prize of 10,000 francs in 1825 to the first European traveler to prove that he had reached the city of Timbuktu, which was ultimately claimed by René Caillié three years later.[33]

Rationalizing Africa

Leo Africanus's *Descrittione* ushered in a new era during which authors supplemented first-hand accounts of black Africa with more comprehensive attempts at understanding and interpreting difference on its own terms. Such was the case with Duarte Lopes's 1591 *Report of the Kingdom of Congo and the Surrounding Countries*.[34] Like Leo Africanus's text on North Africa and the Niger Delta, Lopes's *Report*—which was based on its author's experience in the region as both slave trader and, toward the end of his stay, as ambassador for the king of the Kongo—was a marked departure from episodic travelogues on the Kongo region. Transcending general assessments of topography and vague ethnographic descriptions, it provided broad considerations of the region's flora, fauna, peoples and customs, currency, royalty, regional alliances and disputes; many of the chapters dedicated to these subjects are filled with enough Kongolese vocabulary to constitute a lexicon of sorts.

Much of what is contained in the *Report* must be seen in the context of Lopes's political and diplomatic mission. In his capacity as ambassador for Kongolese King Alvaro I, Lopes came to Rome in order to solicit help for more extensive conversion efforts in the region and perhaps even to mitigate Portuguese influence in the Kongo by forging a direct alliance with the Vatican.[35] Given the tenor of the mission, it is not surprising that Lopes's text paints the Kongo as an appealing land for Europeans. As Willy Bal has pointed out, Lopes and his Italian translator Filippo Pigafetta purposely minimized the inconveniences of the climate and exaggerated the fertility of the soil and the region's mineral riches."[36] Most tellingly, the *Report* portrays the inhabitants of the Kongo as eminently assimilable. Depicted as either converts or potential converts, the Kongolese are described as even looking like the Portuguese:

> The men and women are black, some approaching olive color, with black curly hair, and others with red. The men are of middle height, and, excepting the black skin, are like the Portuguese. . . . Their lips are not large like the negroes, and

> their countenances vary, like those of people in our countries, for some are stout, others thin, and they are quite unlike the negroes of Nubia and Guinea, who are hideous.[37]

This tendency to render the region's inhabitants as European-like or familiar in appearance was carried to an extreme in the description of the fierce group of supposedly anthropophagic humans known as the Anziques. Despite Lopes's assertion that this particular group frequented human butcher shops as a matter of course, he and his translator Pigafetta far from condemned them. Rather, they represented these alleged cannibals' psychological makeup as potentially well suited to Christian martyrdom:

> These people are wonderfully active and nimble, leaping up and down the mountains like goats, very hardy, without fear of death, simple, sincere, and loyal, and, indeed, the Portuguese have greater confidence in them than in any other tribes. So that Duarte Lopes well says, if these Anziquez became Christians (being thus faithful, truthful, loyal, and simple, giving themselves to death for the glory of the world and their flesh to their princes for food, if it would please them), how much more from their hearts would they suffer martyrdom for the name of our Redeemer, Jesus Christ, and nobly defend our faith and religion.[38]

The illustrators for the first Latin edition of the *Report* echoed this optimistic (albeit reductive) view of the region's inhabitants.[39] Indeed, the overall visual image of Africa in this text, which features beautifully rendered orientalized or Europeanized men engaged in hunting, warfare, and the burning of fetishes, faithfully conveys the overarching ideology of a travelogue that sought to facilitate a new type of "cooperation" between Africans and Europeans.

Lopes and Pigafetta buttressed this overall optimistic assessment of the kingdom of the Kongo with an especially positive historical account of the Portuguese discovery of this land. Typical of the era's historiography, Pigafetta portrayed Portugal's King João II as an exceptionally gifted ruler who had sent his sailors beyond the treacherous Cape Bojador in order to discover the sea route to the East Indies. Along similar lines, the authors depicted the Kongolese royalty and population as universally happy to receive European explorers. Not only had King João's emissaries supposedly met with a warm and gracious welcome from the Kongolese ruler, but the latter had immediately converted and dictated a letter to the Portuguese king "begging for priests to be sent him to propagate Christianity."[40] Much like the sanguine presentation

Congo animals depicted in Duarte Lopes, *Regnum Congo, hoc est, Vera descriptio regni Africani* (1598). The plates in the De Bry edition include a variety of real and imagined creatures from different regions of Africa. Despite their somewhat fantastic rendering, these images generally diffused a more realistic understanding of the continent. Courtesy Beinecke Rare Book and Manuscript Library, Yale University.

of the flora, fauna, and economic opportunities in the Kongo kingdom, Lopes and Pigafetta's view of the past overlaps perfectly with a belief in a better, Christian, future for the entire region. This was, after all, a world where Africans were purportedly as desirous of Christian help as Europeans were eager to extend it.

Despite the fact that the historical episodes recounted by Pigafetta took place approximately a hundred years before his "co-author" Lopes set foot in Africa, they are related in the exact same narrative style as the rest of the text. This seamlessness is one of the foundational characteristics of Pigafetta's prose. Whether pronouncing on first-hand ethnographic and naturalist details, second-hand rumors regarding the anthropophagic excesses of the Anziques tribe, or on regions of the continent (e.g., Sofala) to which Lopes had never ventured, Pigafetta's text inconspicuously shifts from carefully rendered personal experiences to pure fantasy.[41]

Not surprisingly, it is the mythical aspect of Pigafetta's account that stands out in many a contemporary reader's mind, particularly when the *Report* is read alongside examples of the work's stunning iconography, most famously the beautiful Amazon-like warriors who supposedly lived in Monomotapa. Pigafetta describes these "extremely agile and rapid" female warriors, who were supposedly feared "on account of their wiles and cunning," with great admiration. Nonetheless, in spite of the imaginative renderings and ideological underpinnings of this text, the fundamental *telos* of Pigafetta's account of the Kongo was to portray a knowable African reality that was very different from that of the terra incognita of the ancients. Far from being an impenetrable territory populated by strange peoples and monsters, Lopes and Pigafetta's Africa is an eminently understandable region.

By 1600, editors and authors reacting to Africanist works such as Lopes and Pigafetta's *Report* began to treat the continent as a potentially discernable whole. John Pory, the English translator of Leo Africanus, played a key role in this regard. As translators often did in this era, Pory supplemented Leo's text with entire sections of ethnography and geography borrowed from other Africanist sources. Drawing from Pigafetta, Livio Sanuto, João de Barros, and Francisco Alvares, among other sources, Pory's *History and Description of Africa* paints a "complete" portrait of the continent. This large *History* is often less sympathetic than Leo's original text, however. Although Pory faithfully translated Leo's admiring views of some of the peoples of the Niger Delta, his supplemental geography also reflects, perhaps more accurately than Leo's text, the increasing prejudices of his era. Proceeding clockwise from East to West Africa, Pory describes most ethnicities negatively: Nubians embrace the "infinite corruptions of the Jewish and Mahomedan religions,"[42] Cafri idolaters are deadly enemies of Christians,[43] Zanzibaris are "much addicted to sorcery and witchcraft,"[44] Anzichi cannibals purchase human meat in "butcher shops" (a detail borrowed from the Lopes-Pigafetta *Report*),[45] Beninians seem "rude and brutish,"[46] and Senegambians are "Black and barbarous."[47]

Although a montage of stereotypes and hearsay, Pory's adaptation and amplification of Leo's text conveyed the impression that Africa, as strange and enigmatical as it was, could now be fruitfully scrutinized, assessed, and

"Amazons" of Monomotapa depicted in Duarte Lopes, *Regnum Congo, hoc est, Vera descriptio regni Africani* (1598). Courtesy Beinecke Rare Book and Manuscript Library, Yale University.

categorized by region. Much of this increasingly continental view of Africa was made possible by a proliferation of new European texts and knowledge regarding Africa that overlapped, not entirely coincidentally, with the waning of Portugal's virtual monopoly on the continent. Before the early seventeenth century, Portugal alone had enjoyed a deep set of contacts and relationships in sub-Saharan Africa.[48] In addition to its factories and footholds along the two coasts, the Portuguese had founded more substantial colonies in the Congo and Mozambique, which were ostensibly protected by papal law and the Portuguese navy. By the 1620s, however, Portugal's hegemony in Africa and elsewhere was increasingly challenged by the Dutch and the English. As early as 1605, the Dutch had taken over the Spice Islands (Indonesia); Hormuz was lost to Persians with the help of the English in 1622, and in 1637, the famous fort of Elmina (Accra, Ghana), a symbol of Portuguese dominance in Africa since the 1480s, fell to the Dutch.[49]

The loss of Elmina opened the door to further European incursions. In 1641, the Dutch also attacked and took possession (for seven years) of the sugar-producing island of São Tomé, while simultaneously waging war against the Portuguese in Angola. The most significant and long-lasting Dutch foray into Africa commenced in 1652, when the Dutch East India Company established a military-commercial base at the Cape of Good Hope. While certainly the main rivals of the Portuguese in sub-Saharan Africa, the Dutch were not the only Europeans to establish new footholds on the continent. By the early 1660s, the Danes and Swedes had established several trading posts on the coast of what is present-day Ghana, the English had constructed Fort Saint James in Gambia, and the French had also created their first fortified enclave at the mouth of the Senegal River.

Although Portugal continued to maintain a grip on the cities of Luanda and Benguela and areas of Sofala (in Mozambique) during this era, by mid-century, sub-Saharan Africa had become much more accessible to European traders from other countries. If the actual number of seventeenth-century voyages from France and England to the region was small by eighteenth-century standards, the effect in terms of the diffusion of knowledge was not. Not only were the illiterate populations of port cities like Liverpool and Nantes now hearing first-hand accounts of idolatrous peoples and African "cannibals," but European elites in Paris, London, Rome, and Amsterdam were reading a growing amount of European travel literature related to Africa and its inhabitants.

What people wrote (and read) about the African continent and its peoples during this era reflected both continuity and rupture with earlier ideas and myths. On the one hand, authors who continued to look to antiquity or the Renaissance for their information on Africa affirmed that there was a litany of things "new," monstrous, or unexpected on the continent. Along the same lines, many geographically oriented authors, such as Pierre d'Avity in *Les Estats, empires, royaumes, et principautez du monde* (1625) and Gerhard Mercator in his *Atlas or Geographical Description of the Regions, Countries and Kingdoms of the World* (1636), persisted in describing the mores and lands of the mythical Ethiopian king Prester John two centuries after he was first "discovered."[50] On the other hand, the most fanciful views of Africa, including the range of monstrous African races that had been passed down from Pliny the Elder to medieval bestiaries, began to disappear from most writings about the continent.

This desire for a more rational view of Africa (not to mention the rest of the world) became a cause célèbre for authors like Richard Hakluyt, who argued compellingly that "history should rely not on descriptions of marvelous and fabulous origins but on authenticated records, acts, and monuments."[51]

This rational view of natural phenomena was increasingly reflected in the works of botanically and zoologically oriented Africanists who wrote "from" Africa itself. Although the magical diversity of African life continued to fascinate European audiences, the simple fact that authors were increasingly subjecting the continent's flora and fauna to naturalistic methods of scrutiny demystified much of what had been perfectly inexplicable a few years earlier.[52] In certain cases, this tendency to reconcile African "data" with European "knowledge" was pushed to an extreme. Edward Tyson asserted in his *Orang-outang, sive Homo Sylvestris, or, The Anatomy of a Pygmie* (1699), for example, that the monstrous races of Cynocephali, Satyrs, and Sphinges alluded to by the ancients were nothing more than "Apes or Monkeys, not *Men*, as they have been represented." In a similar effort to demystify the continent, Tyson also mistakenly proclaimed that the Pygmies evoked by Homer, Strabo, and Pliny were simply chimpanzees.[53]

The trend toward denying Africa a preternatural status was also furthered by the progressive commoditization of the continent's animals. Although Western authors continued to describe the West African manatee as a "fish woman," the terminology of wonder increasingly faded from other descriptions of the continent's beasts, particularly those, like elephants, that had now become articles of trade on the international market.[54] To a certain degree, the same might be said of the at-one-time supposedly "monstrous" Africans themselves. Once they had become European commodities, *pièces d'Inde*, they too had been completely naturalized, if not completely normalized.

While a number of European nations were involved in the ongoing rationalization of Africa by 1600, it was the Dutch—whose commanding position in the era's publishing industry overlapped with their interest in Africa—who led the charge to explain and decipher the African continent. As Ernst van den Boogaart's detailed research on this subject has demonstrated, Dutch writers seized on the European reading public's fascination with the "massive idolatry" outside of Christianity during the eighteenth century and produced the vast majority of "comparative studies of [African and European] beliefs and customs."[55] Approximately fifty titles on the subject of Africa, two dozen of which

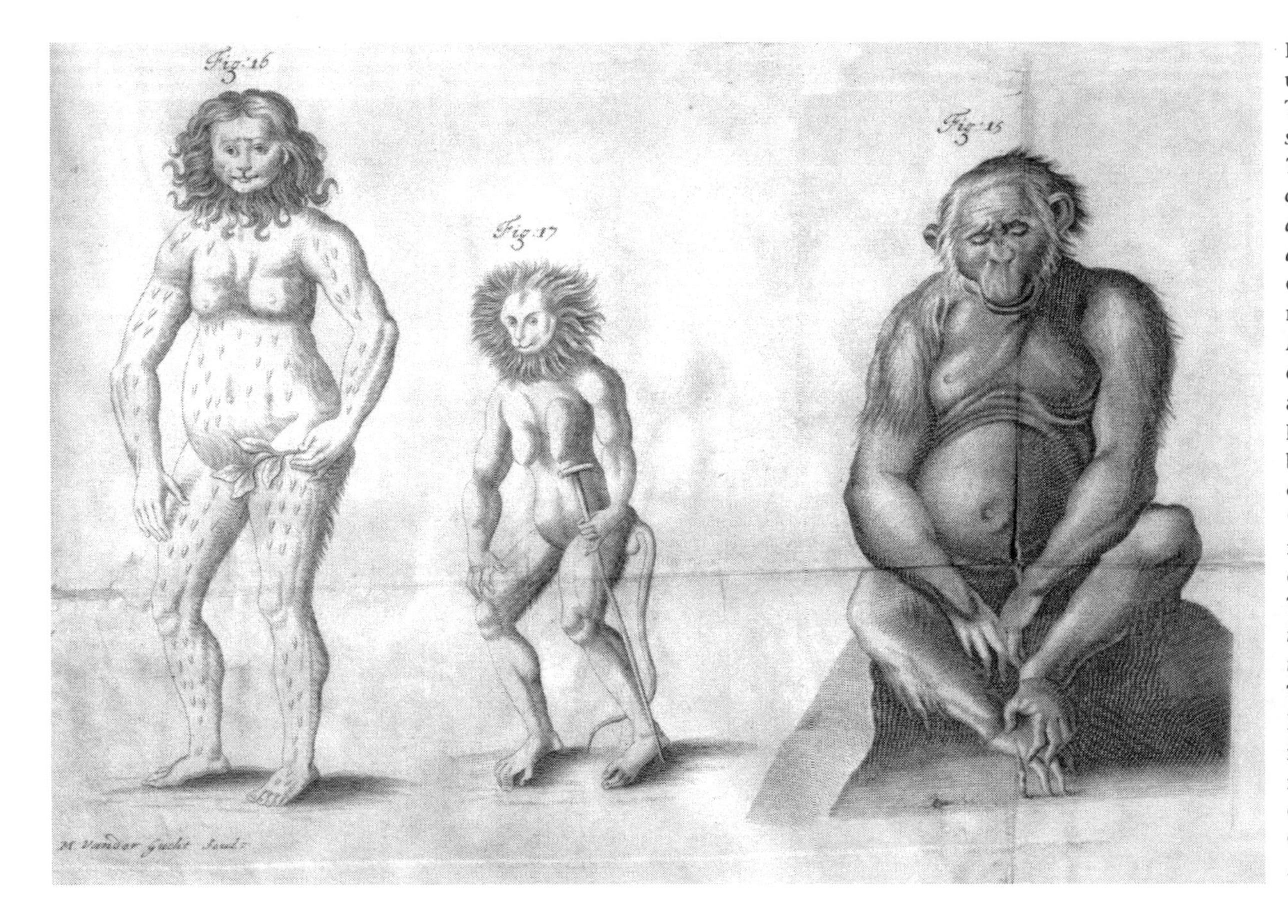

Pygmies and an orangutan depicted in Edward Tyson, *Orang-outang, sive, Homo sylvestris, or, The Anatomy of a Pygmie compared with that of a Monkey, an Ape, and a Man* (1699). Tyson claimed that the "Pygmies" described during Antiquity were actually orangutans (at the time, a generic term for any large primate). Tyson's book depicts several different visions of a "Pongo pygmaeus," the most famous of them copied from Nicholaus Tulipus's 1658 *Historiae naturalis* (drawing on *right,* labeled *fig. 15*). Tulipus's seated female Pongo—she hides her nakedness—was often reproduced during the eighteenth century. Courtesy New York Academy of Medicine.

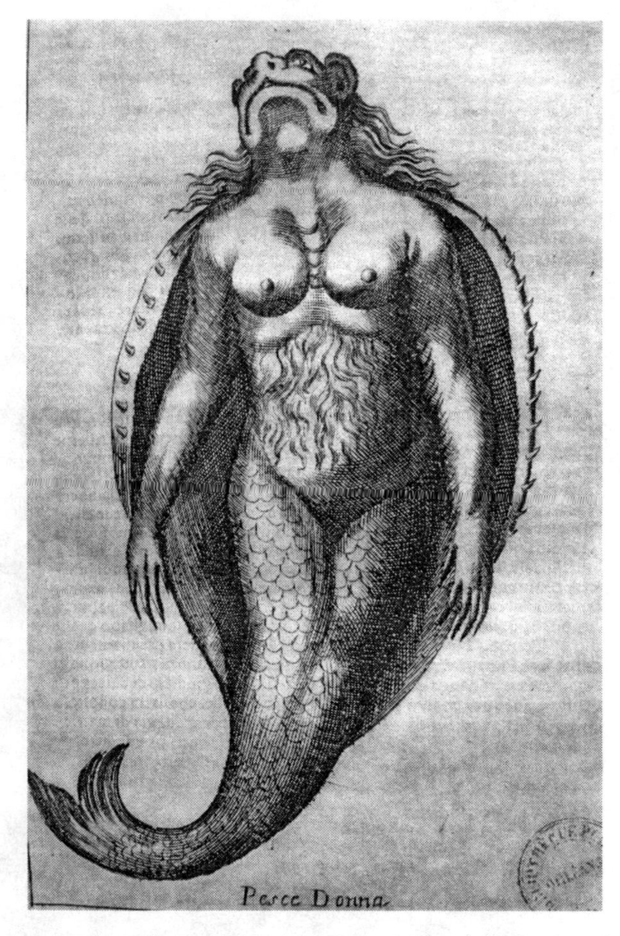

Fish woman depicted in Giovanni Antonio Cavazzi, *Relation historique de l'Ethiopie occidentale* (1732), a translation by Jean-Baptiste Labat of a work originally published in Italian in 1687. Although representations of Africa were increasingly naturalistic, this outlandish rendering of a manatee (*Trichechus senegalensis*) shows the persistent European tendency to depict African wildlife as fantastical. Courtesy Bibliothèque d'Orléans.

were dedicated solely to the "black" portion of this continent, appeared during the century. The most important of these was Olfert Dapper's monumental *Naukeurige Beschrijvingen der Afrikaensche gewesten* (Exact Descriptions of the African Lands) (1668), which was quickly translated into German (1670–71), English (1670), and eventually into French (1686). This was a watershed in the European representation of Africa. In contrast with previous publications on

Africa, Dapper's *Naukeurige Beschrijvingen* provided a synthetic understanding of Africa drawn from a huge body of writings on the continent.[56]

Assessing Dapper's legacy is not an easy undertaking. On the one hand, his text understandably contains a number of negative views reflective of his era. This is particularly true when, much like Leo Africanus, the author had access to limited textual sources. Dapper's short assessment of the Senegalese, for example, is unrelentingly pessimistic. Likewise, his portrayal of the Mandinkas is as reductive as it is unenthusiastic, although he did affirm that they were supposedly an attractive people: "These *nègres* have the reputation for being the best looking in all of *Guinée*, but they are treacherous and barbaric."[57]

Despite the shortcomings of Dapper's text, it is generally agreed that he may be the first Africanist worthy of the name.[58] Not only did Dapper present an exhaustive survey of existing knowledge regarding the continent in the mid-1600s; he proceeded methodically and often non-judgmentally for

African boat building depicted in Olfert Dapper's *Description de l'Afrique* (1686). Courtesy Bibliothèque d'Orléans.

each ethnicity he discussed, touching on subjects such as agriculture, hunting methods, and the intellectual and moral "nature" of the inhabitants, as well as birth, death, and marriage rituals, customs, government, and religion. Indeed, Dapper's treatment of many African ethnicities seems to speak to a desire to go beyond simple cultural indictments: to explicate and to understand difference on its own terms. This moderately relativistic stance, which, during Dapper's era, was more commonly reserved for Arabs, Asians, or Amerindians, often comes to the fore in sections for which the geographer had access to a wide range of documents provided by Dutch merchants.[59] In such cases, Dapper's treatment of African beliefs is more often descriptive than judgmental. While his extended assessment of the kingdom of Luongo, for example, characterizes the inhabitants' religion as both "ridiculous" and composed of remarkable "superstitions," Dapper nonetheless interpreted African talismans in the framework of African beliefs.[60] A similar ambiguity is found in Dapper's treatment of the much-maligned Hottentots (Khoikhoi). Although he certainly depicts them as more savage than noble, he nevertheless emphasizes their decency and loyalty, comparing them favorably in this respect to the greed of the Christians who were increasingly occupying their lands.[61]

The Birth of the Caribbean African

As a narrator or a literary voice, Dapper is simultaneously observant and deliberate. Among his many qualities, it was perhaps his detachment that prompted an impressive number of Enlightenment-era thinkers, including Diderot, to cite him as an authority. And yet the seemingly neutral and careful tone of the *Naukeurige Beschrijvingen* belies the seventeenth-century geopolitical context that had generated the Dutch travelogues that are at the heart of his work.[62] Indeed, behind this relatively detached study of proto-ethnography lay not only the numerous outposts, settlements, and factories in Africa of the Geoctroyeerde West-Indische Compagnie (Chartered West India Company) or WIC—which included (or had included) Axim, Gorée Island, Elmina, São Tomé, São Paulo de Luanda, and Benguela—, but also the Dutch East India Company (Vereenigde Oost-Indische Compagnie, or VOC) establishment at the Cape of Good Hope (Caap de Goede Hoop). Tied, as many of these outposts were, to Dutch interests in the New World, Dapper's Africa was thus the product of the dominance of the Netherlands in the Atlantic economy before the 1650s.

By the time Dapper's *Naukeurige Beschrijvingen* had appeared in print in the late 1660s, however, the preeminence of the Dutch and the WIC was increasingly being challenged on both sides of the Atlantic. In addition to the fact that the Portuguese had ousted the Dutch from Luanda in Angola in 1648, the WIC's primary investment in the New World, its colony at Mauritsstad (Recife) in Brazil, had also fallen back into Portuguese hands in 1654.[63] Losing Recife was not the most important setback; nor was ceding the colony of New Netherland (New York) to the British in 1664. Rather, the Dutch Republic's most substantial loss in the seventeenth century came about slowly and imperceptibly. During the thirty-three years that the Dutch Republic (and the WIC) dedicated significant amounts of capital, time, and effort to their ill-fated Brazilian enterprise, the Dutch missed out on one of the most profitable land-grabs of the seventeenth century. As Pieter Emmer writes, "While the Dutch were keeping the Iberians at bay in the South Atlantic, France and England took advantage of Dutch preoccupation to conquer a sizable part of the Caribbean," effectively dividing up the majority of the large islands.[64]

The English were the first to annex what were ostensibly Spanish territories in the Caribbean, securing Montserrat (1625), Saint Kitts (1728), Antigua (1632), Barbados (1627), and Jamaica (1655). During this same general era, the French also began settling their colonies in Martinique (1626), Guadeloupe (1635), and, finally, Saint-Domingue (1697). While the Dutch did establish colonies in Aruba, Bonaire, and Curaçao (1634), and effectively controlled the vast majority of trade in the Caribbean region until the 1660s, the overall geopolitical realignment of this region during this era actually locked the Dutch merchants and Republic out of what would become the most profitable colonies in the world during the late seventeenth and eighteenth centuries.[65]

The first major success story in the Caribbean was the English island of Barbados. During the third quarter of the seventeenth century, this English territory underwent a spectacular transformation. Once an impoverished isle known for its foul-smelling tobacco, Barbados became the envy of all the colonial powers thanks to a new monoculture, sugarcane. Sugar utterly transformed life on the island; it also revolutionized Europe's view of the potential profits to be reaped in the Caribbean basin. As James Walvin has summarized, the sugar and the rum of Barbados

> transformed tastes in Britain and created wealth "beyond the dreams of avarice" for a fortunate few. By the mid-seventeenth century, thanks to Barbados, slave-grown sugar had entered the popular and mercantile imagination as the source

> of legendary wealth: a cornucopia which disgorged prosperity to all concerned (except for the slaves of course).[66]

Several factors assisted in the conversion of Barbados to its highly profitable monoculture. In addition to the fact the English government was committed to investing in and settling its Caribbean territories—many of its "financial backers were MPs"—the market for sugar itself had also changed dramatically by the 1650s.[67] War in Brazil between the Portuguese and the Dutch had severely disrupted Europe's supply of sugar, making the creation of new plantation complexes much more worthwhile. Always the entrepreneurs, the now-displaced Dutch played a crucial role in this transformation. In addition to financially backing the initial settlement by smallholders on Barbados,[68] Dutch merchants and traders, many of whom were Jewish and new Christian refugees from Brazil, provided critical equipment, expertise, and food supplies to early English settlements and plantations.[69] The final critical element in this success story was obviously slaves: a commodity that Dutch slave traders who had lost their market in Recife were more than happy to supply.

African chattel slavery would, of course, become the defining character of islands such as Barbados. During the early years of the colony, thousands of white indentured servants and prisoners were sent to the island, but this changed dramatically over time. In 1630, there were fewer than a thousand black slaves living on Barbados;[70] by contrast, in 1700, some 170,900 Africans had been shipped to the island.[71] Barbados's conversion to sugarcane and African slave labor was soon reproduced elsewhere in the Caribbean with varying degrees of success. In the British Caribbean, this narrative would play out most fruitfully and dramatically in Jamaica, whose sugar exports alone accrued profits of £800,000 annually by 1770.[72] Not surprisingly, the conversion to this monoculture necessitated an enormous slave labor population; Jamaica imported over one million slaves between 1650 and 1800.[73]

The French were slower to establish themselves as major exporters of sugar. While improved methods of sugarcane production had arrived on both Guadeloupe and Martinique by the 1660s, the French government's 1664 decision to exclude Dutch ships—and their ability to deliver cheap African labor—from French islands initially put French planters in an unfavorable position in comparison to their English competitors. By the mid eighteenth century, however, the French had more than caught up with the English. Between 1676 and 1725, for example, only 49,300 slaves arrived in Saint-Domingue; between 1750 and 1800, however, approximately 593,300 Africans were imported to the island.[74]

The slavery and slave systems that came to characterize the Caribbean did not start with the English and French, of course; various forms of chattel slavery and indentured servitude had long existed in this region. Columbus, who had presumably participated in the purchase of slaves when he sailed to the Elmina region of what is now Ghana in the 1480s, unequivocally advocated the enslavement of the indigenous Arawaks and Caribs of Hispaniola and Cuba.[75] Forced labor was not limited to indigenous peoples, however. In 1504, Spain also began sending white convict slaves to Puerto Rico and, several years later, the colonial authorities even deported Spanish women criminals to the island to improve morale and quell revolts among the enslaved men.[76] During this same general era, Hispaniola also became the first Caribbean island to receive a cargo of black African slaves.

French and English involvement in the forced migration of Africans to the Caribbean came decades later, well after the Spanish had abandoned many of their Caribbean possessions. The English, who would ultimately carry more slaves than any other nation to the New World, began sporadically participating in the trade in the 1560s after John Hawkins, who flagrantly disobeyed English laws regarding the taking and selling of human cargo, exchanged a cargo of slaves for pearls and silver on Hispaniola. Historians generally agree that the French probably began participating in the trade on an infrequent basis around the same time, although the first documented French slave expedition was that of the *L'Espérance* in 1594.[77] For France and England, however, slave expeditions remained irregular until the 1670s.[78] The same might be said regarding English and French encounters with—and representations of—Africans; these, too, were intermittent until the advent of their respective Caribbean colonies.[79] This new chapter in the European colonial venture would not only change the number and nature of African-European encounters; it would privilege a new type of Africanist discourse derived primarily from the Caribbean context.

Among the variables that influenced the description of the African in French eighteenth-century thought, none is as important as the specific geographical context in which the so-called *nègre* was evoked, be it in Africa or on a Caribbean sugar island. If a work such as Dapper's *Naukeurige Beschrijvingen* was generally constructed against a backdrop of African autonomy (in West Africa), this was hardly the case for Caribbean texts, which sought both to explain and, sometimes, to justify the human bondage of the *nègre* in the New World. The resulting overall ethnographic portrait of the black African that European readers accessed was thus a curious hybrid of differing evaluative criteria. Properly Africa-oriented texts dedicated most of their pages to African morphology, occupations, institutions, tools and instruments, pastimes, "religions," trade practices, and sexual mores; although Caribbean writers, too, wrote about many of these "traits," they were also more generally interested in the utility of particular ethnicities, and how to get the most work out of their labor force.

This increasingly utilitarian view of the African was not produced overnight; nor was it invented by one particular constituency of Europeans. Rather, this revised understanding of the "African of the Caribbean" would come from a variety of sources, among them mercantilist treatments of the *nègre*. One of the most influential and widely translated examples of this type of text is Jacques Savary's *Parfait negociant* (1675). A compilation of reports that Savary, an eminent authority on trade, had presented to a French commission in the early 1670s, this book judiciously codified the practices associated with the exchange of money, services, and goods, among them black African slaves.

Like many other advocates of international trade, Savary began his discussion of what he deemed to be the legal and logical business of human bondage by putting forward a *stipulative* definition of the *nègre*. By discussing this variety of human solely within the context of existing trade practices, Savary forced a more restricted understanding of the *nègre* on his reader, one that categorically associated the black African with an assortment of other African goods: "In addition to the previously mentioned merchandise . . . , *nègres* purchased in Africa on the coasts of Guinée are imported: this commerce is considerably more profitable [than others], since the French Islands cannot do without *nègres* to work in the production of sugar and tobacco and at other tasks."[80] While Savary's overdetermined redefinition of the *nègre* in this passage may seem logical given the emerging priorities of France's Caribbean colonies in the 1670s, this strict equivalence between *nègre* and slavery nonetheless signaled a change in the way in which the term was being used in other contexts. In properly African travelogues, for example, the word *nègre*, while often associated with a series of "vices," had functioned more generally as a geo-ethnic marker. In other words, under other circumstances and in an African context, the term *nègre* remained polysemous and as such could connote king, cannibal, warrior, or slave, whereas, in Savary, the term had become monological, reflecting the black African's specific ethno-political status as the highest form of commodity, one that produced profit for European *armateurs* and planters alike.

Curiously enough, it is only after having positioned the *nègre* as a de facto slave that Savary ultimately discussed the justification and the process of enslavement. Echoing hackneyed views that had circulated since the fifteenth century, Savary began by citing a common religious rationale: "This trade may appear inhuman to those who do not know that these poor people are idolatrous or Mahometans."[81] This Christian-inspired justification, which implicitly condemned Africans' belief systems or religions as spurious, was then coupled with a series of pragmatic arguments ostensibly in the *nègres'* own interest:

> Christian merchants, in buying *nègres* from their enemies, take them away from a cruel slavery and allow them to achieve a more gentle servitude on the islands to which they are taken; moreover, [this process provides them] with knowledge of the real God and a path to salvation, [which they obtain] through religious teachings dispensed by priests and religious men who take great care to make them into good Christians"[82]

According to Savary, those Africans who had been rescued from idolatry, Islam, and/or despotic rule in Africa benefited from a better life in European colonies; in addition to avoiding the brutal slavery that they might endure in Africa, their safe passage overseas theoretically had the advantage of guaranteeing them redemption.

Not surprisingly, Savary's assertion that the *nègre* somehow fared better (both physically and metaphysically) in the Caribbean contrasted markedly with other views of life flowing from the colonies themselves. This is particularly true of those Caribbean texts that underscore the supposed ability of the African to adapt to life as a slave. As the Jesuit priest Pierre-François de Charlevoix made clear in his 1730–34 *Histoire de l'isle Espagnole ou de Saint-Domingue,*[83] the reason for the relative "contentment" of the Caribbean slave was not the supposedly improved life he led in the colonies, but his uncanny ability to endure almost unimaginable amounts of suffering and torture:[84]

> Indeed, in addition to the fact that a *nègre* performs as much labor as six Indians, he becomes more easily accustomed to slavery, [a condition] for which he seems to have been born; he does not easily become upset; he is happy with very few

> things with which to live, and he never loses his strength and vigor even if he is poorly nourished. He naturally has a certain amount of pride, but all it takes to dominate him is to show him more [pride], and make him feel with a few lashes that he has masters. What is surprising is that punishment, even when sometimes pushed to [the point of] cruelty, does not cause him to lose his good attitude, and he generally does not hold much of a grudge.[85]

Charlevoix's sanguine assessment of the almost *innate* capacity of Africans to suffer speaks to the self-justificatory logic of chattel slavery during this era. Although his *Histoire de l'isle* was written before the notion of race had crystalized in European thought, the priest's view of the *nègre* nonetheless took an important step in associating this group of humans with a particular corporeal logic, a logic that destined them to feel "the entire weight of servitude."[86]

Much like Savary's assessment, the majority of Charlevoix's presentation of the *nègre* is reductive and monolithic; this type of human, implied the priest, could be theorized, and explained as the product of observable and predicable behaviors. And yet, while Charlevoix put forward an essentialist portrait of the black African, he, like many other writers, also diffused a more "empirically grounded" breakdown of ethnographic knowledge that evaluated various ethnicities against the needs of white planters. The result, in stark contrast to his overall unbending view of the African, is one of the most complete morally based inventories of Africans provided during the era:

> Of all the *nègres,* these Senegalese are built the best, the easiest to discipline, and the most suitable to domestic service. The Bambaras are the biggest, but they are thieves; the Aradas are those who best understand agriculture, but are the most proud; the Congos, the smallest, and the best fishermen, but they run off easily; the Nagos, the most human; the Mondongos, the most cruel; the Mines, the most determined, the most capricious, and the most likely to despair. Finally, *creole nègres,* whatever their origin, inherit only their color and slavish mentality from their fathers. They have more of a love for freedom, however, though they are born into slavery; they are also more intelligent, more reasonable, more skillful, but more lazy, more arrogant, more dissolute than the Dandas, which is the common name for all those who came from Africa.[87]

Although authors of earlier travelogues had mentioned the preferences that plantation owners had expressed for certain types of Africans, Charlevoix's assessment of African types reflected a new era in the interpretation of ethnographic data, an era where the colonial space was increasingly and effectively used as a laboratory within which the "liabilities" or "advantages" of given African ethnicities could be observed, categorized, and subsequently given a relative value. While such a stance may seem to invalidate the justificatory power of more global views of the *nègre*, this labor-based breakdown of Africans into categories of ethnic utility actually had the opposite effect; it further buttressed—through circular reasoning—the ultimate rationality of slavery itself. Here, after all, was a category of human that was literally defined according to its ability to serve.[88]

Parallel to the era's mercantile, "historical," or planter-based views of slaves and slavery, English and French governments (or colonial governments) increasingly legislated specific rules governing the keeping of African slaves in the West Indies during the second half of the seventeenth century. Such edicts had the effect of codifying and normalizing a series of white beliefs and behaviors toward the Africans who worked for them in European colonies. One of the first sets of these laws was issued in 1661 by the acting governor of Barbados, Humphrey Walrond. This was the Barbados Act, which was ostensibly promulgated on the island in order to standardize policies concerning the management of the island's growing slave population. Describing Negro slaves as "an heathenish brutish and . . . uncertain dangerous pride of people," the Barbados Act reflected the growing gap between slaves and freemen, as well as the mounting perception that Africans had a different conceptual status from Europeans.[89] Comprised of a matter-of-fact list of twenty-three clauses, the act enumerated various undesirable behaviors along with detailed and sometimes incremental corporeal penalties that were to be expeditiously carried out without the "legal trial of twelve Men."[90] A slave found guilty of violence to "any Christian," for example, was on first offense, to be "severely" whipped; the Negro who committed a second offense, in addition to a subsequent rigorous whipping, was also to have his "nose slit" and was to be "burned in face."[91] A third offense merited an even greater, unnamed corporeal punishment to be determined by the governor. Like many such documents, the Barbados Act treads a fine line between its stated desire to protect slaves from "[a]rbitrary, cruel, and outrageous wills of every evil disposed person" and its tangible goal of controlling and disciplining a slave population "prone to escape and revolt."[92]

A little more than two decades after the Barbados Act was read in all the parishes of Barbados, Louis XIV's chief minister, Jean-Baptiste Colbert, formulated the most comprehensive set of regulations concerning slave matters during the era, the *Code noir* (1685). This now-infamous document's objectives went beyond the simple list of punishments described in the Barbados Act. The goal of this document, as stated explicitly in the *préambule*, was to maintain "the discipline of the Catholic, Apostolic, and Roman Church on our islands in America, to regulate state issues there and to ensure the quality of the slaves on said islands."[93]

Issued only a few months before the revocation of the Edict of Nantes—the culmination of Louis XIV's campaign of intolerance against French Protestants—the *Code noir* addressed geo-religious preoccupations as much as it did more material concerns regarding the treatment of slaves in French colonies.[94] Not only did its first article effectively expel Jews from the islands; it also sought to eradicate Protestant elements from positions of authority, particularly in situations where they had control over slaves. This clause was followed by another important proviso. In stark contrast to what was occurring in the British West Indies, but close to earlier Iberian practice, the Sun King proclaimed that slaves of French landholders had to be baptized as soon as they arrived on the islands.[95]

To these religious principles were adjoined a series of more slavery-related directives. In addition to those articles written with the benefit of the slave in mind—Article 22 required minimum food and clothing allotments to slaves; Article 47 forbade the breakup of slave families; Article 59 conferred, in theory, the same rights on freedmen (former slaves) as the king's other subjects—much of the *Code noir* codified hierarchical racial relations between a society of masters and a society of slaves. Strict sexual codes were put into place: masters were forbidden sexual congress with their slaves under penalty of large fines and confiscation of the slave involved (along with any offspring). Colbert also addressed the potentially problematic presence of interracial offspring in a slave society. The solution was quite ingenious given the reality of life on the islands: following the Roman precept *partus sequitur ventrem* ("the issue follows the womb"), the *Code* made slave descent matrilineal, or uterine, that is, the offspring of a male slave and a free woman were free, but those of a female slave and a free man were considered slaves.[96] While such rules were blatantly disregarded in French colonies—a class of mixed race and non-slave children became one of the noteworthy characteristics of the French Caribbean—this gendered logic was designed to further secure the colonial enterprise by separating the categories of white and black, freeman and slave.

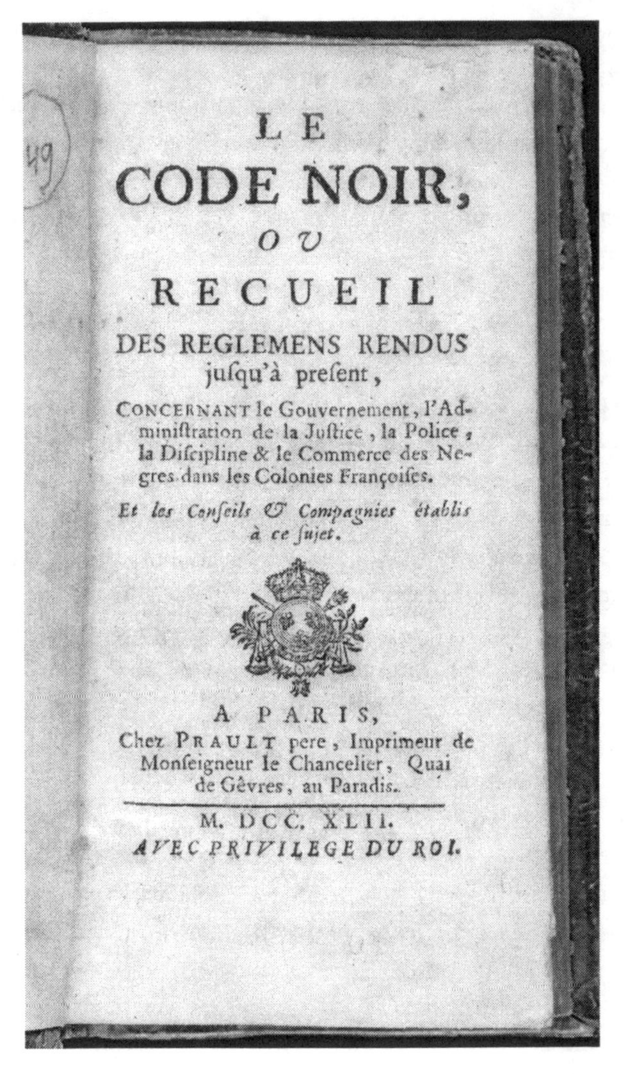
LE
CODE NOIR,
OU
RECUEIL
DES REGLEMENS RENDUS
jusqu'à present,
CONCERNANT le Gouvernement, l'Administration de la Justice, la Police, la Discipline & le Commerce des Negres dans les Colonies Françoises.
Et les Conseils & Compagnies établis à ce sujet.
A PARIS,
Chez PRAULT pere, Imprimeur de Monseigneur le Chancelier, Quai de Gêvres, au Paradis.
M. DCC. XLII.
AVEC PRIVILEGE DU ROI.

Title page of a 1742 edition of the *Code noir* (1685). Courtesy Beinecke Rare Book and Manuscript Library, Yale University.

Finally, and not unexpectedly, the *Code* enacted a series of controls and punishments. Slaves could never carry weapons; they could not gather at night; if they attempted to escape, they were to have their ears cut off, be branded, or be put to death. Attacks on masters were theoretically punished even more severely than in English islands. Any slave who struck his or her master was to be executed.[97] While the *Code*'s documentation of the day-to-day functioning of slave societies does not describe the Africans at the heart of French Caribbean agriculture, the sanctions enumerated in the document testify, by way of inference, to the resistance of African slaves. The same can be said for similar legislation in all slave societies. Whether written for Jamaica, Louisiana, or, most infamously, for the virtual penal colony that was the Danish West Indies, slave laws reflected colonists' fears as much as the real need to manage African populations.

Jean-Baptiste Labat

While slave laws such as the *Code noir* offer valuable insight into how colonial islands institutionalized both whiteness and blackness, individual writers' texts provide more complex and complete assessments of European conceptions of black Africans. The Dominican Jean-Baptiste Labat's best-selling *Nouveau voyage aux isles de l'Amérique* (1722), for example, not only provides a day-to-day account of the routine of the slaves working on Labat's struggling sugar plantation on Martinique; it also describes the perceived mentality of the slaves that he had come to know on the island, as well as what was regarded as the most "suitable" mind-set for a plantation master.[98]

Labat's reactions to the reality of slavery evolved considerably during his time in the Caribbean. According to his own chronicle, he first came into contact with slavery before he even set foot on Martinique. While moored in Saint Pierre harbor, the Dominican priest had the opportunity to contemplate slaves who had come aboard, presumably to unload cargo: "Many *nègres* came aboard; they were wearing only simple cloth undergarments; some had a bonnet or a nasty hat. Many carried the marks of whippings they had received on their backs; this initially provoked compassion among those who were not yet accustomed to this, but one gets used to it."[99]

Labat's description of how "one" became inured to African suffering—he was talking obliquely about himself—speaks volumes about the era's ability to rationalize the plight of a different category of human. This is borne out in the rest of the text. Managing, purchasing, and meting out punishment on his mission's plantation, Labat shows himself to be a clear advocate of progress, the colonial mission, and the *Code noir*. While there are moments in the *Voyage* where the priest seems to have been genuinely concerned with the souls of his slaves—he demonstrates palpable concern for two suicidal Mina slaves, as well as for a slave boy bitten by a snake—Labat is absolutely pitiless in other

Frontispiece in Jean-Baptiste Labat's *Nouveau voyage aux isles de l'Amérique* (1742), with portrait of the late author. The poem translates as: "A writer curious about lands and *moeurs*; he adorns his books with the graces of his style; corrects men's errors while amusing, and knows how to mix the pleasant and the useful throughout." The holder of the portrait of Père Labat is a clearly an African slave, but his feathered skirt evokes the indigenous Amerindians. Courtesy Wesleyan University Library, Special Collections and Archives.

situations. A ceremony involving an African talisman, for example, incited a terrible punishment from the Dominican:

> I had the witch tied up and ordered him to receive approximately three hundred lashes of the whip, which flayed him from the shoulders to the knees. He cried in desperation and our *nègres* begged for mercy for him, but I told them that witches do not feel any pain and that his cries were only meant to mock me. . . . Later, I had the witch doctor put in irons after having him washed with a *pimentade,* which is to say a brine into which we crushed chilis and small lemons. This causes horrible pain in the person who has been whipped, but it is a proven remedy against gangrene, which would otherwise find its way into the wounds.[100]

As gruesome as this incident is, perhaps as significant is the fact that it is recounted without a trace of regret.

Labat's first-hand accounts of his encounters with Caribbean slaves overlap with his other, more sweeping, representations of the destiny of Africans in the French islands. In the chapter dedicated to the *nègres* that the French "use" in the colonies, Labat supplied a multi-level examination of human bondage in the French Caribbean that begins with the often-related story of Louis XIII's initial reluctance to allow slavery in the colonies. According to Labat, slavery only became possible after the pious king realized that this was the only "unfailing way. . . to inspire the Africans to worship the true God, to save them from idolatry, and to make them embrace and continue in the Christian faith until death."[101] Labat followed up on this anecdote with a series of more practical observations. Writing from the perspective of the planter he was, he bemoaned the failure of the major French trade companies to supply enough slaves, and tacitly sanctioned the resulting clandestine traffic that islanders engaged in with the English, Dutch, and Danish. More significantly, Labat also shared how he derived the most edifying behavior from what he thought of as the African slave mentality. Although he described the Africans who came to the islands as initially "idolatrous" and prone to all kinds of sorcery, magic, and dealings in "poison," he also suggested that, after indoctrination into the Christian faith, the African was eminently tractable.[102] Indeed, as Charlevoix would write eight years later, the ethnography of the African slave seemingly corresponded perfectly to a life of servitude, a life that had the added benefit of being closer to God.

Unsurprisingly, Labat's recommendations reflected his belief that European Christian paternalism suited the slave's supposedly childlike mind-set. "All *nègres* have great respect for the aged," he pronounced, and he counseled his fellow Caribbean masters to position themselves as benevolent but firm patriarchs.[103] This process, he suggested, should begin upon the slaves' arrival; during the seasoning process, overseers were told to make life as pleasant as possible for the slave and not fall prey to the "avarice" and "harshness" of planters who put their Africans to work immediately. Based on his own experience, Labat affirmed that "good treatment," "clothes," and any other "tenderness that we show them, render them affectionate and make them forget their country, and the unfortunate state to which their servitude reduces them."[104] The gratitude of the slave could be boundless according to Labat: "Should one do good by them, and in good grace, they love their master infinitely, and will do anything, even at their own peril, to save his life."[105] This pragmatic and theologically justified system of slavery was far superior, according to Labat, to what he had witnessed on English Barbados, where the slaves were "poorly nourished," "treated worse than horses," and where the English neither instructed nor baptized their slaves. In contrast, African slaves on Martinique were, in the Dominican's opinion, much closer to paradise.[106]

Labat's representation of black Africans and slave ethnology is a conflation of slavers' anecdotes, information derived presumably from other Caribbeanist texts, and his own experience. Absorbed into his compelling first-person chronicle, his *Nouveau voyage* effectively initiated eighteenth-century French readers into the complexities of human bondage. Moving, as it does, from Labat's comparatively naïve understanding of colonial practices in the early chapters to a time when he presided over an expanding plantation of, on the whole, purportedly appreciative slaves, this chronicle tells a tale of theologically guaranteed progress. Despite the text's presentation of the undeniable brutality of the slave system, Labat ultimately provided a comforting early-eighteenth-century view of slavery that combined an impassive assessment of the day-to-day economic realities of the plantation system with anecdotal examples of the potentially humane psychology of the slave owner. Perhaps most mollifying for his readership, however, was the fact that the swashbuckling priest supplied a comparative view of slavery that assured the French that their own form of human bondage was both intrinsically and extrinsically superior to that practiced by the English brutes on the islands of Barbados or Jamaica. It was conceivably this opinion that prevented this extremely readable and best-selling book from being translated into English.

Labat on Africa

At the end of Labat's *Nouveau voyage*, the Dominican tells of how he was called back, against his will, to France, where he was ordered to lobby in the interests of his order. Despite his many requests to return to the Caribbean, he never again saw the French West Indies. This was far from the end of Labat's career as an "ethnographer" of the black African, however. In the late 1720s and early 1730s, he published three substantial multi-volume works on West and Central Africa. While he had never set foot on the African continent—as he himself admits in the first sentence of his 1728 *Nouvelle relation de l'Afrique occidentale*—Labat soon became the most prolific "authority" on Africa and its inhabitants during the eighteenth century. Expectedly, both eighteenth-century writers and contemporary critics have upbraided him for the methodological shortcomings of his works on Africa. And yet, the fact that Labat saw no reason to refrain from writing about West Africans is actually one of the most interesting aspects of his Africanist project. As a best-selling expert on these peoples in the colonial context, Labat clearly believed that his own experience gave him the right to be the intermediary between "trustworthy" Africanist sources and the wider public.

Labat announced the scale of this project in the first of his works on Africa, the 1728 *Nouvelle relation de l'Afrique occidentale*. Hinting that he wanted to become something of a French Dapper, Labat promised to provide a survey of the entire continent that would extend well beyond the four volumes on Senegambia he was introducing.[107] This proposed tour of the African continent and its peoples stalled after he had surveyed Senegambia, the Slave Coast, and the Kongo-Angola region, but the three major works that he did publish on the continent offer a veritable typology of the type of Africanist sources circulating in Europe during the eighteenth century. For the *Nouvelle relation*, Labat adapted the memoirs of the directors of the Compagnie royale du Sénégal, in particular those of André de Brüe.[108] For his second Africanist work, the *Voyage du Chevalier Des Marchais* (1730), Labat actually met with, interviewed, and ultimately synthesized the written observations of a French slave trader who had come back from the "Côte des esclaves." And finally, for his *Relation historique de l'Ethiopie occidentale* (1732), which treats the Congo-Angola-Matamba kingdoms, the Dominican drew primarily from a 1687 Italian work written by a Capuchin missionary, Giovanni Antonio Cavazzi.

While one can only speculate as to how Labat himself felt about the "data" that he processed for his works on Africa, it is evident that his basic orientation continued to echo that of his *Nouveau voyage aux isles*. As he had done in his signature travelogue, Labat organized his material into chapters on history, geography, and flora and fauna, as well as on the activities of the always-threatening English. Most notably, he also continued to demonstrate a combination of revulsion and fascination vis-à-vis those accounts of African life that violated a Dominican worldview. This was particularly true in his translation-rendering of Cavazzi's gruesome vision of Africa, a text in which cannibalism, infanticide, and bloodthirsty mercenaries set the tone.

As Labat himself described his relationship to this touchy subject, it was only in his quality as "historian," and not as a religious missionary that he could bring himself to expose the many "defects" of the black Africans portrayed in Cavazzi's text.[109] In addition to the "fact" that the inhabitants of this region were unrepentant thieves living in the "shadows of paganism," the "barbarous climate" in which Africans were born apparently made them so insensitive to the plight of others, according to Labat, that African parents readily abandoned their children to lions or sold their family members into slavery for a necklace or a bottle of wine.[110] The most striking episodes of supposed African inhumanity that Labat recounted came from Cavazzi's lurid account of the Giagues/Jaga, however. According to this Italian missionary, this group of anthropophagic warriors led their lives according to a series of *quixilles*, a code of cruelty that included curious forms of infanticide, most notoriously the crushing of an infant in a mortar in order to make a paste which was then applied to a soldier's skin before battle. It was such a salve that the famous Zingha (Nzinga Mbandi), queen of the Jaga, allegedly applied to herself during her revolt against the Portuguese in the early seventeenth century.

Among the three sources from which Labat drew, Cavazzi's text, more than the two others, confirmed the priest's worst fears regarding the consequences of what he often referred to as a pagan life. Indeed, in many ways, this Capuchin-oriented view of Africa provided the perfect counterpart to the Caribbean project as Labat had described it; clearly, the African excesses of the Kongo supplied an a priori justification for religious indoctrination and the brutality of chattel slavery in the Caribbean. This was not entirely the case for the two other sources that Labat had adapted previously. Despite the fact that both de Brüe and Reynaud Des Marchais's occupations were much more overtly linked to the slave trade itself, their assessments of African reality provided a less sensationalist vision of Africans. Although Labat himself remained true to

Preparing *mangi sami*: anthropophagic Giagues, or Jaga, pound an infant in a mortar to make a paste supposedly applied to soldiers' skin before battle. From the Capuchin Giovanni Antonio Cavazzi's *Istorica descrizione de' tre' regni, Congo, Matamba, et Angola* (1687). Of the three Africanist sources that Jean-Baptiste Labat adapted, Cavazzi's best confirmed his belief in the pitfalls of paganism. Courtesy Bibliothèque d'Orléans.

his own particular orientation as he adapted their ideas—in the *Nouvelle relation,* for example, he reacts violently against African idolatry, digresses on the biblical explanation for black skin, and interjects Caribbean-derived, labor-based typologies of African ethnicities into his text—the overall ethnographic portrait of black Africans in these two texts is much more measured.[111] This was especially true in Labat's adaptation of the experiences of Chevalier Des Marchais, an experienced trader on the Slave Coast. Contrary to the incendiary tone that we find in Labat's translation of Cavazzi, the Dominican priest's adaptation of Des Marchais often gives rise to relatively open, nonjudgmental ethnographic remarks. This can be seen quite clearly in the section devoted to the people of Bouré in Sierra Leone. Praising the local architecture, mores, and most notably, the healthy relationship between the evenhanded local monarch and his people, Labat here paints an uncharacteristically optimistic view of the black African in his homeland. In stark contrast to what he wrote elsewhere, he states that the tall, good-looking inhabitants of Bouré are intelligent and well-mannered, to the point of *politesse.*

While it may seem counterintuitive that among the three sources from which Labat drew to produce his Africanist texts, it was the slave trader Des Marchais who supplied the Dominican with his most "human" understanding of African reality, this is actually quite logical. If individual French slave trader–writers certainly reflected the prejudices of their particular milieus, they often provided the most useful proto-ethnographic assessments of the Africans with whom they came into contact. Catherine Gallagher has best summarized this paradoxical phenomenon in her edition of Aphra Behn's *Oroonoko, or, The Royal Slave* (1668), maintaining that, whatever their personal beliefs regarding the Africans themselves, slave traders often supplied complex views of African customs and rituals. The reasons for this "double vision," Gallagher writes, are self-evident:

> [Slave traders] were intimately involved in African politics on the Ivory, Gold, and Slave Coasts, jockeying for monopolies and making alliances with some nations and leaders against others. For all of these reasons, the trade in Africa forced [them] to observe status and national differences between Africans, pay court to rulers, display marks of respect for the great men of the region, their wives, families, and attendants, and maintain a reputation of trustworthiness. Englishmen who indiscriminately lumped all Africans together as barbarians, showed signs of racial contempt, or behaved treacherously could cause considerable damage to English interests in the region. Hence, even in the works of British slave traders, there is a palpable tension between regarding Africans as just so many potential slaves and responding appropriately to their ethnic and status differences.[112]

Although the ultimate intent of these traders—be they English, French, or Portuguese—was clear, and while their particular ethnographic knowledge allowed them, above all, to profit from the trade in human lives, slave traders' accounts are thus often the era's most complete views of the African. Not only did slavers' texts need to reflect real cultural knowledge, but they necessarily featured *autonomous* Africans functioning within their own systems and institutions, "backward" though they might be. As Labat discovered in adapting

Houses at Cape Mezurade (in present-day Liberia): *A*, House of red clay mortar. *B*, Kitchen. *C*, Storage hut for millet and rice walled with red clay. *D*, Platform for doing business and chatting during the day. *E*, Courtyard. *F*, Public Area. From Jean Baptiste Labat, *Voyage du Chevalier Des Marchais en Guinée, isles voisines, et à Cayenne, fait en 1725, 1726 et 1727* (1730). Labat's adaptation of Des Marchais sometimes offers a more measured view of Africans' lives than his portrayal of plantation slaves on Martinique. Courtesy of Dartmouth College Library.

Des Marchais, this was a very different world from the one that he had known and described on Martinique.

Processing the African Travelogue: Prévost's *Histoire générale des voyages*

Eighteenth-century thinkers interested in the question of the *nègre* had access to a wide range of ethnographic data, including Caribbeanist texts such as Labat's, synthetic Africanist compilations such as Dapper's, stand-alone African travelogues such as Jacques-Joseph Le Maire's, alphabetically organized *géographies*, and natural history treatments such as Buffon's *Histoire naturelle*. The most complete French source of African ethnography, however, was Abbé Prévost's voluminous *Histoire générale des voyages*. Based, originally, on a translation of John Green's four-volume *A New General Collection of Voyages and Travels* (1745–47), Prévost's *Histoire* was commissioned in order to disseminate a broad, international selection of travel literature (much of which was hard to find or disappearing) to a French public for the first time during the eighteenth century.[113] Published initially in a luxury *in quarto* edition (15 vols., 1747–59) filled with sumptuous engravings supplied by Jacques-Nicolas Bellin, the *Histoire* provided a synthetic assessment of African ethnography for writers including Diderot, Voltaire, Helvétius, Rousseau, Raynal, and Buffon, not to mention many participants in the *Encyclopédie* project.[114] Dedicating almost five volumes—some three thousand pages—to Africa itself, Prévost's (and Green's) Africa embodies the Enlightenment era's simultaneously skeptical and speculative relationship vis-à-vis the ethnography of the African.[115]

Prévost's treatment of Africa, as he himself admits in the introduction to the third volume, evolved considerably over time.[116] Although this project was supposed to provide a new "system of modern geography," the abbé recognized that the first tome dedicated to Africa remained reminiscent of the earliest compilations on the subject, filled as it was with a rather uncritical and chronological assemblage of early Portuguese and English voyages.[117] Worse yet, according to Prévost, this early treatment of Africa lacked a standard methodology for the work's so-called *réductions*, the promised syntheses and critical engagement with travelogues on particular geographical areas.[118]

Prévost believed that he (and the English authors he was translating) had addressed many of these problems by the fifth book of the second volume. And, it is true that, at this point, we begin to see the full measure of Prévost's contribution to French *africanisme*. In the first place, Prévost was consciously raising the awareness of his readership by pointing out the inadequacies, factual errors, and diverse orientations that colored an objective assessment of various aspects of African reality. In his introduction to Ca' da Mosto's voyage to the Cape Verde Islands, for example, Prévost finally exposed a long-standing misunderstanding regarding the dates of the voyage itself.[119] Similarly, he openly criticized Labat's adaptation of André de Brüe's recollections of the lands, peoples, and business of Senegambia, taking the Dominican to task for indiscriminately mixing de Brüe's observations with digressions plucked from antiquity or other unnamed sources. At other times, Prévost lauded particular writers over others, saying, for example, "Moore appears to be more accurate in his observations of the Fulani."[120] But it was in bringing together diverse texts that Prévost made the greatest contribution to Africanist discourse. By structuring what were often lengthy, first-person travelogues into multi-author, third-person accounts of a given region or people, the *Histoire* diffused a detailed and useful survey of the peoples and regions of Africa.

The compilation that is the *Histoire générale des voyages* played a critical role in the mid-century genealogy of ideas on Africa. Distilled from multiple perspectives, the *Histoire*'s authoritative synthesis created ethnographic truth where there was often uncertainty or inconsistency. Africans were not, of course, allowed an opportunity to question these European views. Indeed, in the few instances in which the "native informant" is given the floor, it is to confirm the European's superiority over these *naturels*. Such is the case in Prévost's uncritical account of how the inhabitants of Senegal envisioned their relationship to the whites who purchased slaves in the area. Citing this "African" view (reported by George Roberts), Prévost asserted that Africans knew full well that they were inferior to whites, and were thus "destined to be their servants," an idea supposedly confirmed by the fact that whites came every year and bought "thousands of slaves in Guinée."[121]

Recent critics, who have understandably denounced the deleterious effect of negative representations of Africans in the era's thought, have tended to seek out such moments in the *Histoire des voyages*; they have also actively cited more sensationalist accounts of African "ethnography." This is, of course, rich hunting ground. For example, Prévost's amalgamation of several travelogues allowed him to provide "reliable" confirmation of what would become an obsession in eighteenth-century thought, the supposedly furious cannibalism of the Giague/Jaga:

> All of our travelers agree on the anthropophagic traits of the Jagas. Lopes asserts that they feed on human flesh. Battel says that they prefer it to beef and to goats, despite the fact that they have both of these in abundance. Merolla often repeats that they eat men; and . . . does not hesitate to describe the Jagas as having the most barbarous nature in the universe.[122]

Like other writers of his generation, Prévost freely cites the standard inventory of vices commonly attributed to the *nègre*: hypersexuality, dishonesty, and a stunning lack of either common sense or intelligence.[123] And yet, while it is undeniable that Prévost's ostensibly authoritative assessment of African travelogues did sanction some of the worst stereotypes regarding the "barbaric" customs of certain ethnicities or groups, much of the *Histoire* diffused a more complete ethnographic portrait of African ethnicities that flew in the face of more reductive views of the African. In particular, Prévost's assessments of those regions that had been visited by a large number of French and English traders evoked a tableau of African life that allowed European thinkers to envision the people of this continent living freely without the interference of Europeans. These episodic vignettes include, for example, details about a particular livelihood or a practice as intimate as how an infant's name was chosen:

> There are a large number of fishermen in Rufisco and other places along the Senegalese coast. They generally fit three men in an Almadie or a Canot, with two small masts each with its own two sails.[124]

> The infant is given his name one month after his birth during a ceremony where his head is shaved and rubbed with oil in front of five or six witnesses. The most common names are taken from the Mahomedans. Thus the boys are named Omar, Guiah, Dimbi, Maliel, etc., and the girls, Fatima, Alimata, Komba, Komegain, Warsel, Hengay.[125]

While the methodological failings of the *Histoire* are evident to contemporary readers—be it the text's uncritical acceptance of European a prioris or the conflation of recycled accounts into a timeless African ethnography—the publication of Prévost's *Africa* was nonetheless received as the most refined, synthetic, and comprehensive picture of this continent and its peoples available at the time. Enlightenment authors' interpretations of the *Histoire*'s ethnography would, however, often exceed the bounds of Prévost's ethnographic portraits.

Rousseau's *Afrique*

The reaction to Prévost's five-volume assessment of Africa was almost immediate; while Prévost's first volume appeared after Montesquieu had finished his geographically oriented *De L'Esprit des lois* (1748), information from the early volumes of the *Histoire* found its way into Buffon's *Histoire naturelle* (1749) as well as the nascent *Encyclopédie* (1751). But it is Jean-Jacques Rousseau's curious relationship to Prévost's text that typifies how many Enlightenment thinkers engaged with the *Histoire générale des voyages*. In his *Discours sur l'origine et les fondements de l'inégalité parmi les hommes* (1755), Rousseau made very clear that he understood both the advantages and the grave methodological problems of a compilation like the *Histoire des voyages*. Although he, too, drew conclusions based on the ethnography contained therein, Rousseau seemingly sought to beat Prévost at his own game by highlighting the contradictory "facts" found in this supposedly synthetic text. Drawing attention, for example, to Prévost's inability to synthesize the divergent opinions on the African orangutan (by discussing Dapper, Battel, Purchas, and Merolla), Rousseau not only castigated the entire enterprise of travel writing as it existed in its current form, he also indicted his predecessors and contemporaries for their inability to understand foreign lands, fauna, and peoples. Rousseau attributed this last failing to the myopic and solipsistic orientation of the typical European voyager-author: "In the three or four centuries since the inhabitants of Europe have flooded into the rest of the world, ceaselessly publishing new collections of voyages and reports, I am convinced that we have come to know no other men except Europeans."[126]

Not surprisingly, Africa and Africans figured prominently among the subjects that had yet to be understood according to Rousseau. Prescient about what would be affirmed some thirty years later by the Association for Promoting the Discovery of the Interior Parts of Africa—the so-called African Association—Rousseau saw Europe's attempt at comprehending Africa as a dismal failure.[127] To underscore his point, he compared the relative dearth of information on Africa to that generated by explorers of other lands:

> The jeweler Chardin, who traveled like Plato, has said all there is to say about Persia. China appears to have been well observed by the Jesuits. Kaempfer gives us a tolerable idea of the little that he saw in Japan. . . . All of Africa with its numerous inhabitants, as distinctive in character as they are in color, is still to be examined;

> the entire world is covered by nations of which we know only the names—yet we allow ourselves to judge the human race![128]

Taken out of context, Rousseau's indictment has a Voltairian ring to it. Yet, where Voltaire might have left off after this skeptical assertion, Rousseau's appraisal of his century's methodological shortcomings paradoxically co-existed with another entirely speculative view of the African. This arose as part of a larger critique of the century's tendency to send *terre-à-terre* geometers as opposed to philosophes abroad to study foreign cultures. Indeed, in the same paragraph where Rousseau evoked unknown lands, chief among them Africa, he also dreamt of dispatching a true philosophical traveler to such places, a "Montesquieu, Buffon, Diderot, Duclos, d'Alembert, or Condillac," a person whose speculative insight might somehow transcend the lack of information on the continent. According to Rousseau, the advantages of such a venture were manifest. Not only could such a person provide the "natural, moral, and political history of what they had seen," but those Europeans who benefited from such insights into the foreign culture would "see a new world arise from their pens, and . . . we would thus learn to know our own."[129] From Rousseau's perspective, the problem was not a place like Africa itself, but rather, the lack of a suitable occidental and Enlightenment mind in situ able to interpret the seeming paradoxes of such a foreign land.

Rousseau's faith in the capacity of Enlightenment luminaries to "decipher" Africa is more than a banal trait of Africanist discourse; it is a defining feature that reflects a more general conviction; namely, that African behaviors, customs, and culture could only become real knowledge if they were processed by superior European minds. This dynamic is very much at work in the *Second Discours*. If, in this text, Rousseau lamented the fact that Europe had only produced an imperfect and often contradictory assessment of Africa, he also positioned himself as a European capable of making sense of this rough data, in particular his era's reports on the Hottentots (Khoikhoi), supposedly the most backward variety of humans in existence.

In evaluating the Hottentots (and other *sauvages*), Rousseau asserted that these peoples actually compared quite favorably to Europeans; they were stronger, faster, less effeminate, and had better eyesight. This physical appraisal provided the transition for a much more sweeping account of the human species. Positioning the Hottentots as the valorized denizens of an imaginary past, Rousseau recounted his bleak (and famous) chronicle of man's fall from

To the astonishment of his white companions, an acculturated Hottentot doffs his European clothes and returns to the world of his kin in the South African veld. From Jean-Jacques Rousseau, *Discours sur l'origine et les fondements de l'inégalité parmi les hommes* (1755). Courtesy Wesleyan University Library, Special Collections and Archives.

grace. In leaving the state of nature that the Hottentot still inhabited, humanity not only gave up a happier and simpler past; it sentenced itself to a world of painful and deleterious abstractions. To emphasize this point, Rousseau recounted the celebrated story (taken from Peter Kolb) of the Europeanized Hottentot who, upon his return to South Africa, chose to throw away his European clothes and plunge back into the veld. It was this story that was the source for the *Essai*'s celebrated frontispiece.

With this image and in the text to which it refers, Rousseau sought not only to evoke a savage nobility in the Hottentot, but the possibility of living in peace with nature and oneself. For us, however, Rousseau's Hottentot is symptomatic of the Enlightenment era's tendency to go beyond the bounds of the empiricism that it often preached. Indeed, by giving such a prominent place to the Hottentot of his imaginings, Rousseau was indulging in an ethnological *rêverie*. Such bold conjecture about black Africans was increasingly becoming standard practice by 1750. By the time that the *Second discours* appeared in the mid 1750s, anatomists, naturalists, and philosophes had all begun to provide more comprehensive assessments of the *nègre* that sought to transcend the vagaries and contradictions of the travel literature which, until the early eighteenth century, had been the primary locus of interpretation for the black African.

CHAPTER TWO

Sameness and Science, 1730–1750

Holy Scripture and the unvarying tradition of all of the civilized Nations of the Universe teach us that all men came from the same father and the same mother, and that we should see the *Whites* as the branch from which stem all mankind.

ABBÉ DEMANET, *NOUVELLE HISTOIRE DE L'AFRIQUE FRANÇOISE, ENRICHIE DE CARTES ET D'OBSERVATIONS ASTRONOMIQUES ET GÉOGRAPHIQUES* (1767)

Nothing better demonstrates the degeneration of human nature than the diminutiveness of the savage in the enormity of the desert.

FRANÇOIS-RENÉ DE CHATEAUBRIAND, *DU GÉNIE DU CHRISTIANISME* (1803)

As the keeper of the Jardin du Roi, George-Louis Leclerc, comte de Buffon, had one task that was more critical than his other responsibilities: providing an objective inventory of the content of the king's natural history cabinet. This takes up a large portion of the third volume of Buffon's *Histoire naturelle*.[1]

Within this somewhat tedious enumeration of dried, embalmed, and waxed specimens, one finds the description of two preserved African heads. The first was that of a dead black child, the "preserved head of a two- or three-year-old *nègre*." This was, according to Buffon, an unremarkable piece. Reacting clinically to this specimen, as he did to all the objects he was describing, Buffon wrote: "This head was prepared like preceding ones, one could not distinguish it from the head of a white if one did not know otherwise that it comes from a *nègre*."[2] A second black head drew a slightly different comment. After noting its overall physical condition, Buffon highlighted the specific physical characteristics of this black *tête*: "This item is almost in the same condition as the last one, but one can better recognize the traits of the *nègre* physiognomy in this one."[3]

Severed from their bodies, identified by color and morphology, and associated with a particular variety of man, these African specimens reflect the increasing materiality of human categories at mid-century. And yet, while this hint of what would become physical anthropology is indeed significant, it would be an error to associate Buffon's vague identification of these two heads with the rigid categorization of the *nègre* proposed by thinkers including Linnaeus as early as 1735 and Johann Friedrich Blumenbach and Kant in the 1770s. Unlike the Swedish naturalist, Buffon refused to ascribe a scientific label (e.g., *Africanus niger*) to these black African heads and, in contrast to either Blumenbach or Kant, he specifically chose not to assign the *nègre* a specific spot within a schematic breakdown of humankind.

If anything, Buffon's description of these two African specimens more accurately reflects the overall anti-classificatory impulse that we find throughout his thought. In commenting on the specimens in the royal collection, Buffon made no attempt to separate "out" these African heads from similarly embalmed European examples on the basis of physiognomy or color. In fact, the two *têtes de nègres* cited above were actually found among a series of white heads. This was the precise approach that Buffon had endorsed in the preface to the same volume of the *Histoire*: the collection was designed to display "individuals of the same species one after another."[4] To a large degree, the anti-classificatory structuring of human varieties in this section foreshadows the compelling narrative of monogenesis in the subsequent pages of the *Histoire naturelle*.

Scholars who have examined Buffon's single origin theory of humankind have generally overlooked the history and implications of this biological *cousinage*. Whether censors or apologists of Buffon, specialists who have studied the naturalist's monogenist theory have wondered to what extent he was responsible (or not) for the ongoing or subsequent "othering" or racialization of non-Europeans. This book addresses this question as well, but it does so after examining just how the belief in a single human origin evolved from a vague, biblically based notion into a scientific concept grounded on information flowing back from the colonial world. It also demonstrates how the black African became the extreme data point around which an ongoing and much more material discussion of degeneration, degenerative anatomy, and whiteness took place in light of Buffon's theory. Ultimately, Buffon's belief in an original sameness prompted a new generation of anatomists and philosophers to think beyond his version of monogenesis and its emphasis on shared origins, and to concentrate on the specifics of the corporeal transformation that had, according to Buffon himself, separated black Africans from Europeans. This intense investigation into the material and physiological specificity of the black African was a major and unintended consequence of Buffon's environmental theory. While Buffon's intellectual descendants never endorsed polygenist race science, they nonetheless tended to give much more credence than Buffon did to the new anatomical discoveries being made all over Europe. In short, while the generation of thinkers who followed Buffon also envisioned the *nègre* within a human narrative that linked all human varieties, these same thinkers also put forward a much more pessimistic and etiological account of blackness.

The Origin of Shared Origins

The story of human sameness does not begin with Buffon, of course; nor does it begin with the generation of naturalists who began thinking about the origins of human phenotypes during the early-modern era. Well before there was a scientifically derived theory of essential sameness, there was the general, Christian-inspired belief that all humans, even heathen black Africans, were members of the extended family of man. While this vague notion clearly existed alongside a series of forceful and profoundly xenophobic beliefs regarding Africans, both scripture and later Christian teachings emphasized the idea that the human condition was universal and that all men were somehow redeemable. As Colin Kidd has written, Christian civilization, very much unlike either traditional Chinese or Japanese cultures, had "the benign capacity to render racial Otherness as a type of cousinage or remote kinship."[5]

The precise theological sources and multi-tradition history of the concept of human sameness extend far beyond the scope of this book. Suffice it to say here that the foundation for this belief in universal kinship is most clearly articulated in the New Testament when Paul tells the Athenians: "[God has] giveth to all life, and breath, and all things; and hath made of one blood all nations of men for to dwell on all the face of the earth" (Acts 17:26).[6] Paul's view of the human family stems, of course, from the more allegorical explanation of shared bloodlines found in the Old Testament's Book of Genesis. Here it is intimated that the world came to be populated by humans of diverse customs and, presumably, colors, as the result of the breakup of Noah's family after the Flood.

Giovanni Bellini, *The Drunkenness of Noah* (1515). As Japheth, on the right, and Shem, on the left, avert their eyes and help to cover their father's body, Ham (Cham) looks indiscreetly at Noah asleep on the ground. It was this act that supposedly cursed Ham's son Canaan, the father of Cush. Courtesy Musée des beaux-arts et d'archéologie, Besançon. Photo Charles Choffet.

In this well-known account of humanity's origins, Noah steps off the ark with his three sons, Ham, Shem, and Japheth, and their respective families, and "of them was the whole earth overspread."[7] This is followed by an anecdote that would have far-reaching consequences for Africans. According to this Old Testament parable, Noah drinks wine after the floodwaters recede "and [becomes] drunken," and lies "uncovered in his tent."[8] Unlike Noah's other two sons, Shem and Japheth, who conceal their father's body and "saw not" his nakedness, Ham witnesses his father in this undressed state. It is this indiscretion that brings a curse upon Ham's son Canaan, the father of Cush (who would people Ethiopia). Acting as a vengeful patriarch, Noah sentences Canaan to serve forever his now hierarchically superior brothers as "a servant of servants."[9] While there is no mention of the color of Noah's sons in the Old Testament, subsequent exegeses of this episode in both Christian and Jewish theological writings often claim that this was the moment that not only sentenced Africans to a life of base servitude but *darkened* the sinful descendants of Ham.

The myth of Ham, or Canaan, played a key if ambiguous role in the debates regarding the conceptual status of the African. On the one hand, the latent political implications of this parable ultimately allowed certain eighteenth-century pro-slavery writers, including Jean-Baptiste Labat, to cite Scripture itself as justification for the African's sorry lot in life.[10] On the other hand, this biblical structure also provided the foundational idea that all men were part of a shared human genealogy for centuries. This noteworthy belief was clearly articulated by Saint Augustine in book 16 of *The City of God* when he stipulated that all thinking, rational beings were a part of humankind.[11]

The basic belief in human sameness is also present—although not immediately obvious—in the way that many early-modern Europeans envisioned Asians, Amerindians, and even Africans. As indicated in chapter 1, the majority of Renaissance thinkers who wrote about the inhabitants of the *pays des nègres* simply did not have the inclination, or the political need, to exclude the "Ethiopian" from the category of the human. While it is certainly true that one can isolate a strain of European thought that demonized the black African during the late Middle Ages—and, as we have seen, a number of fifteenth-century thinkers, including the chronicler Gomes Eanes de Zurara, did in fact portray black Africans as little more than pagan beasts—such views were tempered by a strong undercurrent in other Africanist "ethnography" that emphasized the knowable and analogous aspects of African life (government, agriculture, justice) as much as their potential differences.[12] This remained a defining feature of the most influential texts on Africa, from Leo Africanus in the sixteenth century through to Dapper in the seventeenth. Although every early-modern text that pretended to evaluate African life also emphasized the supposed cultural superiority of its author (be he Arab, French, English, or Dutch), rare were the sixteenth- or seventeenth-century authors who positioned the black African as wholly *other*. Indeed, many Africanist texts recognized and, in certain cases, lauded those ethnicities closest to Europeans in either habit or morphology. This is significant; the very fact that such comments existed in early "ethnography" signaled a belief in a human continuum that included white, red, yellow, and black humans.

Toward a "Scientific" Monogenesis

Although a limited number of heterodox thinkers challenged the belief in an essential human sameness during the sixteenth and seventeenth centuries, this time-honored principle nonetheless continued to enjoy pride of place in discussions of human differentia.[13] Indeed, in addition to orienting religious thinkers, the conviction that humanity sprang from a single origin also provided the general point of departure for most writers taking up the question of humankind's origins during the eighteenth century.[14] This is clearly at work in Abbé Jean-Baptiste Dubos's ground-breaking work of esthetic theory, *Réflexions critiques sur la poésie et sur la peinture* (1719). In this influential treatise on literature, music, and the importance of "sensibility" in the arts, Dubos, soon to be a member of the Académie française, posed a question that was clearly on the minds of an increasing number of people during his era. Why, he wondered, were "all nations . . . so different among them in body, stature, inclinations, and mind, if they all came from the same father?"[15] Dubos's query and his implied answer to it signal a shift in the way that differences among human varieties would be interpreted during the eighteenth century. Although he echoed the era's widely held belief that humanity's different races all came "from Adam," Dubos hinted that Old Testament genealogy now required a more scientific explanation that could be added to the authority of biblical monogenesis.[16]

Dubos's interest in human varieties, it should be noted, came in the context of a much larger discussion regarding what he deemed to be the undeniable artistic and spiritual differences found not only among the world's many nations, but also between the more or less intelligent generations within a given family tree. To explain such divergences, the esthetic theorist put forward a mechanical explanation, singling out the effect of the air and, to a lesser degree, the "land" on the human body as the main causes. What particular groups of people breathed, according to Dubos, not only had a significant effect on "the physical structure of their organs" but also contributed "to the different qualities of their blood."[17] Indeed, the resulting differences in blood explained even deeper differences in humankind's intellectual diversity. As he put it: "two men who have blood that is so different as to make them dissimilar on the exterior, will be even more dissimilar in terms of their minds. [What is more,] they will be even more unlike in terms of temperament than in terms of coloring or body type."[18] According to Dubos, climatic variables were ultimately responsible—via the blood—for orienting whole nations toward "certain vices" or "certain virtues."[19]

On one level, Dubos's speculative understanding of the forces exerted on the bodies of given populations was hardly a breakthrough. Theories about the effect of the air and/or climate had a long history, beginning with Hippocrates' treatise on "Airs, Waters, and Places" and continuing through Plato, Aristotle, Galen, Ptolemy, Lucretius, and others, among the ancients, and Jean Bodin, Ambroise Paré, and Nicolas Malebranche during the seventeenth century.[20] And yet, despite its obvious link to previous thinkers, Dubos's climate (air) theory also signaled a move toward a more naturalistic explanation of biblical monogenesis, an understanding or conceptualization that simultaneously reasserted a foundational sameness while emphasizing the huge discrepancies that the air had wrought on the people in hot climates. Dubos expressed this view of monogenesis in a memorable aphorism: "There are few brains so badly formed that they cannot produce a man of wit, or even a man of imagination, under a certain sky; it is the opposite in another climate."[21] Given the examples he used to emphasize the deleterious power of the climate, there can be little doubt that Dubos was thinking primarily of the *nègre* when he wrote this.

Dubos was not the only thinker to grapple with the problem of human sameness prior to the publication of Buffon's 1749 theory of monogenesis. In a 1737 issue of his London weekly *Le Pour et contre*, Abbé Prévost provided a temporal view of humankind that evaluated particular human groups, especially Africans, against the possibility of moral progress or regression. This meditation on the prospects of certain human varieties came up during a discussion of how angry black slaves from "Antigo" had planned to blow up a ballroom party with stolen gunpowder, take the best women and ships, and sail back to Africa, with the sailors as hostages. In Prévost's opinion, this crime, coupled with the measurable differences between Africans and the superior civilized "nations of Europe,"[22] was such that he entertained a conceptual difference existing between the *nègre* and the European, namely, that these *machines animales* were perhaps "a different species."[23]

Prévost's follow-up to this brief evocation of polygenesis is very revealing. Having momentarily entertained this heretical view of human difference, the abbé then retracted this idea and stated quite clearly that he believed that all races of men in fact belonged to the same species. In disavowing polygenesis, however, Prévost also felt obliged to explain how such a horrifying act could be conceived of by a group of humans that was in theory part of the same species. His consternation generated a comparative history of humankind's moral evolution over time:

> If one also considers how historians have described certain earlier eras, as well as the state in which the people who now pass for the most advanced languished for a long time before reaching the stage of enlightenment of which they are now so proud, one begins to think, despite oneself, that since the mind and manners are subject to the same changes as anything else . . . , it is hardly surprising that the *nègres* have fallen to the lowest degree of brutishness.[24]

Like many others who grappled with the long-standing belief in the original unity of the human species, Prévost simultaneously conceded and qualified the notion of human sameness. Africans were clearly of the same species as Europeans, he maintained, but they had either degenerated into or languished in a pitiful state of moral animality.

Prévost's preoccupation with the origins and conceptual status of black Africans was not an isolated phenomenon at this time. Two years after he grappled with the question in *Le Pour et contre*, as we saw in the Introduction, the members of the Académie royale des sciences de Bordeaux announced a prize for the best essay explaining the origin and the significance of blackness in a 1739 issue of the *Journal des sçavans*. The question posed by the Académie sought answers to two questions: what was the physical cause of blackness and African hair, and what was the cause of their degeneration?[25] By 1741, sixteen essays had been submitted. No clear winner was ever named—a note affixed to the manuscripts indicates that the contest was never "adjudicated"—but these unpublished papers (never before analyzed to my knowledge) nonetheless bear witness to a significant shift in the interpretation of the *nègre*: the subject of *blackness* had become an international and widely publicized subject of debate.[26]

On one level, the Bordeaux Académie's interest in African pigmentation is a clear example of what Roxann Wheeler has described as the "reification of skin color as a discrete item of analysis" during the eighteenth century.[27] And yet, while this is certainly the case here, the *académiciens* who posed this question also believed that advances in anatomy might generate answers that were more than skin-deep, as it were. This is evident in the way in which the members of the Académie constructed the question. By soliciting theories of degeneration, they were tacitly seeking responses that explained, not only Africans' pigmentation, but the deterioration of earlier, presumably white, humans into blacks. Put simply, the Académie was seeking to move beyond simple questions of skin color and provide a scientific genealogy of blackness.

In looking back at such moments in the early history of theorizing about "race," scholars understandably tend to seek out the antecedents of later trends, the most important being the split between monogenist and polygenist theories.[28] This particular tension is rarely a factor in these 1741 essays, however. Although there are several moments in which the possibility of the African's essential conceptual difference is raised, none of the sixteen thinkers really articulated a coherent polygenist explanation for the African. To the extent that the possibility of a separate origin for blacks was brought up, it was generally evoked in order to be refuted. Author #4's explanation of human origins is typical in this respect. Although this ardent monogenist declared that it was not worth his time to rebut other explanations of human varieties, he nonetheless attacked the belief that Africans had different sperm or blood. More significantly, he also discredited the seventeenth-century polygenist, Isaac de la Peyrère, who had heretically postulated the existence of pre-Adamite races that included a black race that "did not descend from Adam, but from another man who was black, and consequently, [was] of a different species . . . from our own."[29]

Author #4 was among the few to acknowledge the polygenetic theory of the human species circulating in certain circles. More commonly, essayists sought to reconcile the measurable differences of *nègres* with the generally accepted belief in the common origin of humankind.[30] Like Author #4, many of the essayists still based their belief in monogenesis on the Old Testament account of human origins. This was the case among the majority of the authors who submitted their essays in Latin, many of whom recounted their own versions of the moment when Noah's indiscreet son Ham (Cham) laid eyes on his naked father, thereby bringing about a curse on the descendants of his own son Canaan.[31] This conviction—that Africans' roots could be traced to the "sons of Ham"—was also taken for granted by a number of essayists who added a naturalist explanation to this basic theological paradigm. Author #7, for one, based his belief in a common human origin on, among other things, the fact that different human varieties had similar moral weaknesses: "If *blancs* and *nègres* are different in color, they resemble one another in their capacity for pride. When considered from this angle, it is clearly the same species."[32] Author #5, however, chose to emphasize the importance of interfecundity in determining the category of species. This submission (which I have discovered is a manuscript version of Pierre Barrère's 1741 *Dissertation sur la cause physique de la couleur des nègres*)[33] asserted that the ability of blacks and whites to produce fertile offspring together proved the fundamental unity of the human species:

> There are doubtless many people who imagine that *nègres* are a different, degenerate or, as some say, bastardized species of man and, consequently, different from other [varieties of humankind], and that the products of their blood mixed with another, different blood could be compared to male or female mules. But what contradicts this notion is that, if this were the case, mulattos . . . should be sterile and completely unable to produce offspring, just like male and female mules . . . , which is not consistent with what we see on a daily basis.[34]

While more naturalistic than many other theories, Barrère's belief in interfecundity and the cousinage of the human species overlapped with the ideas found in other essays. There was less agreement among the essay writers, however, as to just why Africans had broken off from the rest of humanity. This, of course, was the crux of the matter in 1741: determining if this degenerative change was an "accident" or an act of Providence.

Not surprisingly, a number of essayists put forward a vague providential explanation for the African's blackness. This was the case with Author #1: "We cannot seek the cause of the origin of this blackness either in sacred or profane history, except by reflecting on the state of the known world and Providence's daily benevolence toward all men."[35] According to this essayist, a munificent Providence had changed people as they spread out over the globe to correspond to the particular climates in which they found themselves. This was most evident in the case of Africans, of course, whom Providence had rendered black to enable them to "bear the heat of Africa."[36]

Other essayists, including Author #2, attributed the morphology and color of Africans more directly to the African climate itself. While the net result of Author #2's theory may be the same as that of the first author—a change in human pigmentation—the latter's theory is anything but providential. Drawing from a long history of climate theory, Author #2 provided an explanation of the black branch of the family tree based on the notion of the *accident*: or a random variable acting upon the fundamental unpredictability of matter. Although the intent of Author #2's accidentalist (and environmentalist) explanation was clearly to give a "scientific" account of the biblical unity of the human species, his emphasis on chaos over more providential notions of order signifies a real shift—soon to become dominant—in the overall view of the human species at this time. In addition to the fact that this essayist implicitly asserts that the characteristics of the earth's human varieties were the product of certain physical variables, he also argues that their present physical form was no longer immutably fixed. Much like the "African Portuguese" who had supposedly been transformed into *nègres* after a certain number of generations, the human species as a whole was now being interpreted as the result of a natural, if not a preternatural, history.

While a number of the submissions to the Académie posited an almost mechanical relationship between a very hot climate and a pathological physiology, this explanation varied in interesting ways among authors. Author #5, for example, inflected this belief with a profoundly non-essentialist (and botanical) metaphor: "*nègres* who are transplanted into the countries of true whites lose their blackness and whites transplanted to the country of the blacks lose their whiteness."[37] More commonly, however, this transition from human variety to variety was qualified with certain value judgments. As Author #4 put it, the conversion from black to white was clearly an enhancement, accompanied by physiological and intellectual improvements, including a lightening of the *reticulum mucosum,* or Malpighian layer:

> Take away the causes [of blackness, viz., the sun and heat] and the effects will cease. Transport black men and women to temperate climates, give them healthy food and appropriate clothing; the friction of this environment will no longer take place; the glands will retract; the filtration process will work in a different

> way; the nature of the blood and the bodily humors will change; the children that will be born from these people, or at least their grandchildren, will have the reticular membranes a few shades less black than those of the people from whom they came."[38]

In short, while the climate theory endorsed by many of these thinkers often generated a belief in racial reversibility, it also generally explained blackness less favorably than it did whiteness. Author #2, for example, maintained that the climate effectively created human pathologies that could be explained by the imbalance of various humors. Echoing aspects of classical humoral theory, he affirmed that the heat and sun of Africa adversely affected "the quality of the blood," which in turn "can explain the languidness of the mind and the phlegmatic passions of [African] peoples."[39] This vestigial humoral belief from antiquity remained a vital part of the overall explanation of the African's pigmentation during the eighteenth century. If African skin remained the most telltale sign of the *nègre*'s degeneration, African blood and vital bodily fluids were also seen as the cause of this blackness and the potential source of other problems (e.g., assumed stupidity) as well.

Of the various contributors who submitted manuscripts to the Académie royale des sciences de Bordeaux in 1741, Author #14 best encapsulates the astonishing ambiguities of mid-century proto-raciology. Like his fellow essayists, this writer was a staunch believer in the original unity of the human species. And yet, in contrast to most of the other contenders for the prize, he maintained that the best explanation for human unity—climate theory—was unrealistically optimistic regarding racial reversibility and other mutations. After all, he argued, it was clear that black Africans who traveled to different climates nonetheless remained *nègres*.[40] This "fact" led Author #14 to suggest new ways of interpreting blackness. The first had to do with the African's classification: despite his own belief in a shared lineage, Author #14 nonetheless classified the *nègre* as a fixed variety and as conceptually distinct. This taxonomical impulse was accompanied by a more speculative move. Looking back (quite vaguely) to a time when the races had somehow split or divided, this author wrote rather dogmatically that the varieties of the human species that had the greatest number of people must be the least degenerate and thus members of the original prototype group. This group could only be, in his opinion, the white race: "There is no doubt that the first species (which is to say, that of which we are the members) is the original, and thus the legitimate; and that all the others have degenerated."[41] Although his methods and rationale were admittedly quite unsubstantiated, Author #14 nonetheless offered a speculative understanding of humankind's origins that asserted both the genealogical primacy and *legitimacy* of the white variety.

More than the other essays submitted to the Bordeaux Académie, essay #14 hints at the full impact of a new degenerative genealogy for the human species. In addition to positioning the *nègre* as a degenerative subset of prototypical whites, this essayist hinted that the *nègre* could be understood in terms of the Lockean-inspired belief that the mind at birth was a tabula rasa, a "white paper, void of all characters, without any ideas" and with only the potential for "reason and knowledge."[42] How Author #14 used this concept to inflect the overall conception of the African was really quite simple. If the ability and functioning of the human mind were being increasingly interpreted as the product of experience, then different (and brutal) environments (which suggested very disparate experiences) must likewise produce different types of minds (varied potentials for reason and knowledge) in the same way that different environments produced a range of human physiologies.

Author #14's essay presents but one of the many divergent theories of blackness circulating in the early 1740s, albeit perhaps the most prescient of them. Indeed, the diversity of opinion among the sixteen essayists who submitted their essays to the Bordeaux Académie reflects the immense uncertainty underlying a question that elicited a range of answers from biblical scholars, climate theorists, and anatomists. As Author #4 described it, anyone who wrote on the cause of blackness in 1741 was necessarily entering into a "land of conjectures" where "it was permissible for each person to speculate as much as he wanted."[43] And yet, while these essays certainly underscore the hesitation regarding the African's blackness and conceptual status just before mid-century, this same competition also reveals several important facts. First of all, for many writers at this time, the question of human difference sometimes necessitated a hybrid answer, one in which arcane interpretation of the Book of Genesis could be combined both with beliefs from antiquity (climate theory and four-element theory) as well as with more recent discoveries in the area of human anatomy. Secondly (and despite the era's blending of theology and physiology), the orientation that was clearly in the ascendency at this time was one based on ostensibly theology-free explanations of blackness produced within a naturalist frame of reference; it was only in this sphere, after all, that the various "liabilities" of the African could be effectively explained as the pathological outgrowth of an original whiteness. Third, and perhaps most significant, was the fact that the essayists who pontificated on the source of Africanness in 1741 saw this as a profoundly *apolitical* question. While several of the authors who submitted essays briefly mentioned the slavery of black Africans, in general, the essayists avoided any reference to the colonial backdrop that was clearly generating both the "data" and the increasing interest in the *nègre* in the first place. This conscious or unconscious ellipsis—in an era where French ships carried approximately ten thousand Africans to the colonies every year—reflects the political myopia of most of those writing on the black African during the first three-quarters of the eighteenth century.

Historicizing the Human in an Era of Empiricism: The Role of the Albino

Taken as a whole, the essays submitted to the Bordeaux Académie royale in 1741 contain the three essential elements needed to put forward a new, synthetic narrative of human sameness. The first was a mechanistic understanding of how various forces (climate, temperature, mores) could affect human bodies (pigmentation, humors, intelligence). The second was a new non-biblical narrative that explained these changes as having taken place in deep time. The third and final element was the conviction that white had been the original color of humankind.

This monogenetic temporalization of humankind's history would become the most important explicative paradigm for blackness and black Africans during the next four decades. From approximately the 1730s up to and after the French Revolution, countless thinkers cited African climate, food, and mores as causal agents in humankind's monogenetic narrative. What was initially missing, however, was a key concept or link that proved that the two most divergent actors in this narrative (whites and blacks) possessed the same genealogy. Curiously enough, the solution was found in the very emblem of racial sameness/difference: the human albino.

Long before the Enlightenment, Europeans had read numerous accounts of pale, night-dwelling people living on the African continent—the so-called *Leucoaethiopes*—in classical authors such as Pliny the Elder, Ptolemy, and Pomponius Mela.[44] The earliest European report of people suffering from hypopigmentation came in 1519, however, when Hernán Cortés and his soldiers discovered a roomful of people with "white faces," "white bodies," and "white eyelashes" living among other curiosities in Montezuma's palace.[45] The first published account of the African version of this phenomenon—"white children" born of "black parents" living with the king of Loango—was described approximately a century later by Andrew Battel.[46] These *dondos*, which were soon to be known as *nègres blancs* in France, implicitly raised a series of questions that would preoccupy Europeans for centuries.[47] What was the exact nature of this monstrosity? What was the *dondos'* relationship to both black Africans and white Europeans? And, finally, what kind of lives did these strange humans lead?

Some fifty years after Battel's African account was published, the Dutch compiler Olfert Dapper attempted to address these concerns by synthesizing the opinions of a variety of travel writers. His work provided two key points of information relative to the white African. Of most concern were the inopportune and unnatural symptoms of hypopigmentation:

> [In the court of the king of Loango], there are also certain White Men . . . , with Skins on their heads, and at a distance seem like our *Europeans*, having not only gray Eyes, but red or yellow Hair; yet coming nearer, the discovery grows easy; For they have not a lively colour, but white, like the Skin of a dead Corps, and their Eyes as it were fixed in their Heads, like people that lie a dying: the sight they have is but weak and dim, turning the Eye like such as look asquint, but at night they see strongly, especially by Moon-shine.[48]

In addition to this physical appraisal of albinism, Dapper also produced a sort of "ethnography" of the albino, the echoes of which were heard in many eighteenth- and even nineteenth-century texts. Although some of these *dondos* were presumably lucky enough to be subjects of an African monarch, Dapper asserted that the majority lived in a state of constant war with black Africans, so much so that the latter did "not allow [the albino] to multiply [or procreate]." In sum, the life of the albino was one of misery by Dapper's account; subjugated at court, hunted like animals, and isolated in small groups, the creature was the black Africans' own "monster."

Dapper's view of the question of albinism reflects the widespread belief that the monstrosity was particular to Africa: no one had yet made the link between the "white populations" of Mexico and those found in Africa. Albinism was, in short, a geo-ethnic condition that seemed to confirm a long-standing belief that nature escaped its conventional boundaries in Africa. "Ex Africa

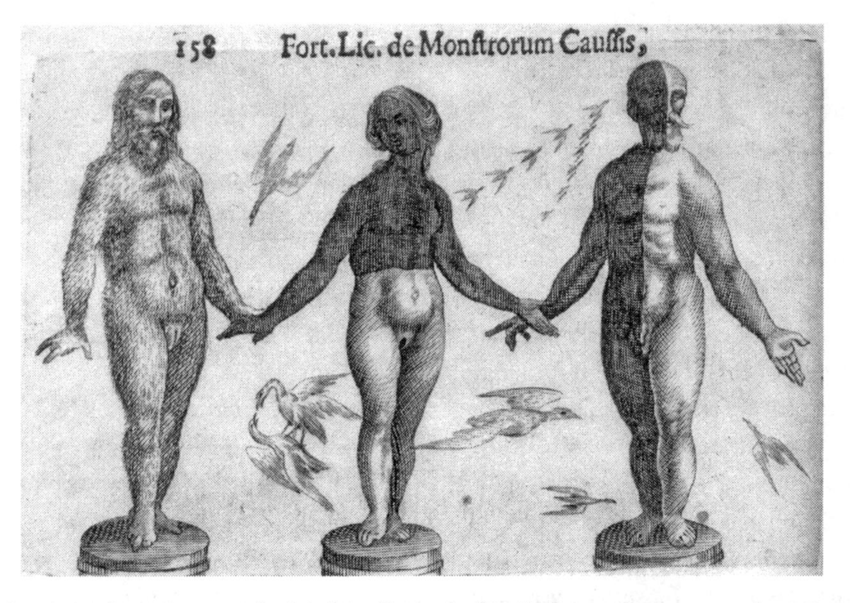

Monstrous human types depicted in Fortunio Liceti, *De monstrorum caussis, natura, et differentiis libri duo* (1616, 1634). Liceti studied people with unusual skin disorders, which he often affirmed were the result of natural developmental aberrations, as opposed to either divine or infernal intervention. Courtesy Division of Rare and Manuscript Collections, Cornell University Library.

semper aliquid novi" (Always something new out of Africa), as Pliny the Elder famously put it.[49]

This reduction of the *nègre blanc* to an African *caprice of nature* had an insulating effect throughout the seventeenth and early eighteenth centuries; although evoked occasionally in the era's travel writing, this strange being was simply another example of an African semi-monstrosity, like Pygmies or long-tailed humans. By the 1730s, however, natural philosophers and naturalists no longer contented themselves with contemplating the potential wonder of the *nègre blanc*. Instead of placing humans afflicted with such skin disorders on pedestals, as Fortunio Liceti had done in his famous *De monstrorum caussis*, Enlightenment-era thinkers studied the *nègre blanc* with the specific intention of reconciling the phenomenon with the basic paradigms of the burgeoning life sciences. As has been noted, this inquiry was much more than an investigation into a particular pathology: it was an attempt to understand how these "white negroes" simultaneously presented the racial or varietal traits of two different groups.

The effort to decipher these pale-skinned Africans began in 1733, when Jean-Claude Helvétius took the floor at the Paris Académie royale des sciences, alerting the assembled body that he had been contacted by the governor of Surinam regarding the existence of a young white-skinned boy born of black parents in the South American colony who "did not appear very intelligent, and was destined to be an imbecile."[50] Reporting on this for the Académie's annals, its secretary, Bernard de Fontenelle, wondered aloud about the "life science" implications of what Helvétius had related. How was it possible that this creature—who had both black parents and white skin—could defy his racial origins? How could it be that this *white* human had "all the distinguishing traits of the *nègre*, [such as] wool instead of hair?"[51] Although these queries may seem uninformed to us now, they indicate a major reorientation in the study of the *nègre blanc*. By implicitly inviting his peers to investigate the phenomenon from a scientific perspective, Fontenelle freed the *nègre blanc* from the confines—both generic and epistemological—of the Renaissance travelogue. No longer a semi-mythological figure existing on the same ontological plane as the antipodal races of antiquity, the albino needed to be brought into line with new notions of variety, race, and category.[52]

Parisian *académiciens*, naturalists, and philosophers waited ten years before they had an opportunity to speculate on the source and implications of a real *nègre blanc*. This occurred in 1744 when a small four-year-old albino boy was taken to Paris. Born in South America and shipped to Europe for exposition at venues that would come to include the Hôtel de Bretagne and the Académie royale des sciences, this intriguing piece of colonial merchandise was an immediate sensation, and inspired a number of the era's philosophers and naturalists to take up the pen. Among those thinkers to have their say on the *nègre blanc*, Pierre-Louis Moreau de Maupertuis was the first to understand the huge influence that the matter would have on the era's thought.

An insightful observer of mid-century French life sciences (as well as a pioneer in astronomy and geography), Maupertuis first used the advent of the *nègre blanc* as a springboard for a series of reflections on the fundamental beliefs held by contemporary naturalists and philosophers in his 1744 *Dissertation physique à l'occasion du nègre blanc,* although curiously enough, this text hardly mentions the *nègre blanc* at all.[53] Maupertuis's real "examination" of albinism actually came a year later, however, when he combined the *Dissertation* with a series of further questions related to the life sciences in *Vénus physique* (1745).[54] This anonymously published bestseller was divided into two parts.

The first is an introduction to the initial moments of conception, including a summary and critique of reigning embryological theories. The second part contains a meditation on the classification and origins of the human species as well as a section dedicated to the quandary of the *nègre blanc*.

Although the white negro plays a foundational role in the overall architecture of *Vénus physique*, Maupertuis carefully placed his remarks regarding this curious phenomenon at the end of his revised text. This was a strategic decision. Before Maupertuis could explain the existence of the *nègre blanc* (and of the world's many varieties of humans), he had first to refute the era's metaphysically inclined theory of embryo preformation, which held that conception simply activated a preexisting or preformed egg or animacule.[55] To do so, Maupertuis drew from a variety of countervailing evidence, not only from the writings of the English epigenesist William Harvey—who had scrupulously examined deer embryos culled from the does of Charles I's forests—but also from "the study of monsters, observations of racial differences and family resemblances, animal breeding, microscopy, the natural history of insects, and chemistry."[56] The cumulative effect of these "data" was indisputable for Maupertuis: to his mind, the fetus did not come from a preformed entity residing in an egg (as the ovists maintained) or in the sperm (as the animaculists affirmed); rather, the fetus was the product of mechanical forces set into action by the coming together of two seminal fluids.

The implications of Maupertuis's epigenesist embryology went far beyond a refutation of preformationism, however; epigenesis was the epistemological precondition for his generation-based examination of what he called the African variety and the *nègre blanc*.[57] To explain or classify either of these phenomena, Maupertuis believed, it was necessary to understand the organic processes that produced such beings in the first place. This was nothing short of a revolution in 1745. In contrast to most naturalists, who assumed that pigmentation, humors, or geography alone delineated the specific categories of the human, Maupertuis looked to his new and dynamic understanding of reproduction in order to characterize the limits and the origins of human varieties, particularly those of the *nègre*.[58]

Not surprisingly, miscegenation played a critical role in Maupertuis's overall theory of humankind. In an era when race mixing was considered a serious moral violation in France, and was even a punishable act under the 1685 *Code noir*, Maupertuis reveled in the philosophical implications of this indisputable biological phenomenon.[59] In addition to conjuring up a fantasylike thought experiment involving "bored Sultans" who oversaw the creation of new beauties in a multi-ethnic seraglio, Maupertuis also looked to the concept of racial mixing in order to invalidate his era's trenchant categories for white and black humans.[60] In *Vénus physique,* where, as Mary Terrall writes, "individuals emerged not from the hand of God, but from the physical congress of two profane creatures," Maupertuis played the role of libertine Prometheus, putting dark and light bodies into play in order to create, entertain, instruct, and perhaps even to transgress on a certain level.[61]

Although Maupertuis's miscegenation scenarios are presumably designed to amuse more than to arouse, these "literary" moments are of critical import for the text as a whole. By evoking an occurrence—miscegenation—that was playing out on a daily basis in France's Caribbean colonies, Maupertuis asserted forcefully that, despite the obvious political separation of black from white, these two groups were very closely related. Even more important, Maupertuis argued that *métissage* undermined essentialist classifications that had reduced the human species to three or four markedly different races.[62] As Maupertuis memorably stated: "the varieties of the human species cannot be reduced to white or black; there are a thousand others."[63] This is a significant moment both in *Vénus physique* and in the history of the life sciences. Unlike most thinkers writing on phenotypes during the eighteenth-century, Maupertuis's understanding of miscegenation allowed him to use the term *variété* as a descriptive rather than a stipulative marker; far from denoting a fixed and separate category—as "race" would later—*variété* simply suggested a subgroup of humans related by morphology and color.[64]

Maupertuis's conception of *métissage, variété,* and the uncertainty of generation shaped his understanding of Africans in various ways. Most optimistically, his belief in the interrelated nature of the human species led to a provocative history of ancient "race relations" between blacks and whites:

> Had the first white men who saw black men encountered them in forests, they might not have bestowed upon them the name of man. But those [black people] that they found in large cities, governed by wise queens, under whom the arts and sciences flourished at a time when almost all other peoples were barbarians, might well not have wished to accept the whites as their brothers.[65]

These views were tempered elsewhere, however, by an assessment of Africans more characteristic of the 1740s:

> From the Tropic of Cancer to the Tropic of Capricorn, Africa has only black inhabitants. They are distinguished by their color and their facial features. The nose is broad and flat, the lips are thick, and there is wool instead of hair, all of which seems to indicate a new type [*espèce*] of men. Moving from the Equator toward the Antarctic Pole the color lightens, but the ugly features remain [and] on the southernmost point of Africa one finds the hideous people [known as the Hottentots].[66]

Despite its seeming straightforwardness, this section of the *Vénus physique* has produced misreadings on the part of both eighteenth-century and contemporary readers.[67] Frequently cited out of context, Maupertuis's declaration that Africans were members of a new *espèce* might seem to indicate a polygenetic theory when his intent was precisely the opposite, namely, to explain African morphology and color in terms of a particularly vigorous monogenetic view of racial bifurcation. Put simply, Maupertuis believed that the differences between blacks and whites had actually grown out of a shared sameness or original root species.

This unified human genealogy—belief in "one mother," as Maupertuis himself put it—raised two critical questions.[68] Among humankind's varieties, which was the original race? And what exactly had caused the species to branch off into different varieties? Answering the latter of these two questions first, Maupertuis theorized rather generally that what we might call genetic modifications within an original group perpetuated themselves over time; specific traits—such as polydactylism—arose in "a common but unknown ancestor" and were passed on to descendants who, in time, constituted separate groups.[69] In short, nature's fluctuations and improvisational power had produced new life forms that, while only "accidental individuals" at first, became self-reproducing and statistically significant, if not entirely normal, in due course.[70]

The most telling of these "new" life forms for Maupertuis's theory was clearly the occasional birth of a *nègre blanc,* or "white negro," the supposedly cross-racial product of black parents.[71] In addition to being a living illustration of how the generation process could be affected by chance or mutation, the *nègre blanc* also proved to Maupertuis that white and black races shared a common, but hidden, history. As it turned out, the most revealing aspect of the *nègre blanc* had nothing to do with the condition itself. It was, rather, the fact that the opposite phenomenon—blacks being spontaneously born among whites—had never been documented. And here is where Maupertuis moved beyond the dynamic and unpredictable Epicureanism from which he had drawn many of his ideas. Whereas an epicurean thinker like Lucretius was fascinated by a history of regeneration in which Nature yielded every kind of creature imaginable, Maupertuis was most interested in the "fact" that a particular monstrosity did *not* occur.[72] This constitutes a significant moment in *Vénus physique*. By drawing attention to the predictability of certain aspects of the generation process, Maupertuis was able to speculate about the actual origins of humankind and its various categories.[73] The conclusion that he drew from this modified understanding of race was heard throughout the eighteenth and nineteenth centuries; since there were only *nègres blancs* and no *blancs nègres*, he argued, the white variety was clearly the forerunner variety from which the black race originated, and to which it sometimes reverted when an albino was born.[74] According to Maupertuis, "blackness [is] only a variety that became hereditary after many centuries, but that has not entirely eradicated its [vestigial] whiteness, which tends to reappear."[75]

In the albino, Maupertuis had seemingly discovered the missing link between white and black; indeed, the *nègre blanc* was a curious indication of the black African's whiter and perhaps *brighter* past. While this explanation of albinism and race is obviously more temporal than taxonomic—Maupertuis did not assign either the albino or the African to rigid categories—the speculative chronicle of both the *nègre blanc*'s and the *nègre*'s respective origins nonetheless produced an unaffected, white-centered genealogy of the human species. If, following Maupertuis's own understanding of human generation, the white race, too, was something of an accident—the product of matter and its inherent determinants—it was only the black *variété* that was subjected to the full power of Maupertuis's anti-teleological understanding of certain natural processes. In short, only the black race appeared entirely accidental, whereas the white race, as primary, seemed to retain a vestige of *telos* or intent.[76] This white-oriented understanding of humanity did not limit itself to the realm of natural history; it spilled out into other areas of Maupertuis's thought as well, particularly in his social explanation of the location of these "new" *variétés* around the globe. According to Maupertuis, this human geography was the result of a conscious exclusion: the "more numerous [white] species" must have cast out these "deformed races" to the least favorable zones of the world, whether hot or frigid.[77] This social segregation relegated peoples perceived as ugly or backward, be they blacks or Eskimos, to lands where, presumably, their esthetic appearance corresponded perfectly to the unpleasant climate.[78]

Creating the *Blafard*

Like many life science concepts in the eighteenth century, Maupertuis's understanding of human categories was at the crossroads of two competing views of nature. Although the author of *Vénus physique* had certainly emphasized the stability and predictability of generation when he spoke of family resemblances and reproducible patterns in animal and human reproduction, his writings on albinism had also accorded a preponderant role to chance in the generation process.[79] This latter characteristic is exemplified in the title *Vénus physique*. Like the Roman goddess of fertility, love, and spring, humankind's generative force was supposedly capricious and, in limited circumstances, capable of producing new and strange beings as a result of the sex act. This accidentalist understanding of human generation soon became the most influential explanation of the *nègre blanc*. After the publication of *Vénus physique,* most naturalists accepted the "fact" that these "white negroes" were accidents, mistakes, or lapses in nature. But Maupertuis's treatment of the question of albinism also left many questions unanswered. What was the etiology of the *nègre blanc*? Was this phenomenon, which was entirely naturalized in Maupertuis's work, still a monstrosity? And, finally, to what degree could one trust the lineage between whites, blacks, and albinos that Maupertuis had posited as part of his overall understanding of the question?

Four years after Maupertuis published *Vénus physique,* Buffon provided answers to many of these queries in the *Histoire naturelle*. Unlike Maupertuis, however, Buffon did not initially take up the question of the *nègre blanc* in a separate chapter or section. Rather, Buffon's insights regarding both the *nègre blanc* and albinism took place incrementally as the naturalist synthesized hundreds of travelogues from around the globe for his 1749 "Variétés dans l'espèce humaine." As I made clear in the introduction, this 150-page inventory of the human species appears, at first, to be a simple catalogue of racial phenotypes. But behind this flowing map of human varieties lies Buffon's conviction that humankind had had a shared origin, but had since undergone a series of degenerations or mutations into the different groups of humans living on the planet in 1749.

Buffon's climate-based catalogue of humanity's many colors had very few seeming inconsistencies. In fact, when Buffon encountered reports of pigmentation seemingly out of synch with geographical location, the naturalist confidently attributed such anomalies to cooler sea or mountain breezes. And yet Buffon also realized, more problematically, that travel writers sometimes reported the existence of small groups of very white humans living among much larger dark-skinned populations in various parts of the tropical world. Three such counterexamples to his climate theory appear in his inventory of humankind. The first, found in Java, was a nation called the Chacrelas, about which Buffon wrote: "[they are] completely distinct, not only from the other inhabitants of this island, but from all the other Indians. These Chacrelas are white and blond, they have weak eyes and cannot tolerate bright daylight; on the contrary, they see well at night, [whereas] during the day they walk with their eyes lowered and almost closed."[80]

Buffon's initial inability to recognize this population for what it was—a group of exiled humans suffering from the same condition as the *nègre blanc*—underscores an important point. Despite the fact that the Chacrelas group bore a striking resemblance to the *nègre blanc* that Buffon had himself examined in 1744, the famous naturalist could not envision this pathological state occurring simultaneously in isolated, non-African populations across the globe. Buffon as yet knew nothing of albinism as a generalized phenomenon. Instead, he took a race-specific view of the occurrence; *nègres blancs,* by definition, came only from Africa and occurred among Africans. It was precisely this myopia that led Buffon to attribute European descent to the Chacrelas: "The inhabitants of Malacca, of Sumatra, and the Nicobar [islands] seem to originate from the Indians of the peninsula of India; those of Java, from the Chinese, with the exception of these white and blond men called the *Chacrelas,* who must come from Europeans."[81]

Several pages later in the "Variétés," Buffon deploys the same theory to account for a group of white-skinned humans found in Sri Lanka:

> The inhabitants of Ceylon resemble those of the Malabar coast quite a bit . . . [but] there are . . . certain types of savages that are called *Bedas*. They live on the northern part of the island, and only occupy a small district; these *Bedas* seem to be a type [*espèce*] of men that is very different from those of these climates. . . . They are white like Europeans, and there are even some that are red-headed. . . . It appears to me that these *Bedas* from Ceylon as well as the *Chacrelas* from Java, could well be of European stock, particularly since these white and blond people

exist in such small numbers. It is quite possible that a few European men and women were abandoned at some point in time on these islands, or that they came from a shipwreck.[82]

One can detect a crack in this shipwreck theory by the time that Buffon's inventory of humankind leads him to Central America and Lionel Wafer's detailed description of the white-skinned humans living among the Indians on the Isthmus of Darien (Panama). Of particular import to Buffon was the fact that Wafer had affirmed that these white humans had darker parents:[83]

> One finds among [these Indians] men who are completely different and, although they exist in small numbers, they merit comment; these men are white, but this white is not the white of Europeans, rather, it is a milky white close to the color of the hair of a white horse. . . . [T]heir eyebrows are milky white, as well as their hair, which is quite beautiful. . . . These men do not constitute a particular and distinct race, but it does sometimes happen that yellow-copper skinned [parents] do produce a child such as the one we have just described.[84]

Although Buffon admitted that he was impressed by the detail in Wafer's account of this albino population, the naturalist was also puzzled by the potential implications of the discovery:

> [If it is indeed the case that these children come from dark-skinned parents, then] this color and this singular tendency of the white Indian body would only be a type of malady that they contracted from their fathers and their mothers; but if we suppose that this last fact did not prove to be true, which is to say that instead of coming from yellow Indians they constituted a separate race, then they would resemble the *Chacrelas* of Java and the *Bedas* of Ceylon, both of whom we have already discussed; or if [Wafer is correct], and these whites are born in fact of fathers and mothers who are copper-colored, one could believe that the *Chacrelas* and the *Bedas* come also from dark-skinned fathers and mothers, and that all these white men that we find at such great distances from each other are individuals who have degenerated from their race by some accidental cause. I must admit that this last opinion appears to me to be the most credible.[85]

After briefly considering two ideas, namely (1) that the albinos of Panama could be the pathological result of a disease afflicting their parents, and (2) that these humans might actually be members of a distinct biological group, Buffon arrived at what would be his final position regarding albinism: these white humans were in all likelihood suffering from a widespread, cross-racial, pathological condition that affected certain populations across the globe including the so-called Chacrelas, the Bedas, and *nègres blancs*. While he had certainly not set out to do so, Buffon had established a new de facto category within his "Variétés dans l'espèce humaine" for humans with albinism. Given the fact that these humans did not "respect" the parameters of his climate-determined understanding of human populations, they were clearly examples of a singularly degenerate group, the result of some as-yet undetermined pathological cause. As Buffon put it himself regarding the *nègre blanc*: "Independent of what travelers have said, what I have seen [of *nègres blancs*] leaves no doubt regarding their origin: these [beings] are Negroes that have degenerated from their race, they are not a regular and consistent species of man; they are singular individuals that are only an accidental variety."[86]

Although Buffon did not treat the question of albinism in detail in 1749, he had come to the conclusion that this peculiar subset of white beings hovered awkwardly between the human and the monstrous. The *nègre blanc* was clearly an example of how the usually stable reproductive process could sometimes produce category-defying puzzles within a "succession of individuals."[87]

Twenty-eight years later, Buffon again grappled with the conceptual status of albinos in his "Addition à l'article qui a pour titre, Variétés dans l'espèce humaine" (1777). Inserting his discussion of the question between sections dedicated to *terres australes* and *monstres*, Buffon began, first, by asserting that this condition, although somewhat bewildering, was nonetheless an expected feature of humankind. Having received confirmation from around the globe that these white beings were much more widespread than he had previously indicated in 1749, Buffon now affirmed that albinism was not specific to the world's dark-skinned peoples, but was found among all humans, even the French. Although he stressed his belief that people living in torrid zones were more often affected by the condition, Buffon endorsed a more general term to describe this global malady, *blafard*.[88] Originally an adjective meaning sickly pale or wan, the noun *blafard* now indicated the sum of the world's pigmentless humans, whether albinos, Bedas, Chacrelas, Dondos, or *nègres-blancs*.[89] Despite Buffon's tendency to separate such creatures from the rest of humankind, he had taken the first step in both naturalizing and, on a certain level, normalizing the condition; henceforth, it was recognized that one could be born a *blafard* anywhere, a novel idea that underscored the statistical predictability of the condition.

Buffon's dialectical understanding of the *blafard*—partially normalized yet conceptually ostracized—is at the heart of his writings on the subject. Nowhere else is this truer than in his description of a young *négresse blanche* whom he examined in April 1777, several months before his "Additions" went to press.

Buffon stressed that his inquiry was aimed at comparing the "measurements" of the *négresse blanche* with the dimensions of other *blafards* reported by his correspondents across the globe. True to his proposed methodology, he began his examination by measuring the young woman's body against a series of both esthetic and physiological norms. Not surprisingly, he was able to enumerate a significant number of supposed defects during this inspection: over and above her impure "suet-like" complexion, her eyes were disproportionately spaced; her head was "too fat" and "too round"; her neck "too short and too fat"; her arms "too long"; and her ears placed "too high."[90] In contrast with these features, however, the portions of the girl's body having to do with sex were described with more sympathy. Buffon's description of the young girl from the island of Dominica often "reveals a voyeuristic eye probing every inch of her firm, sweet-smelling young body," Julia Douthwaite writes.[91] Evoking the naked girl and her observer (himself) in a mise en scène that recalls all the power politics of a libertine novel, Buffon portrayed Geneviève as a "virgin" who was "moved by the shame of . . . being seen nude." Not only did the naturalist ascertain that this sugary-breathed girl had perfect teeth; he also acknowledged the quality of the skin on her thighs and buttocks as well as her "plump, round, very firm, and well placed" breasts.[92]

While it may be going too far to suggest that Buffon himself felt a sexual desire for this young girl, his description of Geneviève nonetheless demonstrates what he deemed to be a pertinent fact: this *blafarde*'s body was not without appeal.[93] And here is where the seventy-year-old Buffon's examination of this *blafarde* may overlap with Maupertuis's more scabrous suggestions regarding racial hybridization in the closed seraglios of Asia. As in *Vénus physique*, it was not only morphology or beauty that incited curiosity; it was the potential knowledge that could be derived from the sex act and its possible outcomes. In the case of Geneviève, or any other *négresse blanche* for that matter, Buffon believed that sexual experimentation could resolve two major questions. First, since the naturalist believed that a true conceptual link between the *blafard* and other subgroups (viz., races) was determined not by morphology or skin color, but by an ability to reproduce with members of other categories, "cross-

The white *négresse* Geneviève, an albino Buffon examined in May 1777. Proclaiming her to be a *blafarde,* Buffon theorized that her lack of pigmentation was a degenerative fluke. From Georges-Louis Leclerc de Buffon, *Histoire naturelle, générale et particulière: Supplément* (1777). Courtesy of the Watkinson Library, Trinity College, Hartford, Connecticut.

ing" a "specimen" like Geneviève with men of different races could determine if the *blafard*'s obvious pathology interfered with reproduction and, therefore, its potential status as a member of the human species. Contiguous to this question of species was the question of race, as Buffon defined it. To be a member of a real race, he argued, a given human had to be able not only to procreate with another member of this same group, but to produce offspring whose traits would, generation after generation, mirror those of the progenitors in question.[94] Clearly, the best way to confirm the dubious conceptual status of someone like Geneviève was, as Buffon put it discreetly, to increase the number of "observations" in this domain in order to understand the "nuances" and "limits" of humankind's "different varieties."[95]

This understated longing to "experiment" with the category of the *négresse blanche* (and the limits of human sameness) is clearly reflected in the *Histoire naturelle*'s plate of Geneviève.[96] Supposedly executed *d'après nature*, this illustration represents her leaning against a piece of furniture surrounded by various "other" colonial trophies, her folded arm resting on draped fabric. Although Geneviève is rendered somewhat disproportionately, she is posed rather gracefully, with one foot placed behind the other, staring enigmatically away from the artist. In addition to the erotic value of this staging, the most significant aspect of this plate is its very appearance in the *Histoire naturelle*. Although Buffon's text has literally hundreds of illustrations dedicated to the world's fauna, the naturalist purposely did not commission corresponding plates for the human species. As Thierry Hoquet has written, Buffon believed that humankind was not reducible to specimens, and certainly not to ethnographic types that could be forever preserved in a series of oversimplified illustrations.[97] Geneviève, however, did not represent humankind as a whole, as far as Buffon was concerned; she was a fluke, akin to the monstrous humans that were engraved for the *Histoire naturelle*. But more important, she was a living emblem (and enigma) representing the limits of humankind's procreative abilities.

Establishing the fecundity of the *négresse blanche* was an unrealized priority for Buffon. Since this research was impossible to undertake in Paris, he had to rely on "data" flowing from "the colonial laboratory." His reaction to these secondary sources is quite telling. In consulting Johann Christian Daniel von Schreber's *Histoire naturelle des quadrupèdes* (1775), Buffon was surprised to find that the German naturalist had affirmed that, when albinos were crossed with non-albinos, they reverted to the "primitive color from which they had degenerated."[98] As von Schreber put it: "the Dondos [*nègres blancs*] produce black children with Negroes" and "the Albinos of America produce mulattos with Europeans."[99] Since albinism is in fact carried on a recessive gene, von Schreber was correct that humans with hypopigmentation would (unless they mated with another carrier) produce normal offspring (including mulattos). But Buffon quickly dismissed such assertions as inconceivable, apparently convinced that it would be impossible for the degenerate *blafard* to reenter the category of humanity so easily.[100]

Not surprisingly, Buffon was much more receptive to reports suggesting that the *blafard*'s physiology was incompatible with, or contrary to, a procreative teleology. Although Buffon readily admitted that *négresses blanches* like Geneviève were clearly fertile, he was convinced, thanks to an unnamed source, that the more sickly and short-lived *nègre blanc* was not.[101] Were the *blafarde* Geneviève to mate with a *nègre blanc*, Buffon argued, "she would produce nothing, because in general the male *nègres blancs* are not fertile"; they were a dead branch on the human family tree.[102] This left open the question of what would happen in other cases, the most obvious being what would happen were Geneviève to attempt to have a child with a *nègre noir*. The answer came from an unlikely source. Contrary to the information provided by his colonial correspondents, the naturalist's best "proof" of the *nègre blanc*'s degenerative tendencies came from a man from Dunkirk, a certain Monsieur Taverne. In addition to being a retired borough master, Taverne had been an investor in a French corsair that had captured an English vessel in 1746 and had obtained, presumably as part of the booty, a portrait of a white and black child born with vitiligo (leukoderma) in Columbia in 1736. The engraving that Buffon had made from this painting was more harmonious and, dare I say, cheerful than the portrait he had commissioned of the understandably unsmiling Geneviève. Unlike the *négresse blanche,* the small but elegant *négresse pie,* or piebald negress, is carefully posed in front of an open window, surrounded by a range of colonial icons; she looks sweetly toward the reader while one of the Spaniards' most prized birds, a parrot, perches lightly on her right hand. Buffon did not comment at all on the composition of this interesting colonial moment. What fascinated him was the *négresse pie* herself, living proof in his eyes of what would happen were one to cross a *nègresse blanche* and a *nègre noir.*

When Taverne sent this portrait to Buffon in 1772, he speculated somewhat mechanically in the accompanying letter that this blotchy human had resulted from the union of a white and a *nègresse.*[103] Replying to this hypothesis in the *Histoire naturelle,* Buffon gently reminded Taverne that, contrary to what the latter supposed, race mixing between white and black generally produced mulattos. "[O]ne would be on much solider ground to attribute the origin of this child [i.e., the piebald negress] to *nègres* in which there are white individuals or *blafards,*" he wrote.[104] While arbitrarily attributing the birth of

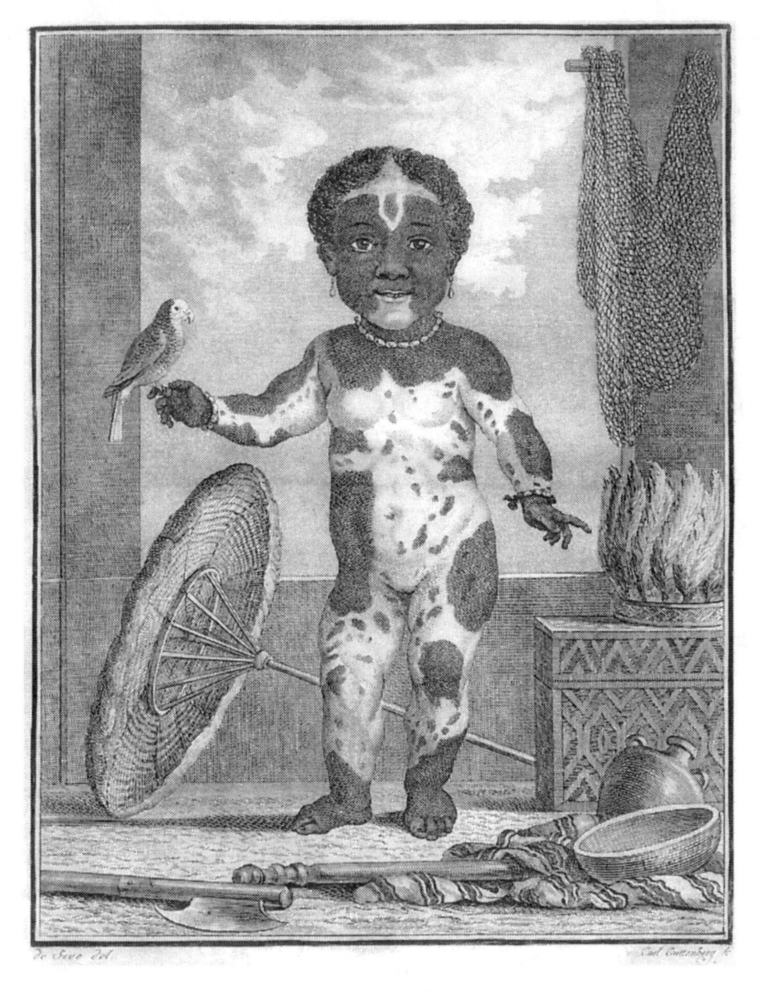

Piebald *nègre*. Based on his understanding of the albino, Buffon judged this child to be the offspring of a *blafarde*, or *négresse blanche*, and a *nègre noir*. From Georges-Louis Leclerc de Buffon, *Histoire naturelle, générale et particulière: Supplément* (1777). Courtesy of the Watkinson Library, Trinity College, Hartford, Connecticut.

this spotted human to the crossing of a *nègresse blanche* and a *nègre noir* was, to say the least, methodologically risky, it nonetheless buttressed Buffon's overall dubious view of the *blafard*. The curiously piebald human was just another confirmation that these *blafards* were not a *race constante*, but rather a collection of individuals who were "treated with disfavor by nature."[105]

By the time that Buffon had finished writing about the *blafard* in his "Additions," he had substantiated his 1749 intuition that these creatures were a degenerate variety of humans. He had, in essence, scientifically ejected the *blafard* from the normal functioning of nature with a "study" of procreative boundaries, thereby preserving and codifying more stable categories such as those of the *nègre* and the *blanc*. Additionally, Buffon had also put forward a new race- and climate-influenced understanding of the *blafard* that demonstrated that the condition, while found all over the world, manifested itself in a particular fashion as a function of environment. As such, there are, he wrote, *blafards* with "with wool, others with hair, and . . . others have neither wool nor hair, but a simple down; some have red irises and others have pale blue." As he himself put it, the geographically diverse *blafards* came in "many types."[106]

This proliferation of albino types, a grouping that came to include European *blafards*, should have posed a major problem for Buffon's overall conception of humankind. Indeed, his discovery that white people suffered from albinism, just like Africans, contradicted the entire theoretical underpinning of his 1749 "Variétés dans l'espèce humaine." In this earlier text, Buffon had positioned the *nègre blanc* as *the* key concept in his monogenetic understanding of human varieties. According to the "Variétés," the fact that dark-skinned races living along the equator sometimes produced white children confirmed the genealogical primacy of the white race. As he had written, "white appears to be the primitive color of nature . . . that reappears under certain circumstances, but with such a great alteration, that it hardly resembles the primitive white, which was effectively denatured by the causes that we have just indicated [a hot climate along the equator]."[107] In the 1777 "Additions," however, Buffon could no longer make any such claims regarding either the geographic specificity of albinism or the implications of this supposedly race-specific phenomenon; in sum, the nègre blanc *could no longer function as a symbol of lost whiteness within the chronicle of humankind's origins*. While Buffon never admitted it, he had forfeited the major element of "empirical proof" in his white-based chronology of monogenesis.

Notwithstanding this substantial (and unnoted) ellipsis, Buffon's 1777 assessment of the *blafard* rationalized this phenomenon more completely than had ever been done before. Conceptually and anthropologically, *blafardisme* had come to be seen as a global, yet insignificant, occurrence within (or contiguous to) the human species. This was taken as a given by the naturalists who followed in Buffon's footsteps. While the exact cause of this degeneration remained a subject of debate—Johann Friedrich Meckel the Elder maintained that the cause was smallpox, while Johann Friedrich Blumenbach attributed it to a change in body chemistry—it was generally believed that the *blafard* was the result of some immediate, albeit unknown cause. And yet, despite the general consensus regarding the pathological status of the *blafard*, the primary uncertainty that loomed over the condition, namely, the ontology of white-blackness or black-whiteness, was never really resolved during the eighteenth century. In fact, albinism continued to raise two troubling possibilities. The first was that the indissoluble link between pigmentation and race (and its inherent political implications) had come undone, leaving an uneasy blending of what were supposed to be fixed categories. The second was that the human race had either originated from, or could possibly degenerate into, the feared and denied category of the semi-monstrous.[108]

Buffonian Monogenesis: The *Nègre* as Same

The conceptual ambiguity of the albino was both bothersome and useful; bothersome because there was no easy way of rationalizing it, useful because the marginality and supposed infertility of the albino allowed Buffon to assert the existence of more solid categories for the notion of species and, to a lesser extent, its subcategory, race. To the extent that this latter category even existed in the naturalist's worldview, the question of "variety" or "race" demanded particular attention. This was particularly true for the sub-Saharan or black African, the most vexing group of humans on the planet as far as Buffon was concerned.

Much like the "grubby" Laplanders, the Africans that Buffon read about in his travelogues and correspondence were almost universally seen as occupying a marginal space in the human family. In fact, ethnographical information about the black African often tended to give credence to the idea that there was an essential difference between whites and the dark-skinned peoples of Africa. After all, here was a "group" of people that seemed both extreme and illogical in their comportment, a group that observed strange practices, such as the excision of one testicle by the Hottentots, or, like the inhabitants of Sierra Leone, seemed backward and lazy with no ambition and who lived in deserts and "miserable huts . . . while nothing prevented them from residing in beautiful valleys."[109] Irrational and dim-witted, Africans seemingly functioned as living counterexamples to things civilized or European in his era's thought.

Buffon's reaction to the ethnographic data he processed for his discussion of the black African was twofold. In the first place, he examined the *nègre* on a conceptual level. Once again defining human categories not by color or anatomy but by an ability to procreate together, the naturalist forced his readers to recognize blacks' cousinhood to whites.

> If a *nègre* and a white person were unable to reproduce together, even if their offspring were sterile, if the Mulatto were a real mule, there would indeed be two truly distinct species; the *nègre* would be to the man what the donkey is to the horse; or rather, if the white person were a man, the *nègre* would no longer be a man, it would be an entirely different animal like the monkey, and we would be right to think that the white and the *nègre* in no way had a common origin; but this presumption itself is refuted by reality, and since all humans can mate and reproduce, all humans derive from the same stock and are of the same family.[110]

As was the case in Barrère's 1741 *Dissertation sur la cause physique de la couleur des nègres* and Maupertuis's 1745 *Vénus physique*, humankind's ability to produce mixed-raced children in the present implied a communal biological past; in short, miscegenation provided the proof of monogenesis.

Having seemingly established humankind's common origins, Buffon's next step was to speculate on the details of the species' divergence into separate varieties over time:

> As soon as mankind began to move around the world and spread from climate to climate, its nature was subject to various alterations; these changes were minimal in temperate regions, [lands] that we presume to be the place of its origin; but these changes increased as man moved farther and farther away and, once centuries had passed, continents had been crossed, offspring had degenerated due to the influence of different lands, and many [people] had decided to settle in extreme climates and populate the desert sands of southern lands and the frozen regions of the north, these changes became so significant and so apparent that it would have been understandable to believe that the *nègre*, the Laplander, and the white constituted different species. . . . [But] these markings are in no way original [or distinct]; these natural alterations, these differences, being only on the exterior are only superficial. It is [in fact] certain that all humans are nothing more than the same man who has been adorned with black in the torrid zone and who has become tanned and shriveled by the glacial cold at the Earth's pole.[111]

During the most positive and heady moments of the *Histoire naturelle*, monogenesis, miscegenation, and movement seem to fully explain the human species. While scholars have argued quite convincingly that this explanation of the world's human varieties is an overwhelmingly negative history—in which a series of unfortunate human degenerations from a white prototype produce the maligned and marginalized peoples of Africa and other antipodal regions—the treatment of the world's "races" in the *Histoire naturelle* is, above all, a story of communal ancestry. As Thierry Hoquet has written, "Buffon's enterprise is designed to treat the notion of the Linnaean subspecies as nothing more than accidents of history."[112] Indeed, one might add that Buffon's understanding of humankind was implicitly *horizontal* whereas Linnaeus's understanding of the genus *Homo* was *vertical* and hierarchical.

What this reading of the *Histoire naturelle* glosses over, however, is the significance of the "accidental" transformation of white humans into *nègres*. This change is treated as inconsequential when Buffon evokes the black African within a sweeping biological narrative of essential sameness, but that is far from the case when he examines the contemporary physicality and mentality of the African.

Blackness Qualified: Breaking down the *Nègre*

Although rife with stereotypes, the many Africanist sources that Buffon consulted during the writing of the *Histoire naturelle* painted a diverse portrait of the sub-Saharan or black African. In addition to the divergent cultural practices that Buffon encountered, he also confronted a range of skin complexions that began with the light-skinned Hottentots and ended with the very dark Wolof of Senegal. The significance of African diversity was not lost on Buffon. It was obvious to him that the all-encompassing rubrics of *nègre*, *noir*, or, more rarely, *Africain* inadequately reflected the multiplicity of human phenotypes in Africa. As he put it: "It seems . . . that there are as many varieties in the race of blacks as there are in that of whites; the blacks have, like the whites, their Tartars and their Circassians."[113]

Faced with this range of complexions and peoples, Buffon found it necessary to venture into the realms of nomenclature and classification (how Linnaean!) for the large group of *noirs* he sought to describe: "It is necessary to divide the blacks into different races, and it seems to me that we can reduce these to two principal races, that of the *Nègres* and that of the Caffres."[114] This uncharacteristic categorization (of lineages) seemed to be based on the criteria of body odor and "beauty" as much as morphology or mores:

> In this first [race of blacks] I include those of Nubia, of Senegal, of Cap-verd, of Gambia, of Sierra Leone, of the Ivory Coast, of the Gold Coast, of the Coast of Ouidah, of Benin, of Gabon, of Luongo, of the Congo, of Angola, and of Benguela as far as Cap-nègre; in the second [race of Caffres], I put the people who inhabit the region beyond Cap-nègre to the tip of Africa, where they take the name of Hottentots, and also the peoples of the eastern coast of Africa, like those of Natal, of Sofala, of Monomotapa, of Mozambique, of Malindi; the blacks of Madagascar and its neighboring islands are also Caffres and not *nègres*. These two types of black men resemble one another more by the color of their skin than by the features of their faces; their hair, their skin, the odor of their body, their mores, and their nature are also very different.[115]

In the subsequent treatment of the second of these groups, the Caffres, Buffon mixes restraint and sensationalism. In describing the so-called Hottentots, he eschews the scandalized tone found in many of the writings from which he had culled his information (e.g., Prévost, Gui Tachard, Joris van Spilbergen, John Ovington, and Courlai), and exhibits a measured understanding of human difference, contrary to frequent assertions that the Hottentots were often bestial itinerant semi-humans who lived in the most abject, miserable ignorance. Although he did qualify their hygiene as "of the most dreadful uncleanliness," Buffon also depicted their peripatetic tendencies more tolerantly, explaining that they were "independent and very jealous of their freedom."[116]

Despite his comparatively understated asides on the Hottentots, it is clear that Buffon was fascinated, as was his era, with certain lurid aspects of Hottentot sexuality and morphology. Although he would mistakenly come to deny the existence of the famous Hottentot *tablier* in 1766, in 1749 Buffon saw this "outgrowth or wide and stiff skin that grows over the pubic bone" as one of the defining morphological features of this group.[117] Even more inflated than his description of what he called a *monstrueuse difformité*, however, is Buffon's detailed, two-page account of the testicle excision ceremony that young men underwent in Hottentot society. Citing Peter Kolb (and perhaps drawing from the graphic illustration of such rituals in Bernard Picart's beautifully illustrated *Cérémonies et coutumes religieuses de tous les peuples du monde*), Buffon emphasized the suffering, isolation, and seeming inhumanity of this rite of passage.[118]

Considered from the perspective of the history of ideas, Buffon's view of the Hottentot clearly played a significant role in recycling, synthesizing, and diffusing ethnographic "data" on the Khoisan peoples. Not only was his au-

Hottentot initiation. From Bernard Picart, *Cérémonies et coutumes religieuses de tous les peuples du monde* (1723–37, 1783). Courtesy Wesleyan University Library, Special Collections and Archives.

thoritative view published in multiple international editions, authors shamelessly poached his paragraphs on the Hottentots in countless follow-up ethnographies. Jaucourt, for one, drew exclusively from Buffon's portrayal of the Hottentots for his unflattering *Encyclopédie* article on the subject. These second-generation uses of Buffon's ethnography were perhaps as important as the great naturalist's views themselves. Conjured up outside of Buffon's overall assessment of humankind, these abridged and decontextualized views of the black African often seem to suggest that the Hottentots and Caffres were conceptually different from the rest of humankind. And yet, if we return to the *Histoire naturelle* itself, it is clear that the most critical aspect of Buffon's treatment of the Hottentot is not the ethnographic details themselves, but the way in which the naturalist processed this information into a conjectural history:

> From all of these accounts, it is easy to see that the Hottentots are not *true nègres*, but men who, within the race of Blacks, are beginning to move closer to white. . . . The Hottentots, who can only trace their origins to black nations, are still the whitest of all the peoples of Africa, because, in effect, they are in the coldest climate of this part of the world.[119]

Although Buffon never stated this explicitly, he believed that the Hottentot had undergone two major shifts in color. According to the overall scheme of the "Variéties," this former member of the black race had originally degenerated from a white prototype before it got its dark skin. Given the Hottentot's light skin in the present, however, Buffon had no choice but to admit that the dynamism of the human race was now working in the opposite direction; the Hottentot was becoming *white* again as a function of the mild South African climate. This movement toward whiteness is significant. In addition to the fact that Buffon was once again rejecting any trenchant categories for race, the naturalist was also—unintentionally—undermining the seeming synchronization between whiteness and civilization. In stark contrast to the logic of the era's stage-theory ideas regarding progress and ethnicity, the move toward whiteness in the Hottentot far from produced a positive moral development; it had, in fact, clearly been accompanied by a remarkable *moral* deterioration, even more so than in the *nègre*'s case.[120]

Such was the paradox of Buffon's interpretation and systematization of his era's ethnography. While far from unprejudiced, aspects of his monogenetic and racially reversible scheme seemingly transcend his unflattering assessment of the world's peoples. According to the naturalist, a black race could move to white, and a white race toward black. This latter scenario was, in fact, the case for the Ethiopians, whom Buffon declared to be moving toward an increased blackness.

> The Ethiopians . . . the Abyssinians, and even the people of Malindi, [the latter group] whose origin can be traced to whites because they have the same religion and the same practices as the Arabs and because their color resembles that of whites, are, in fact, more dark-skinned than the southern Arabs, but even that proves that within the same race of man, different degrees of blackness depend on the intensity of the climate; it may take several centuries and a great number of generations for a white race to progressively develop a brown color and to become finally completely black, but there is evidence that with time, a white people transported from the north to the Equator could become brown, and even completely black, especially if this same people were to change their customs and ate only the food produced in the torrid region into which they were transported.[121]

The primary reason that Buffon could assert racial *reversibility* was that he saw such transformations in primarily physical terms. While a given race's degeneration could presumably result in a whole host of new liabilities (including black blood and bile), Buffon never stated explicitly that such changes necessarily involved an explicit irreversible decline in intellectual capacities. This is one of the most intriguing and puzzling areas in Buffon's thought. While the naturalist wrote extensively on the changes that humankind underwent when it bifurcated and degenerated, Buffon carefully avoided discussing the category of mind within this environmental determinism: "Men [now] differ from white to black in terms of color, . . . size, strength, and in terms of their mind; *but this last characteristic, which does not belong to the material world, should not be considered.*"[122] This was a significant pronouncement: according to Buffon, humanity's *esprit,* its defining essence, could not be reduced to the same base material variables that affected traits such as skin color.

Despite this declaration, the ethnographic descriptions that Buffon supplied elsewhere in the *Histoire naturelle* sometimes seem to reflect the belief that the capacity to think *like a European* had either disappeared or, more likely, never developed in the African. In describing the *nègre* as a type, Buffon affirmed in the most general terms that they were "tall, large, well-proportioned, but simple-minded and lacking in intelligence."[123] Similarly, Buffon seemed to mock the one literate group of Africans whom he encountered in his tour of the continent, the Ethiopians, who supposedly took days to write a letter.[124] Although Buffon did not state it categorically, he clearly believed that the African lacked an aptitude for intellectual pursuits or the high arts of civilization. Unable to perform on a cerebral level, the *nègre* was not only associated with but also defined by the physical realm.

Indeed, Buffon's ethnography echoed the era's belief that the value and the prominence of the African body overlapped somehow with its moral potential and comportment. This perspective is an integral part of Buffon's description of Senegalese women, for example:

> [Senegalese women] are, with the exception of their color, as beautiful as those of any other country of the world. They are very well proportioned, very joyous, very vibrant, and very drawn to lovemaking; they have a taste for all men, and particularly whites, whom they pursue with eagerness, as much as for their own pleasure as for the hope of obtaining some gift.[125]

In reviewing Senegalese "ethnography," Buffon was clearly fascinated by portraits of beautiful hypersexual African women whose lubricity was only surpassed by their desire for material gain and white men. A more extreme and less alluring version of the African female body can also be found in his description of the women of Sierra Leone, whom he described as being "depraved" and whorish.[126]

On the one hand, Buffon suggested that the African woman's body was closer than the European woman's body to its primal function. On constant display and prone to a heightened sexuality, the female African body supposedly benefitted from a heightened fecundity and an enviable ability to produce offspring: "*Négresses* are incredibly fertile and give birth easily and with no aid; the aftermath of giving birth for them is hardly troublesome, and they need only one or two days of rest to recover."[127] On the other hand, Buffon believed that the excessive debauchery that supposedly defined an African woman's life left an indelible mark on her body:

> The untimely wear on women is possibly the cause of the brevity of their life; children are so debauched and so minimally constrained by their fathers and mothers that, from a very tender, young age, they indulge in anything that Nature may suggest; nothing is as rare as to find a girl who can remember the moment at which she ceased to be a virgin among these people.[128]

Drawn almost mechanically to the undeniable pleasures of corporeal existence, the African girl/woman seemed to be pushing the limits of her own body. This sexual excess, Buffon asserted, was surely being censured by nature. While wantonness seemingly did not to lead to infertility, it did appear to shorten African lives, thereby guaranteeing some kind of poetic justice in this physico-moral ecosystem.

The Colonial African and the Rare Buffonian *Je*

One of the things that Buffon acknowledged while discussing the *nègre* was that this group of humans could not be reduced to a single concept. This is reflected in the overall portrait of the black African in the *Histoire naturelle*. A composite if there ever was one, Buffon's black African is the product of (1) the naturalist's monogenetic worldview and (2) the varied geographical contexts from which Buffon derived his information on the *nègre*.

In chapter 1, I emphasized the importance of geography in the overall representation of the black African. This is also the case in the *Histoire naturelle*. When Buffon cited strictly Africanist sources [on Africa], he echoed a portrait of the African in situ; although full of prejudice and misunderstanding, first-person narratives regarding West Africa invariably conjure up an image of free, albeit supposedly backward, Africans living in their own land. As a thorough naturalist, however, Buffon also consulted the ethnography that had flowed back to Paris from French sugar plantations. This view of the *nègre* generally asserted, all evidence to the contrary notwithstanding, that Africans "become easily accustomed to the yoke of servitude."[129]

Among the Caribbean "specialists" that Buffon consulted were the prolific Jean-Baptiste Labat, the relatively moderate Jean-Baptiste du Tertre, and Pierre-François de Charlevoix. Rather than engaging separately with what each author had to say about the "African of the Caribbean," Buffon initially bundled their various remarks into a revealing framework: a comprehensive labor-based typology based on the ultimate *utility* of the various ethnicities working in the colonial world:[130]

> On our islands, we prefer the *nègres* from Angola to those of Cap-verd [Senegal] for the strength of their bodies, but they smell so awful when they become hot, that the air in the places where they pass is contaminated for over fifteen minutes; those from Cap-verd do not have an odor nearly as horrible as those from Angola, and they also have the more beautiful and darker black skin, a better proportioned body, less harsh facial features, a better disposition, and a more

advantageous height [taken from du Tertre]. Those from Guinea are also very good for farm labor and for other large undertakings; those from Senegal are not as strong, but they are the better suited to domestic service, and more capable of learning trades [taken from Labat]. Father Charlevoix says that of all the *nègres*, the Senegalese are the best looking, the easiest to discipline, and the best suited to domestic service; that the Bambaras are the biggest, but that they are thieves; that the Aradas understand agriculture the best; that the Congos are the smallest in stature, that they are good fishermen, but that they run off easily; that the Nagos are the most human; that the Mondongos are the most cruel, and that the Mines are the most determined, the most capricious, and the most likely to despair."[131]

While Buffon clearly felt that information on the appropriateness of African ethnicities to certain tasks fit well into his overall assessment of the *nègre*, one can sense his increasing frustration with the accumulation of dogmatic and sweeping generalizations regarding Africans in the Caribbeanist texts, particularly Charlevoix's. More so even than Labat's assessments of the African variety, this Jesuit's brutal, reductive view of the African asserted that the only redeeming feature of the African was, paradoxically, that several of his liabilities might actually make him a good soldier. Buffon distanced himself from such views by taking great pains to point out that he was in fact *citing* the priest's assessments of the African. Such is the case regarding Charlevoix's famous appraisal of the African mind: "[Charlevoix] adds that all the *nègres* from Guinea have very limited intelligences, that some of them seem completely stupid, that one sees some *nègres* who could never count past three, that they do not think about anything, that they have essentially no memory, that their past is as unknown to them as their future."[132]

While we can never know exactly how Buffon situated himself vis-à-vis this ethnography, what is evident is that each time he cited Charlevoix, Buffon countered the priest's views with a more measured take on the variety that was inspired from du Tertre. Conceding the point that Africans had little wit, Buffon nonetheless "replied" to Charlevoix by effectively endorsing a condensed version of Labat's and du Tertre's paternalistic assessments of the *nègre*:

Although the *nègres* lack intelligence, they are far from lacking in emotions; they are gay or melancholy, laborious or lazy, friends or enemies, according to the manner in which one treats them; when one feeds them well and does not mistreat them, they are happy, joyous, ready to do anything, and the satisfaction of their soul is painted on their faces; but when one treats them badly, they take

Indigo plantation. From Jean-Baptiste du Tertre, *Histoire générale des Antilles* (1667). The illustration in Du Tertre's *Histoire* depicts productive *nègres* in appealing landscapes.

> their sorrow to heart and occasionally die of despair: they are thus very sensitive to kind deeds and to affronts, and they carry a fierce hatred toward those who have mistreated them; when, on the contrary, they become fond of a master, there are no longer limits to what they are capable of doing to show their zeal and their devotion to him. They are naturally compassionate and even tender toward their children, their friends, and their compatriots.[133]

By citing du Tertre (in particular), Buffon tempered harsh planter-derived views of African difference with what would become the "enlightened" sentimentalism that defined much of the "progressive" thought on Africans after mid-century. The prescriptive rhetoric of such a view of the *nègre* delivered a unambiguous message: treat these simple, mechanical, yet sensitive creatures well, and they will not only live up to their sentimental potential, they will be devoted to their planter/master.

Buffon's processing of the conflicted Caribbean portrait of the African is a defining moment in the *Histoire naturelle*. In working through the contradictions among various Caribbean authors, many of them as prescriptive as they were descriptive, Buffon obliquely confronted the political status of the *nègre* for the first time in his text. In a work of ethnographic synthesis and systematization, a work that asserted the fundamental sameness of the human species, Buffon found himself writing about members of the human family who were clearly suffering like no other. This realization prompted the rare appearance of an emotional and personal Buffonian *je*: "I cannot write their story without being moved by their situation. Are they not already unhappy enough as slaves, without being required to work every day without the possibility of ever acquiring anything at all? Is it also necessary to overwork them, and treat them like animals?"[134]

Buffon followed this rhetorical question by challenging Charlevoix's contention that the *nègre* could subsist on virtually no food: "How can men with any degree of humanity adopt these maxims, turn them into prejudices, and use them to rationalize the abuses that the thirst for gold makes them commit?"[135] While heartfelt, this indictment of Charlevoix is far from an anti-slavery or anti-colonial diatribe. What Buffon objects to here is both the pitiless treatment of black Africans in the colonial context and, perhaps just as important, the use of falsified or misapplied ethnography in the service of the slave trade. These were clearly distractions from what Buffon called the "object of our study": the more edifying narrative of humankind's fundamental sameness.[136] Indeed, faced with the fact that he himself was inadvertently undermining his own monogenist views by citing these essentialist African stereotypes, Buffon "interrupted" Charlevoix and the other Caribbean authors, declaring: "Laissons ces hommes durs" (Let us leave these harsh men).[137] This is perhaps the most revealing moment in Buffon's treatment of the *nègre*. In conjuring up a particular group of sensible and sensitive people (the *nous* of *laissons*), Buffon posited the existence of an ideal audience: an enlightened readership able to recognize the pitfalls of ethnographic knowledge production and transmission. What the great naturalist perhaps did not anticipate, however, was that his best-selling *Histoire naturelle* would have a huge readership that would extend well beyond these imagined philosophes and their progressive intentions. Perused and interpreted by a new generation of anatomists and more materially oriented pro-slavery writers, Buffon's degeneration-based ethnography would ultimately give rise to a new and more brutal version of the story of human sameness that he first recounted in 1749.

CHAPTER THREE

The Problem of Difference

Philosophes and the Processing of African "Ethnography," 1750–1775

> The effects of heat on the constitution of men under the [equator] are phenomena which have been discovered by dissecting negroes, and analyzing their most essential humours.
>
> "PHILOSOPHICAL ENQUIRIES CONCERNING THE AMERICANS BY M. DE PAUW" (1771)

Throughout the second half of the eighteenth century, the major thinkers associated with the French Enlightenment accessed a wide range of proto-ethnographical information on black Africans. Many perused popular travel narratives such as those by Jacques-Joseph Le Maire, William Bosman, William Snelgrave, Francis Moore, Reynaud des Marchais, and Abbé Proyart. Others, like Diderot, also consulted the synthetic overviews of Africans found in Olfert Dapper's *Description de l'Afrique*. Rousseau, as we have seen, pored over the ethnographic portraits of the *nègre* found in Abbé Prévost's *Histoire générale des voyages*. There were also those philosophes who looked at Africanist texts with specific objectives in mind; the anti-clerical Baron d'Holbach was clearly fascinated with the beautifully rendered illustrations of "senseless" African animism found in Bernard Picart's *Cérémonies et Coutumes religieuses de tous les peuples du monde*.[1] Along similar lines, Helvétius studied Charles de Brosses's portrait of African idolatry in the 1760 *Du culte des dieux fétiches*.[2] In both cases, these materialists researched the "inanity" of black African religion in order to mock the pretension of Christianity.

While eighteenth-century philosophes consulted a wide range of works featuring observations on the *nègre*, no "genre" was more informative to them than that of natural history. As discussed in the previous chapter, virtually every person associated with the High Enlightenment read or referenced scientific views of the *nègre*, particularly those endorsed by Buffon in his 1749 "Variétés dans l'espèce humaine." It would be difficult to exaggerate the impact that this chapter from the *Histoire naturelle* had on the overall understanding of this subject after 1750. Published with the king's imprimatur, Buffon's "Variétés" not only supplanted biblical and/or providentialist explanations of the human species (and the *nègre*); it explained the existence of black humans within a degenerative and pessimistic chronology of humankind. For the vast majority of eighteenth-century thinkers who reflected on the *nègre* after the publication of the *Histoire naturelle*, blackness had become a climate-induced condition measurable in terms of symptoms; what was more, the *nègre* was considered to be an *accident* produced by an unthinking nature.

Deciphering the black African's status within the human species nonetheless remained a vexing problem for a number of Enlightenment-era philosophes. If, as most people assumed, the *nègre*'s degeneration from a white prototype had altered nerves, brains, and the African capacity for reason, what did this mean for human educability, progress, and the destiny of the *nègre* "variety" itself? This chapter examines the rise of an increasingly authoritative and naturalized understanding of the *nègre*, as well as the way in which philosophes processed these ideas within the framework of their own preoccupations. What was at stake in the philosophes' discussion of the *nègre* was not only the significance of human "difference" but the limits of Enlightenment universalism as well.

The "Symptoms" of Blackness: Africanist "Facts," 1750–1770

For the majority of philosophes working around mid-century, the "symptoms" of blackness extended to three overlapping realms: the moral, the intellectual, and the physical. The supposed moral or behavioral liabilities of the *nègre* included sloth, barbarism, paganism, and hypersexuality. For many philosophes, these were natural extensions of the black African mind or *esprit*, which, unlike its white counterpart, supposedly lacked basic intelligence, an ability to reason, and historical memory.

Philosophes writing around 1750 absorbed such notions from a variety of sources. A number of travelers who had visited Africa, including Jacques-Joseph Le Maire, had long cited the supposed shortcomings of the African mind.[3] But a belief in a specific and limited form of African "cognition" was most forcefully and effectively conveyed by authors who wrote from the Caribbean world. Perhaps the most persistent "truisms" about the black African were articulated by the Jesuit missionary Pierre-François de Charlevoix in the *Histoire de l'isle Espagnole ou de Saint-Domingue* (1730–31).[4] Typical of other Caribbean assessments of the *nègre*, Charlevoix asserted that the African variety's constitution prevented him from either thinking or reflecting:

> With respect to the mind, all the *nègres* of Guinea are extremely limited; many seem idiotic, as if stupefied. One sees some that could never count past three, nor could they learn the Lord's Prayer. They think of nothing of their own accord; their past is as little known to them as their future. They are machines that must be rewound whenever one wants to make them move.[5]

This authoritative plantation-generated understanding of the *nègre*'s mind/*esprit*/mentality was endorsed by some of the major figureheads of the Enlightenment. Montesquieu, who knew Charlevoix's work well, was presumably thinking of the Jesuit priest's book when he evoked the strict limits of the warm-climate mind in his 1748 *De l'esprit des lois*. Buffon, too, provided a pessimistic assessment of African intelligence along these lines in his *Histoire naturelle*.[6] Such views showed up in a variety of other lesser-known works as well. Pierre Poivre, author of the *Voyage d'un philosophe* (1754), was also probably drawing from Charlevoix when he explained that owing to an inherent incapability to envision future challenges or opportunities, Africans had not yet achieved an "intelligent" agricultural system: "These stupid men [are content] to live from day to day in an environment where there are few real needs. They cultivate only as much as they must to avoid dying of hunger."[7] Twelve years later, the slave-trading priest Abbé Demanet pushed this "futureless-human" notion even further in his 1767 *Nouvelle histoire de l'Afrique françoise*, affirming that the African's entire automaton-like essence made him like wax that could be molded: "The African seems to be a machine that is wound up and released by springs; [he is] similar to soft wax, with which one can form whatever shape one desires."[8] A comparable remark also shows up in Jean-Gaspard Dubois Fontanelle's popular 1775 *Anecdotes africaines*. In this survey of outlandish African beliefs and behavior—including the Serpent cult of Ouidah, the human sacrifices of the kingdom of Benin, the cannibalism of the Jaga, and the insect "god" of the Hottentots—Fontanelle attributed Africans' barbarism to a lack of historical memory: "The Africans live, for the most part, from day to day. . . . They dream neither of the past nor of the future. Much like their ancestors, who were not inclined to leave historic accounts, they too have no interest in providing such chronicles for future generations."[9]

As critics including Christopher L. Miller have demonstrated, such conjectures about black Africans' customs or mores "tended to repeat each other in a sort of cannibalistic, plagiarizing intertextuality."[10] Indeed, what was written about particular aspects of black African behavior or ethnography in 1800 often directly resembles what was written on the same subject in 1710.

This timeless ethnographical consensus did not hold for the *nègre*'s body, however. Philosophes, in particular, debated several problematic questions regarding the supposed corporeal liabilities of the black African beginning at mid-century: in addition to wondering about the precise causes of blackness, they argued about the *nègre's* kinship (or lack thereof) to the European. The polygenist Voltaire, for one, declared by the 1730s that the *nègre*'s anatomy irrefutably demonstrated that whites and blacks were separate races or species. Maupertuis, however, soon countered this view by contending that the obvious interfecundity of whites and blacks proved that there was an ancestral link between Europeans and black Africans.

Despite such strong divergences, the coterie of philosophes who came of age at mid-century did have one thing in common: they were writing in an era during which the "pious absurdities" of the past—such as the myth of Ham—would have no bearing on their discussion of the *nègre*.[11] Indeed, the basic "data" consulted by philosophes after about 1740 was limited to the physical realm. These sources ranged from classical notions of human difference (e.g., humoral theory as it applied to different human varieties) to much more recent anatomical discoveries that emphasized the corporeal specificity of the *nègre*.[12] This latter source of information was by far the most compelling for many philosophes. Not only was it better than the fantasy of scripture; it was also more reliable than what was found in African travelogues. After all, in contrast to both the Bible and the "unscientific" accounts of Africanness produced by slavers or missionaries, European anatomists were dealing with certifiable facts, not telling stories.

As mentioned in the Introduction, the first key discovery regarding African anatomy occurred in 1618 when Jean Riolan *fils* identified a new method for studying the structure of black skin.[13] Using a chemical agent to provoke a blister on his "Ethiopian" subject, Riolan extracted a skin sample in order to seek out the precise "source" of blackness on an anatomical level. In his published descriptions of this experiment, Riolan asserted that there were two distinct layers in Ethiopian skin: an outer layer that was indeed black, but also an inner layer, the *cutis,* that was white.[14] This white *cutis* in black bodies, he contended, confirmed the explanation of exterior human blackness dating from antiquity: the hot sun had produced the black African.

Although Riolan's discovery is rarely discussed by historians of science, his search for the specific anatomical structures of a particular human variety is a milestone in the history of race. Some 150 years before the era of a real racialized anatomy, Riolan had both identified the African body as a fruitful site for study and grasped skin's potential for the understanding of human varieties.[15] This development was not lost on other members of his profession, chief among them the famous microscopist Marcello Malpighi.[16] In 1665, Malpighi made what was undoubtely the most important skin-related discovery of the early-modern era: he identified a third and separate layer of gelatinous "African" skin located in between the outer "scarf" skin and the inner "true" skin. The detection of this "African" structure—which would come to be known as the *rete mucosum, reticulum mucosum,* or Malpighian layer—changed the way that European thinkers conceived of blackness.[17] Very much in contrast to Riolan's research, which maintained that black and white skin had the same basic anatomy, Malpighi had found evidence of a measurable and readily identifiable "racial" feature.

By the eighteenth century, discussions regarding the exact cause and significance of Malpighi's *rete mucosum* would become an international scientific debate involving Anton van Leeuwenhoek, William Hunter, Samuel Stanhope Smith, Bernard Siegfried Albinus, Voltaire, Petrus Camper, and Johann Friedrich Blumenbach, among others.[18] Judging from French eighteenth-century references to the *rete*, philosophes (and naturalists like Buffon) primarily came to know about the structure by reading Alexis Littré's replication of Malpighi's experiment in 1702. Littré, a prolific member of the Académie royale des sciences de Paris, was the first Frenchman to test Malpighi's hypothesis that the black color of the reticulum was caused by a thick, fatty black liquid.[19] His findings, which were based on repeated attempts to isolate the gelatinous substance of blackness by soaking African skin in various liquids, ultimately refined what Malpighi had affirmed several decades earlier.[20] While he failed to extract the stuff of "blackness" from African skin, Littré concluded that the pigmentation of the *nègre* was "partly due to the *reticulum* but also partly to the effect of hot air."[21] He also asserted, significantly, that a darker bile played a role in the pigmentation of Africans.

Like many of the theories proposed at this time, Littré's "anatomy, bile and climate" explanation of the *reticulum* echoed the classical belief in the "systematic relations between internal (physiological) traits and external (environmental) factors."[22] This biology/climate explanation of blackness achieved a new prominence in 1741 when Pierre Barrère published his influential *Dissertation sur la cause physique de la couleur des nègres, de la qualité de leurs cheveux, et de la dégénération de l'un et de l'autre.*[23] A version of the essay that he had submitted to the Bordeaux Académie royale des sciences during the same year, this treatise drew from the anatomist's dissection studies of African cadavers in Guyana. Like Littré, Barrère asserted that black pigmentation was found in the *corps réticulaire* and that the color itself was derived from a dark bile that tainted the skin and blood alike.[24]

Although Barrère's findings regarding the *nègre* did indeed rely on real experiments, his most influential ideas were actually based on a simple analogy, namely, that since yellow-tinged "white" men had yellow bile, black men necessarily had black bile. While this quasi-logic was noted by some of his contemporaries, for the most part Barrère's findings were very favorably reviewed (in the May 1742 *Journal des sçavans,* for example). Indeed, it is no exaggeration to state that Barrère convinced a generation of thinkers—Buffon and Diderot included—that the African had black blood and black bile, all evidence to the contrary notwithstanding. Frequently reprised in texts including the *Mémoires de l'Académie royale des sciences*, the *Journal de Trévoux*, and the *Encyclopédie*, Barrère's text ushered in a new era, one in which anatomical research was claiming the ability to put forward an all-encompassing explanation of the *nègre*'s fundamental difference.

Eight years after the publication of Barrère's *Dissertation*, Buffon explicitly addressed the anatomist's discoveries in the context of his 1749 theory of degeneration. While he had no reason to disbelieve Barrère's writings on the *nègre*, Buffon downplayed the importance of such "varietal differences" within the overall category of the human. Anatomical details such as the *rete* or black bile were simply ancillary "accidents" produced by the African environment, insignificant in and of themselves, Buffon believed. Examining the African from a purely anatomical point of view was thus entirely wrongheaded. To concentrate only on such anatomical observations was to overlook the more

Frontispiece to Pierre Barrère's *Observations anatomiques, tireés des ouvertures d'un grand nombre de cadavres* (1753). Barrère dissected black Africans while living in Guyana and asserted that he had discovered that black bile and black blood were the source of black skin. This image of the established anatomist hard at work in his dissection room comes from his later career in Perpignan. Courtesy H. Richard Tyler Collection of the American Academy of Neurology Library, Bernard Becker Medical Library, Washington University School of Medicine, St. Louis.

important question of the African's origins. As Buffon put it, "if one asserts that it is the blackness of the blood or the bile that gives this color to the skin, then instead of asking why the *nègres* have black skin, one will ask why they have black bile or black blood; this is thus to dismiss the question, rather than resolving it."[25] The significance of this statement cannot be underestimated. In rejecting this anatomical perspective, Buffon sought not only to orient his discipline toward what he deemed to be the more important environmentalist narrative of human mutation and reversibility; he attempted to avoid a discussion that would underscore the essential differences of the black African.

His own reluctance aside, however, Buffon's degeneration theory ultimately encouraged more, as opposed to less, speculation regarding the corporeal specificities of the African "variety." Indeed, his compelling narrative of racial transformation—from white to tan, tan to brown, and from brown to black—prompted anatomists to "flesh out" the story of degeneration. By the mid 1750s, a new series of significant anatomical discoveries regarding black physiology began to be claimed in print.

None of the resulting "findings" in anatomy were more important than those of the German anatomist Johann Friedrich Meckel. Building on the already sizable corpus of recycled beliefs regarding African blood, bile, phlegm, and skin, Meckel's ground-breaking dissection study of African cadavers in the 1755 *Mémoires de l'Académie royale des sciences et des belles lettres de Berlin* proclaimed that black skin was not the only part of the body that was dark; even interior parts of the African had different pigmentation.[26] Most significant, he claimed, was that his careful comparative study of African and European brains had revealed that Africans had darker, bluish brains: "The color of the medullar substance of the brain . . . differed slightly from that of other brains; this color was not white, as we ordinarily find in brains this fresh, but was bluish; and as soon as a detached piece of the brain was exposed to the air, it immediately became completely white."[27] Meckel supposedly observed a similar phenomenon when examining the pineal gland, which "was not, as one ordinarily finds, of an ashen shade, but of a blackish blue. From its base extended two completely white peduncles [stalks]. . . . [containing] the medullar substance arranged in strips between the cortical substance, which was bluish, or blackish."[28]

Meckel's new "discoveries" had a huge impact on nascent race science. Sanctioned by a range of thinkers and academic journals, including the *Bibliothèque des sciences* (1755), the *Annales typographiques* (1759), the *Journal des sça-*

vans (1766), and the *Journal encyclopédique ou universel* (1780), his findings were treated as fact for decades. Summing up the importance of this article for the *Nouvelle bibliothèque germanique,* Johann Heinrich Samuel Formey wrote that Meckel's important breakthroughs not only suggested a difference in color but also "a characteristic difference between the brains of *nègres* and of whites."[29]

This reorientation toward evoking deeper, organ-based differences (as opposed to simply blood or skin) became widespread in the subsequent decades. Ten years after Meckel's study became widely known, Claude-Nicolas Le Cat took up the question of the origin of human skin (and human varieties) in his 1765 *Traité de la couleur de la peau humaine*. Refuting the earlier belief that the source of black pigmentation was the bile and blood of Africans, Le Cat confirmed Meckel's findings regarding bluish African brains with his dissection studies at the Hôpital de Rouen.[30] But Le Cat went far beyond Meckel as well, putting forward the most comprehensive understanding of blackness to date by positing the existence of an elemental African fluid that he dubbed *oethiops* or *ethiops*. Based on his belief that this ink-like fluid must come from the brain—which he considered the clearinghouse for all nervous activity—Le Cat affirmed that *ethiops* flowed throughout the body and into the interior of the "nerves" of the *nègre*.[31] In addition to asserting that this liquid stained structures such as the *rete mucosum* along the way, Le Cat also confidently declared (although by analogy) that it necessarily darkened African sperm: "This is the ancient theory proposed by [the Greek geographer] Strabo, that the color of man's skin is found in the semen of his ancestors. Here it is, I say, established by observation, because no one doubts that the brain is a spermatic organ, much like the fertile seed that produces the rest of the animal."[32]

Le Cat's revelation had an immediate effect on the understanding of what was fast becoming a more fixed view of racial transmission. Not only did scientific works such as the 1771 *Manuel du naturaliste* cite Le Cat's findings, but a significant number of writers readily incorporated the notion of a degenerate African "seed" into their own theoretical understandings of the African.[33] The most significant and complete interpretation of this view of black physiology was undertaken by the Dutch geographer Cornelius de Pauw in his *Recherches philosophiques sur les Américains,* first published in 1768. De Pauw's environmentalist theory was a marked departure from Buffon's, although his approach to the varieties of the human species was certainly Buffonian.[34] If de Pauw affirmed that the *nègre*'s color was indeed the direct result of the brutal climate in which he lived, the Dutchman also emphasized the specific physical changes

Frontispiece to Claude-Nicolas Le Cat, *Traité de la couleur de la peau humaine* (1765), showing a seated colonial European woman and her pet monkey framed by a yellow Asian, a black African, and a red Amerindian. In this treatise on human skin, Le Cat asserted the existence of an elemental fluid that flowed from the black African's brain, which he dubbed *ethiops*. The Latin reads "Not a single face, not a single color." Courtesy Harvey Cushing/John Hay Whitney Medical Library, Yale University.

that the *nègre* had supposedly undergone as a result of degeneration. Drawing from Meckel and Le Cat, de Pauw asserted that autopsies of black Africans had demonstrated that the soft tissues of the brain were "blackish" and that the pineal gland was almost entirely black. He also asserted that "the inside of the optic nerves [was] brownish," and "the blood is of a much darker red than ours."[35] These characteristics, according to de Pauw, were transmitted on the most essential level, since *nègres*' "spermatic fluid" was "colored by the same process that we find throughout their mucus membrane."[36]

Before de Pauw, traditional environmental theory had generally explained away differences between whites and blacks as insignificant corollaries to environmental shifts in color. De Pauw was convinced that the exact opposite was true: the color of the African was, in fact, symptomatic of a larger, prior set of underlying physiological changes.[37] Indeed, de Pauw argued that the African's tainted sperm was ultimately responsible for the other measurable and demonstrable liabilities of this particular "variety," particularly the race's intellectual deficiency and poor memory.

Black African sperm, which de Pauw called an "active and violent" structure, had become the metonymy for a pathological, gendered, and quasi-permanent transmission of blackness. Anatomically speaking, the liabilities that defined the black race now seemed more than skin deep; they were ingrained, permanent: in a word, hereditary. As de Pauw put it, *nègres* seemed forever stuck in the initial and unfortunate stage of human development; they were ensnared in a "childhood" or "original state" that would prevent them from civilizing themselves "if they remain[ed] perpetually below the equator, exposed to the hottest temperatures that the Earth experiences."[38]

This more stable understanding of the black African category prompted de Pauw to fantasize about the best way to eradicate blackness (and black sperm) despite its "tenacious" nature. His solution involved the purging of the African's *matière colorante* through prolonged race mixing: from white (and black) to mulatto, from white and mulatto to quateron, from white and quateron to octavon and, finally, from white and octavon to a "true" white. While this process might be seen as evoking an optimistic mixture of different human varieties, nothing was farther from Pauw's intent. In the overall context of his *Recherches philosophiques*, it was through this purification of blackness that the unfortunate subcutaneous "black gel" of African physiology might be eradicated, along with a whole host of other unfortunate African physiological and conceptual features. In short, the only way that the nefarious effect of black sperm could be remedied was through the repeated introduction of white sperm into the racial economy.

To a large extent, de Pauw's reference to racial bleaching intersected with his skepticism regarding racial reversibility.[39] Although he did not completely reject Buffon's belief that white or black humans could literally change "races" over several generations if they moved to a different environment, he nonetheless contended that these shifts would be quite different for each race. In the case of a white to black transition, for example, the essential and unblemished white group could easily take on the pigment of the African if exposed to a hot climate. The opposite scenario (black to white) was not nearly as straightforward and effortless according to de Pauw. Given the fact that, as he maintained, the black African race had already suffered a tremendous amount of damage on the level of its anatomy, this "return" to whiteness could only be achieved with great difficulty and over a much greater span of time:

> I have no doubt whatsoever that it would take *longer* for *nègres* migrating to the northern provinces of Europe to lose their blackness than it would for Europeans in the heart of Ethiopia to become *nègres*. [This is because] the spermatic fluid and the soft glandular substance of the Africans, which has become colored and impregnated by this acidic substance that we call *animal ethiops*, would subsist [in the *nègre*] for many years. Passed on from father to son, it would disappear only after a succession of many generations.[40]

Although de Pauw's anatomy-driven understanding of degeneration may not seem to add much more to the evolving set of dissimilarities that supposedly existed between whites and blacks, his recasting of the traditional climate-theory narrative had a huge effect on the overall conception of the *nègre*. Already ostracized from the high-functioning world of European minds by the era's anatomists, the *nègre* was now denied even the possibility of environmental redemption.

For many thinkers of his generation—naturalists and philosophes alike—de Pauw's view of black sperm offered an appealing answer to the uncertainty of climate theory. In England, Tobias Smollett's *Critical Review* gave a very favorable evaluation of this discovery in 1771.[41] Likewise, the German naturalist Johann Reinhold Forster also took time to comment very positively on de Pauw's findings in his 1778 *Observations made during a Voyage Round the World*.[42] In France, it was the second-tier philosophe J.-B. Delisle de Sales who seized most forcefully upon the implications of de Pauw's "spermatic" breakthrough. To de Pauw's assertion that the brutal African climate had literally dimmed the *nègre*'s sperm, organs, and brain marrow in his 1773–74 *Essai philosophique sur le corps humain*, Delisle de Sales added his own speculation that Africans' brains had suffered "fatal degradation," judging from "the weakness of their intellect."[43] According to Delisle de Sales, it was clear that the heated air that the African was forced to breathe on a daily basis made the "balsamic" elements of his blood evaporate, wiped out the "organ of memory," and damaged his overall nervous system or "bundle of sensitive fibers."[44] The overall result, according to Delisle de Sales, was a general "inertia of the mind that differs little from stupidity."[45] Despite his belief in humankind's common origin, Delisle de Sales's understanding of de Pauw's anatomy led him to believe that the black African barely rose above the level of an animal.

Although Delisle de Sales's untamed guesswork was presumably not taken very seriously during his own era, his understanding of the *nègre* was significant in that it was among the first to conflate climatic, humoral, and anatomical causalities in order to provide a coherent portrait of the moral and cognitive failings of the *nègre*. In this respect, Delisle de Sales's view of the *nègre* heralded an increasingly powerful belief in an elemental difference between whites and blacks. This can be seen in several spheres of eighteenth-century thought. Ostensibly forward-thinking members of the medical profession, for example, found it of paramount importance to provide special manuals (aimed primarily at Caribbean planters) that took Africans' particular physiology into consideration. In his 1776 *Observations sur les maladies des nègres*, Jean-Barthélemy Dazille began his study of African diseases by lamenting the poor treatment of slaves, as well as the era's inadequate understanding of specific liabilities of the African body, chief among them "the natural dryness and aridity of the fibers."[46] Such statements regarding the identifiable nature of African physiology were only the beginning of Dazille's meditations on a branch of the "human species" that was "the most unhappy and the most neglected, despite its usefulness."[47] In a discussion where Dazille described the unnecessarily high mortality of slave populations in the Caribbean due to "insufficient nourishment, a shortage of clothing, and tasks well beyond their capacities," the physician chided the colonists for causing unnecessary deaths, in particular, for allowing a significant percentage of the *nègres*' "annual generative production" to perish.[48] While Dazille was criticizing European "greed" on a certain level, the clinical euphemisms he used to refer to the death of African babies certainly speak volumes regarding the growing conceptual difference between Africans and Europeans.

In sum, the overall understanding of the black African body to which mid-century philosophes subscribed differed markedly from earlier flippant assessments of African physical features, whether in regard to "wool-like hair," "ugly, flat" noses, or "thick" lips. By the late 1750s, the more anatomically oriented study of African skin, blood, brains, and sperm suggested something altogether more significant. Indeed, if Europeans had long maintained that black Africans suffered from a series of anatomical and intellectual shortcomings, these long-standing prejudices were increasingly taking on a conceptual significance. In the first place, a number of thinkers in Europe were reintroducing more trenchant classification schemes according to which the *nègre* "occupied the lowest rung among humans on the chain of being."[49] Even more ominously, a vocal group of unapologetically pro-slavery writers were increasingly giving credence to polygenist theory, as well as to the idea of the African's possible consanguineous relationship with apes.[50]

Montesquieu and the "Refutation" of Difference

The "reality" of African anatomy both fascinated and troubled Enlightenment-era thinkers. While the most compelling "data" on the African body were not published until the second half of the eighteenth century, the possibility of some sort of essential difference had vexed philosophes since the 1740s. Among the thinkers associated with the so-called High Enlightenment, it was Montesquieu who was the first to address the clear political implications of his era's beliefs regarding the black African body.

Montesquieu's meditation in *De l'esprit des lois* on the conceptual status of the *nègre* came about during his often-cited and highly ironic justification of African chattel slavery. I have translated the passage in its entirety below, numbering the points for convenience.

> If I were obliged to defend the right that we had to enslave the *nègres*, here is what I would say:
>
> 1. The peoples of Europe, having exterminated those of the Americas, were obliged to make slaves of the Africans in order to clear so much land.

2. Sugar would be too costly if the plants that produce it were not cultivated by slaves.
3. These slaves are black from head to toe; and they have such flat noses that it is almost impossible to pity them.
4. It is hardly to be believed that God, who is such a wise Being, should place a soul, especially a good soul, in so black a body.
5. It is so natural to look upon color as the essence of human nature that the Asians [i.e. Turks and Persians] who make use of eunuchs, deprive blacks of that which they have in common with us in a more marked fashion [i.e., by amputating both testicles and penis].
6. The color of the skin may be determined by that of the hair. This was so important to the Egyptians, the best philosophers in the world, that they put to death all red-haired men who fell into their hands.
7. The *nègres* prefer a glass necklace to one made of gold, which is so valued by civilized nations. Can there be greater proof of their lack of common sense?
8. It is impossible for us to suppose these creatures to be men; if this were truly the case, it would follow that we ourselves are not Christians.
9. Weak minds exaggerate too much the injustice done to Africans. For, were this truly the case, surely the princes of Europe, who [spend their time] negotiating so many useless treaties, would have agreed to a general convention on behalf of mercy and compassion?[51]

Coming into focus against the preceding chapters of the *De l'esprit des lois*, where Montesquieu both explains and often condemns the institution of slavery in a range of political systems, this ostensibly pro-slavery discourse occupies a preeminent position in the history of anti-slavery thought, as well as in the history of the African's "difference."

Of the nine reasons that Montesquieu would supply were he obliged to defend the enslavement of Africans, the first two are basely and self-justificatorily mercantile. Having exterminated the Amerindians, he affirms, Europeans clearly need Africans to toil for them in America. Furthermore, were whites not to make use of Africans, he continues, sugar would be too expensive in Europe. Montesquieu's insincerity in both of these cases is clear. In addition to the fact that neither of these reasons actually justifies slavery, what they *do* demonstrate—through an unthinking circular logic—is that the genocidal "imperatives" that explain the continued existence of human slavery in European colonies can be reduced to nothing more than the importance of European luxuries and pleasure. Eighteenth-century readers, familiar as they were with the mode and biting tone of the 1721 *Lettres persanes*, would have clearly heard the screaming irony and absurdity in such statements.

In the ensuing seven "reasons" rationalizing African slavery, Montesquieu focuses on the conceptual status of the African body, not to mention the African soul. Here his "pro-slavery voice" explicitly takes up the question of *difference* and cites African physical features, if not as a justification, as an *explanation* for Europeans' willing blindness to the plight of this group. In a sentence reminiscent of the *Lettres persanes'* celebrated "comment peut-on être Persan?" remark, Montesquieu proclaims that the utter blackness and flat nose of these Africans makes them difficult to pity.[52] While designed, rhetorically, to ridicule a justification based on nothing more than pigmentation and nose shape, reason number three also highlights the irrefutable reality of the African body in the era's thought. What is obliquely called into question in point number three is not the essential corporeal difference between whites and blacks; it is the race-based moral relativism that permits enslavement in the first place. Montesquieu's next "reason" is the logical corollary to this reading of the African body; if African corporeality is either deficient or different, he asks, how could a supremely intelligent divinity put a soul in such a body?

By the time his pro-slavery "voice" articulates his fifth reason—"that color is the essence of humanity"—Montesquieu has attacked the very foundation of slavery's biopolitics.[53] This comes through most clearly in his ironic praise of those civilizations that make use of color as a determining factor in the way that they view, treat, and potentially enslave others. In reason five, for example, Montesquieu's pro-slavery voice lauds Asians for recognizing the black's separate status and signaling this conceptual inferiority by castrating their slaves. In reason number six, Montesquieu commends the ancient Egyptians for having employed hair type (another "essential" African trait) to distinguish among human varieties and supposedly putting redheads to death. These extreme and absurd examples of foreign cultural practices set up the two final rationalizations of human bondage in this chapter, both of which employ a *modus tollens* argument designed to prove the contrary of their initial assertions. Reason number eight, a typical example of Montesquieu's *humeur noir,* is that "Africans are not men," since if they were, it would mean that we are not Christians, because Christians would never treat *men* in such a manner. Finally, reason number nine is that if Europeans were truly harming Africans,

Europe's rulers would have already done something about it; since they have not, the suffering of African slaves is clearly being exaggerated.

Montesquieu was the first French thinker to criticize the long list of beliefs regarding Africans' fundamental difference (and the implicit smugness of whiteness that such notions generated). He was also the first to show how these same convictions had allowed Europeans to rationalize the mass suffering of black Africans during his era. The targets of his chapter 5 are exhaustive; physiognomy and (white and black) color essentialism, ethnography and conceptions of the African mind, metaphysics and mercantilism, divine teleology and government complicity—all are subjected to ridicule. While eighteenth-century writers including Raynal criticized Montesquieu for the flippant tone he used to treat such a grave subject, this short chapter is a critical moment in the evolving discussion of the link between *Africanness* and slavery.

The Nagging Context of Montesquieu's Anti-slavery Diatribe

As an often-cited, stand-alone example of anti-slavery thought, Montesquieu's chapter 5 undeniably played a critical role in highlighting the spurious justifications of African chattel slavery. When fellow philosophes thought of Montesquieu and the question of human bondage, it was this passage that they evoked. Voltaire, for one, quipped that Montesquieu had brilliantly highlighted the plight of the *nègre* with Molière's paintbrush.[54] Nonetheless, despite Montesquieu's undeniable role in unmasking the specious defense of human slavery, this famously ironic chapter is bookended by other considerations relative to the African and slavery that raise serious questions about the status and political import of human difference in *De l'esprit des lois*.

In its characteristic way, the partitioned text that is *De l'esprit des lois* does not contain a comprehensive or easily accessible theory regarding the conceptual status of the black African. Compared to Buffon, Voltaire, or Diderot, Montesquieu hardly engages with the era's Africanist ethnography.[55] Instead, he scatters his implicit or explicit beliefs regarding the *nègre* in three main areas: in the aforementioned ironic chapter on the justification of slavery, in his discussion of the effect of climate on *des peuples sauvages et des peuples barbares*, and in his overall assessment of the different types of *esprits* that characterize the world's peoples.

In *De l'esprit des lois*, the African generally falls into the category of a *sau-*

vage.[56] Although a descriptive rubric in Montesquieu's thought—referring to small, dispersed groups involved primarily in hunting—the term *sauvage* also plays a key role in Montesquieu's theory of human difference. In book 19, he asserts quite broadly that the minds of the world's *sauvages*, unlike the minds of more sophisticated peoples, are the uncomplicated product of two major influences: (1) the brutal environment in which *sauvages* live and (2) their nature, which serves as a catchall notion encompassing both anatomy and a limited, mechanical psychology. "Men are ruled by many things: climate, religion, laws, rules of government, precedents set by past events, mores, customs; a general spirit results from these. . . . Nature and climate are nearly the only forces that bear upon savages."[57]

In contrast to the European, whose national or general *esprit* is supposedly affected by *both* physical (e.g., climate) and moral (e.g., tradition, manners) influences, the *sauvage* mind is the seemingly ahistorical product of primarily mechanical forces. To explain the specifics of these physical influences in book 14, Montesquieu borrows directly from John Arbuthnot's *An Essay Concerning the Effects of Air on Human Bodies* (1733), a work that updates Hippocrates's understanding of air-based national dispositions with new information on the "mechanical" results of different climates on human fibers and character.[58] Arbuthnot enables Montesquieu to describe the specific effects of cold and hot climates on the functioning of the human body:

> The cold air tighten[s] the exterior fibers of the body; this increases their elasticity, and enhances the flow of blood from the extremities toward the heart. [The cold also] decreases the length of these same fibers; as a result, their strength is further increased. On the contrary, hot air loosens the ends of fibers, and lengthens them; as a result it decreases their elasticity. Thus, one has more vigor in cold climates.[59]

This speculative and deterministic physiology provided a useful typology for Montesquieu's readers.[60] In cool or cold climates, which supposedly have the effect of retracting body fibers and—counterintuitively—increasing blood flow, humans have more confidence, more knowledge of their superiority, more courage, and less desire for vengeance.[61] These are only some of the advantages, according to Montesquieu. In particular, he asserts, cold-weather brains are not subject to the full range of raw sensation that potentially excites the nerves, because cold has the effect of compressing or paralyzing nerve endings in the skin called *mamelons* ("nipples") and *petites houppes* ("little tufts").[62]

While this might seem like a disadvantage, the infinite number of "small sensations" that do in fact filter through to the cold-climate brain, according to Montesquieu, produce the defining characteristics of a high-functioning (viz., European) mind: "imagination, taste, sensitivity, and vivacity."[63] By contrast, those humans who live in hot climates and who suffer from the aforementioned "distended or relaxed fiber endings" become sloth-like, pleasure-driven, uninspired machines: "The heat of the climate can be so excessive that the body becomes absolutely lethargic. In these cases, the reduction of strength extends to the mind itself; no curiosity, no noble enterprise, no generous sentiment; all inclinations become passive; idleness constitutes happiness."[64]

This physio-psychological view of the warm-climate human had decidedly political implications. In a digression strangely reminiscent of Charlevoix and other Caribbeanist authors, Montesquieu asserts that the inhabitants of hot countries—which clearly refers to Africa in particular—have such a hard time thinking and conducting themselves that a life of punishment and slavery is perhaps easier for them to endure than their endless indolence. As Montesquieu puts it: "the majority of punishments [for the inhabitants of hot countries] are easier to endure than mental exertion; servitude is also more tolerable than the force of mind necessary for human affairs."[65] Montesquieu follows this speculative explanation of "servitude" with a more pointed discussion of his own era's so-called cruel slavery.[66] Dialoguing with Aristotle's belief, from the *Politics*, that there are people whom nature destines to become slaves, Montesquieu equivocates, since this seemingly coincides with the climate-based explanation of slavery that he provides elsewhere:

> Aristotle wants to prove that there exist slaves by nature, yet what he says hardly proves this. I believe that, if such slaves do exist, they are the people [in warm climates] of whom I have just spoken. But, as all men are born equal, one must admit that slavery is against nature, even if, in certain countries, it is founded on natural reason; further, we must clearly distinguish these countries [where slavery seems natural] from those where even natural reason rejects slavery, which is fortunately the case in European countries where slavery had been abolished.[67]

Characterized by subordinate clauses and hedging, this paragraph is difficult to parse. Put schematically, Montesquieu asserts that:

1. Aristotle has not proven the existence of natural slavery in the *Politics*.

2. If there is a natural slavery, it is restricted to those indolent, cowardly people who live in hot climates and who can only accomplish difficult work through the threat of punishment.
3. This restriction regarding natural slavery, that is, its justification in particular climates, contradicts a fundamental axiom, "all men are born equal."
4. But if slavery is a violation of the natural equality of humankind [contrary to what is asserted in Aristotle's *Politics*], it does sometimes make sense [via *raison naturelle*] in hot climates.
5. We must therefore distinguish between countries where "natural reason" provides the foundation for human slavery and those countries where this same "natural reason" rejects it.

At the heart of this seemingly inconsistent passage are two competing epistemologies, namely, the pressing "reality" of the indolent black, warm-climate body and the universal notion that, as Montesquieu himself puts it in 15.7: "all men are born equal."[68] In the above quotation, Montesquieu brings these two conflicting notions into play by evoking the labor that must be accomplished in any society, as well as the means that must be used to motivate those performing said work. In hot climates where cowardice and slothfulness reign, the ethnographic reality of the labor pool clearly complicates the idea that men are not *naturally* born into slavery. Indeed, in such places, enslavement "is less offensive to reason."[69]

This physiological and psychological relativism vis-à-vis the question of servitude has not gone unnoticed by Montesquieu scholars. Some have privileged the anti-slavery elements of *De l'esprit des lois* over more problematic presentations of human bondage. Sue Peabody concludes, for example, that despite contradictions in his environmental theory, the most important aspect of Montesquieu's thought is that he opposes slavery.[70] Laurent Estève and others have taken a more aggressive position on this question. While Estève does not accuse Montesquieu of being a *négrier* (slave trader) as Louis Sala-Molins does in *Les misères des Lumières* (1992),[71] he nonetheless has highlighted the most sticky and inconsistent aspects of Montesquieu's thought:

> It is in the context of his reflection on the legal and ideological foundations of slavery—and on the slave trade itself—that we must emphasize the fact that climate theory in Montesquieu's *De l'esprit des lois* is utterly subordinate to the legitimization of natural servitude. We must also stress the semantic slippage of

> the concept of natural order in *De l'esprit des lois*. For some, this notion implies the obligation of being respected as persons because they are beneficiaries of a certain anthropological and political status, for others, it validates the crack of the whip.[72]

In addition to suggesting that Montesquieu's environmental relativism calls into question the status of his anti-slavery diatribe, Estève argues that Montesquieu's understanding of ethnographic *spirits* produces a parallel natural (physical) order for the black African, an order that explains, if not justifies, human bondage. Such a notion was presumably so distasteful to Montesquieu that he quickly accompanied the assertion that slavery might *reasonably* exist in various lands with a prescriptive and restrictive declaration: "It is necessary to limit natural servitude to a few particular countries on Earth."[73] The implications of this sentence are staggering. While Montesquieu clearly found the enslavement of any human to be antithetical to his overall philosophical project, his understanding of climate and the *nègre*'s physiology produced a significant exception to this rule: although unfortunate, under certain conditions, enslaving the black African was a *reasonable* decision.[74]

Voltaire: The Philosophe as Essentialist

Given his speculative views on the acceptability of slavery under certain climatic conditions, it is hardly surprising that Montesquieu sought proof of this type of bondage in the era's travelogues. Such is the context for Montesquieu's claim in book 15 of *De l'esprit des lois* that the inhabitants of Achim (Achem, Sumatra) were engaged in a continuous effort to sell themselves into slavery for hope of a better or at least more secure life:

> In Achim, everyone is looking to sell himself. Some of the chief lords have no fewer than one thousand slaves; these [slaves], who are the principal slave merchants, have a great number of slaves themselves, and these [slaves] are also not without their own slaves as well. People inherit their slaves and trade them. In these states, the free men are too weak to oppose the government and seek to make themselves slaves of the tyrants in office.[75]

Voltaire refuted this assertion in his *Dictionnaire philosophique* (1764) in the strongest of terms:

> The author of *l'Esprit des lois* [writes] that, according to William Dampier's account, "everyone is looking to sell themselves in the kingdom of Achim." What

> a strange trade that would be. I have never once seen anything in the *Voyage de Dampier* that even approaches such an idea. It is unfortunate that a man with such a great mind [Montesquieu] speculated on so many subjects, and cited [authors] incorrectly so many times.[76]

Although Voltaire takes Montesquieu to task for a supposed falsehood in this passage, it was the right to enslave that was the real point at issue. Montesquieu, as we have seen, explained chattel slavery as the logical result of a certain set of conditions. Voltaire, on the other hand, believed that such factors should not be used to rationalize human bondage even if there was a clear difference between whites and blacks (or any other groups); he generally rejected the institution outright, most famously in *Candide* (1759).

This aspect of Voltaire's thought seemingly presents a striking inconsistency. After all, there was no other major contemporary French thinker who spent as much time as Voltaire did arguing that the African was *essentially* different from the European; Montesquieu's views of the *nègre*, by way of comparison, are extremely mild. And yet, if Voltaire is one of the earliest eighteenth-century writers to put forward a brutally trenchant view of the human species, he generally seems to have carefully withheld his remarks about the African's problematic physiology and intelligence from political discussions, perhaps to avoid undermining his anti-slavery diatribes. Thus, while Voltaire avoided generating an ideology of race, what he did put forward was one of the first pseudo-scientific views of the African's fundamental dissimilarity from Europeans.

Voltaire developed his views on the African race at a relatively early stage in his career. While most other philosophes of his generation did not engage with the question of the African body before the 1750s, the young Voltaire set out his understanding of the *race nègre* as early as his 1734 *Traité de philosophie*. In this unpublished philosophical primer/*conte philosphique* written for Madame du Châtelet, two main questions relative to the black African preoccupy Voltaire: the origin of the *nègre* and the generally accepted idea that the *nègre* was part of a shared lineage with Europeans.

Voltaire's treatment of these particular questions begins with what he deemed to be the most popular and most unempirical theory of human origins, the Old Testament account of a shared genesis.[77] Whereas naturalists including Buffon debunked such theories implicitly—by simply avoiding any mention of spiritual causality—Voltaire preferred to ridicule the Bible directly, presenting it as a series of far-fetched myths.[78] To convey this message effectively to Madame du Châtelet, Voltaire conjured up an extraterrestrial character who arrives on Earth, as he did in many of his short stories.

The "narrative" in this part of the *Traité* is quite simple. During his travels, an alien from "Mars or Jupiter" arrives in Goa, where he meets a priest dressed in a long black cassock, who explains the existence of the different varieties of men living in this city: "All of these different men that you see, he tells me, are born from the same father; and he then proceeds to tell a long story [i.e., Genesis]."[79] The alien then poses a "naive" question to the priest, which is designed to highlight the inadequacies of the cleric's beliefs: "I enquire whether a *nègre* and a *négresse*, with their black wool and flattened noses, occasionally create white children with blond hair, aquiline noses, and blue eyes; if some people without beards have come from bearded people, and whether the whites have ever produced yellow people."[80] As anticipated, the answer that the alien receives explicitly refutes the priest's biblical view of humankind: "The response is no; *nègres* who have been transplanted to Germany, for example, only produce *nègres*."[81] While anything but decisive for the modern reader, Voltaire's staging of this "fact"—that parents engender members of their own race—reflects his belief that white and black races are, barring any *métissage*, fixed, independent categories that remain constant regardless of climate.

While such a stand may seem directed against the Buffonian view of race that became widely known after 1749, Voltaire was actually rebelling, in this 1734 text, against the extremely influential belief diffused by Abbé Dubos that environment played a determinant role in the propagation not only of genius, but also in the forming of racial characteristics. This target clearly explains why Voltaire constantly evoked botanical metaphors such as "transplanting" Africans: he was mocking the tropes that Dubos employed in his very popular 1719 *Réflexions sur la peinture et sur la poésie*, among them the use of botanical images in order to explain the different brains that develop in different climates:

> Just as two seeds from the same plant produce fruits of different qualities when these seeds are sown in different regions, or even when they are sown in the same region in different years, two children who are born with their brains formed in precisely the same manner will become two different men with regard to their mind and their disposition if one of them grows up in Sweden and the other in Andalusia.[82]

As indicated in the previous chapter, Dubos's view of human difference not only explained existing human varieties as a function of climate; his transformist environmentalism evoked the possibility of *category-changing* racial shifts, the most telling example being an African living in England who "would finally lose the natural color of the *nègres*."[83] Voltaire scoffed at such notions in the *Traité*, reducing them to a profound ignorance: "never has an even slightly educated man suggested that a species, without intermixing, would degenerate; the only one to make such a foolish remark is Abbé Dubos."[84]

To underscore this message within this *conte philosophique*, Voltaire has his extraterrestrial set off on a worldwide trip, during which he encounters the different species of men who populate the globe. His first destination is Caffreria, home of the Hottentots, whom the unsuspecting extraterrestrial first associates with the land's other extreme and exotic "fauna." Voltaire's ethnography is a bit hazy at this point and he calls the Hottentots *nègres* in this text.

> I see monkeys, elephants, *nègres*, who all seem to have a bit of reason, albeit deficient. They all have a language that I do not understand at all, and all of their actions seem to relate to a certain end. If I were to judge them according to the first impressions that they made on me, I would have to believe that of all of these beings, the elephant is the reasonable animal.[85]

After making this unintentional race-based joke regarding the reasoning capacities of the *caffre*, Voltaire's extraterrestrial ultimately identifies these *animaux nègres* as the most intelligent creatures in the region because they possess the most fully articulated language.[86] This identification of the *nègre hottentot* as the most reasonable animal on earth is doubly sardonic. In addition to the fact that it is the Hottentot's language that earns him this distinction—Voltaire refers elsewhere to the click language of this group as a gobbling "turkey" speak—the distinctive characteristics of the *nègre hottentot* become the source for the alien's initial definition of humankind: "Man is a black animal with wool on its head, walking on two paws, almost as well-postured as a monkey, less strong than other animals of its size, having slightly more ideas than they, and a greater capacity to express them; otherwise subject to all of the same needs; birth, life, and death just like them."[87]

This working definition of the human species is of course put to the test as the otherworldly traveler moves across the globe. In China, for example, Asian physical features force the alien to alter his notion of man: "I am now forced to change my definition, and to classify human nature under two different species, the yellow one with horse hair, and the black one with wool."[88] This binary classification of humankind is then further challenged by the proliferation of varieties that the alien encounters during the rest of his trip:

> But in Batavia, Goa, and Surat, which are the meeting places of all nations on Earth, I see a great number of Europeans who are white and have neither horse hair nor wool, but loose blond locks on their heads as well as beards on their chins. I am also shown many [Native] Americans who have no beards at all; clearly my definition of man and [the number of] species of men must both be significantly increased.[89]

Reacting to the multiplicity of disparate types of humans, the extraterrestrial concludes that the morphological differences he confronts must indicate genealogical differences. Once again dialoging obliquely with Dubos, Voltaire uses his own botanical metaphor to deride the general intuition that there is but one species of man on earth:

> It seems to me that my belief that there are different kinds of men, just as there are different kinds of trees, is well-founded; pear trees, fir trees, oak trees, and apricot trees do not at all come from the same [kind of] tree, and the bearded whites, the wool-bearing *nègres*, the horse-haired *jaunes*, and the beardless men, do not come from the same [species of] man.[90]

By the time that Voltaire composed this text in 1734, his fixist, polygenetic view of humankind was already set in stone. During the next three decades, however, Voltaire would be forced to think through this stance on multiple occasions as developments in the life sciences explicitly tested his convictions. At no time was this truer than when Voltaire's contemporaries began focusing, almost collectively, on the category-crossing *nègre blanc*, or African *albino*. As chapter 2 has demonstrated, the question of albinism was inextricably linked to nascent theories of race at mid-century, and played a catalytic role in the advancement of a monogenetic theorization of human origins. Not only did Maupertuis and Buffon identify the birth of *nègres blancs* as clear proof of the vestigial links between black and white "races"; they inferred from this phenomenon that the white race was the prototype, to which the black race occasionally reverted when albinos were born. Indeed, by the 1750s, most French naturalists assumed that the albino was the embodied evidence of an essential and past "biological" link between whites and blacks. But Voltaire's polygenetic understanding of human *races* would not allow him to accept such a theory. If Maupertuis and Buffon used the albino to theorize a kinship connection between the "varieties" of the human species, Voltaire's reaction would be the exact opposite: he identified this strange white being as a specific *species of human*.[91]

Voltaire and the Albino of 1744

Like many prominent thinkers residing in Paris, Voltaire had the opportunity to examine the small albino who was displayed at various Parisian venues in 1744; his viewing took place in a private setting, at the Hôtel de Bretagne. Voltaire's *pensées* regarding this strange creature—Voltaire claimed he was a full-grown, but pygmy-sized man—appeared in print the following year in the 1745 "Relation touchant un Maure blanc amené d'Afrique à Paris en 1744." Although no scholar has commented on the title of this work itself, much can be learned by looking at the descriptive heading that Voltaire chose for this short essay. Instead of referring to this small boy as a *nègre blanc*, as was the customarily the case, Voltaire invented the neologism *maure blanc*, or "white Moor." The reason behind this new term becomes clear when one remembers that the very notion of a *nègre blanc* had come to indicate degeneration from the *nègre* "variety" into an ancestral whiteness. In Voltaire's opinion, this was a physical impossibility: unmixed races did not degenerate into each other; they were separate entities.

Voltaire's title also contains another significant albeit spurious detail; according to this short essay's designation, this *maure blanc* was taken from a tribe of his own kind in deepest *Africa* (*amené d'Afrique*), although it was common knowledge that the albino had been born of black parents in South America. This inconsistency flummoxed eighteenth-century readers; it has also puzzled contemporary critics. But Voltaire's "error" actually makes complete sense in the context of his fixist worldview. In the exact same way that the term *maure blanc* served notice that there was no link between this white-skinned being and either the white or *nègre* race, the supposed geographical origin of this so-called *maure blanc*—darkest Africa—allowed Voltaire to assert that this was a geographically specific race. All evidence to the contrary notwithstanding, Voltaire simply could not admit that this freakish and diminutive "man" had been engendered by anyone other than his own monstrous kind in situ. In short, this unnaturally white-skinned human was a member of a separate species (and nation) of men who presumably came from somewhere near the Congo.

Such were the limits of Voltaire's empiricism. While both Maupertuis and Buffon were obliged to rethink their overall conception of race in light of the albino, Voltaire not only forced the newly dubbed *maure blanc* into his own preconceptions, he marshaled his vast repertory of rhetorical skills to drive home his polygenist explanation of this strange race. After brutally dehumanizing this strange pink-eyed "small animal," described as "white as milk," Voltaire suggested that this creature was a curious assemblage, with "fuzzy wool like a negro" and a "muzzle shaped like that of a Laplander."[92] These comparisons, although they drew parallels between the supposedly unpleasant traits of the albino and certain stereotypical aspects of other races, were anything but an attempt to link the creature to other groups through what one might call comparative anatomy. Indeed, Voltaire cited the varied and conflicting points of overlap between this *nouvelle richesse de la nature* and other human categories in order to question the criteria according to which other thinkers (perhaps too freely) assigned the category of the human to phenomena such as the *maure blanc*: "This animal is called a *man*, because he has the gift of speech, memory, a little of what we call reason, and a sort of face."[93] Far from evoking any filiation between this creature and the rest of humankind, Voltaire cited the *maure blanc*'s curious racial markers as a sign of its overall monstrosity, its fundamental contradiction with anatomical and racial norms.[94] Reminiscent of how he had treated the *nègre hottentot* in the *Traité de métaphysique*, Voltaire plainly suggested that this esthetically unappealing creature belonged to a category that was hermetically sealed off from the rest of humanity.

In assessing Voltaire's disquisitions on albinos, Emeka P. Abanime deemed him an anthropologist.[95] Technically speaking, however, Voltaire's writings on the subject reflect neither a great interest in the supposed "culture" of albinos nor a belief that travelogues describing these white Moors could really teach him anything he could not already intuit. Nonetheless, Voltaire's views had shifted slightly by the time that he once again took up the questions of race and the albino in the *Essai sur les moeurs et l'esprit des nations* (1756). In the first place, the albino had made conceptual progress in the interval between Voltaire's former and newer work. Formally relegated to the status of semi-human in the 1745 "Relation," the albino had now become a man like all others, which is to say a member of a human race, albeit separate and defined by its unpleasant morphology and sickly color. This racial specificity was so obvious, according to Voltaire, that only a blind man could not see that "the Whites, the *Nègres*, the Albinos, the Hottentots, the Laplanders, the Chinese, and the [Native] Americans are entirely different races."[96] Furthermore, Voltaire now speculated more openly on the ethnography of albinos (still not calling them *nègres blancs*) by combining elements of Dapper's reports on their pariah status in Africa with his own observations regarding the midget-sized albino "man" he had seen in 1744: "The Albinos are, in truth, a very small and very rare nation: they live in the middle of Africa; their minimal strength barely enables them to make their way out of the caverns where they live, however the *nègres* capture them from time to time, and we purchase them from the *nègres* for curiosity's sake."[97] In a curious evolution of his discourse on albinism, Voltaire asserted that albinos were a coherent, self-perpetuating and insular group of people, a fact that emphasized the relative stability and independence of this "category" from the rest of humanity.

By the 1750s, Voltaire's belief that albinos were conceptually and biologically independent of other races must have seemed terribly rearguard to his contemporaries. Most naturalists of Voltaire's generation agreed that the albino was the diseased or degenerate product of black African parents; many of these same thinkers had also begun to speculate on the specific causes for albinism, such as an improperly stimulated maternal imagination, leprosy, smallpox, or miscegenation. Voltaire ridiculed such ideas. In particular, he mocked the belief that leprosy had brought about this change in certain black individuals: "To claim that these are dwarf *nègres*, whose skin has been bleached by some form of leprosy, is akin to saying that the blacks themselves are whites whom leprosy has darkened. An Albino doesn't resemble a *nègre* from Guinea any more than he does a man from England or Spain."[98]

Out of step with most naturalists and philosophers of his generation, Voltaire provided a unique view of the albino that echoed his belief that this was a conceptually distinct category.[99] The advantages of denying any kinship between other races and the albino were numerous. In addition to preserving Voltaire's overall fixist understanding of generation from contamination, the albino's conceptual distinctiveness allowed him to place the albino into a polygenist typology according to physical and intellectual capacities. Voltaire did precisely this in chapter 143 of the *Essai*: "The wool that covers [albinos'] heads and that forms their eyebrows is like a fine white cotton: they are inferior to *nègres* in terms of bodily strength and ability to comprehend; nature has perhaps placed Albinos after the *nègres* and the Hottentots, and above the monkeys, like one of the gradations that descend from man to animal."[100]

Although this Voltairian "chain of being" may suggest a biological contiguity between racial categories (not to mention between man and beast), nothing could have been further from the philosophe's intentions. In this and all other discussions regarding the albino category, Voltaire maintained that this was an impermeable, measurable, and conceptually distinct "species."[101] Indeed, to shore up the conviction that this was a discrete group whose very being seemed to contradict the larger social and physical orders, Voltaire asserted that this curious race was on the brink of extinction: "the Albinos are so few in number, so weak, and so mistreated by the *nègres*, that it is reasonable to fear that this species will not survive much longer."[102] This is one of the few instances where Voltaire's ideas are in agreement with Lucretius's *De rerum natura*.[103]

Voltaire, the *Nègre*, and Human Merchandise

In the years after he wrote the *Essai sur les moeurs*, Voltaire continued to put faith in his trenchant view of separate human types. As one of the few voices who refuted the more fluid view of human varieties made famous by Buffon's *Histoire naturelle*, Voltaire took great pains to underscore what he believed to be the fundamental difference between Europeans and the so-called *race nègre*, which he regarded as the clearest refutation of monogenetic theory.[104] This comes across most strongly in Voltaire's 1768 *Des singularités de la nature*. In this meditation on a series of life science quandaries including the oyster and the polyp, Voltaire examined the *nègre* in a chapter on "Monsters and Diverse Races." Lashing out against the continued publication of monogenetic theories by "the ignorant" who maintained that, in Voltaire's words, "the blacks are a race of whites blackened by the climate," the philosophe crowed that now that his era had identified the source of Africans' blackness—the *reticulum mucosum*—any informed person had irrefutable proof that they were separate species.[105]

This reticular tissue discovered by Malpighi was an essential element in Voltaire's discourse on the human. *Reticulum mucosum,* in fact, became the philosophe's battle cry in defense of racial difference. Having observed a sample of this tissue—"the obvious cause of the inherent and specific blackness of the *nègres*"—when he visited Frederik Ruysch's famous house in Amsterdam,

Voltaire cited this anatomical feature whenever he sought to emphasize the fundamental difference of the African.[106] While he perhaps exaggerated when he wrote that he had cited the importance of this tissue "twenty times" in the *Dictionnaire philosophique*, he was, in fact, not far from the mark. For Voltaire, the *reticulum* was proof that "in each species of man, as in plants, there exists a differentiating principal."[107] In the *nègre*, this "mucus membrane, [this] network that nature has spread between the muscles and the skin" was an unmistakable indication that "the race of *nègres*" was as different from whites as "the race of spaniels is from that of greyhounds."[108] It was with this "fact" in mind that Voltaire joked horrifyingly in *La défense de mon oncle* that doubters need only check for themselves: "Whoever wants to have a *nègre* dissected (I mean after his death) will find that this mucus membrane is black like ink from head to toe."[109]

Rhetorically speaking, the unmistakable and unimpeachable "specific difference" of the *reticulum mucosum* allowed Voltaire to extend his polygenist views to the other physical traits of the *nègre*; most significantly, it allowed him to bundle physical divergences with a supposed lack of intellectual aptitudes. In an early part of the *Essai*, this conflation of characteristics is still quite vague: "Their round eyes, their flat noses, their always fat lips, their differently figured ears, the wool on their heads, and even the measure of their intelligence, indicate prodigious differences between them and other species of men."[110] Later on in the same text, however, Voltaire refined this view when recounting the discoveries of the Portuguese. The context for this discussion is significant in that it clearly links black Africans' physical and intellectual aptitudes with their expected, albeit unfortunate, role in the colonial project, which was to toil for white masters:

> The shape of their eyes is not at all like our own. Their black wool does not whatsoever resemble our hair; and one could say that if their intelligence is not of a different type of understanding from ours entirely, it is at least greatly inferior. They are not capable of maintaining attention for long; they plan very little in advance, and they seem to be made neither for the advantages nor for the errors of our philosophy. They originate from this part of Africa, like the elephants and the monkeys. . . . they believe themselves to have been born in Guinea to be sold to whites and to serve them.[111]

Voltaire's association of the Africans' morphology and supposed lack of intelligence with a willingness to enter into slavery is reminiscent of some of the most caustic comments made by Caribbeanist writers. In point of fact, Voltaire went even further than many Caribbeanist authors in that he envisioned these perceived liabilities in terms of stages of development and civilization.

Voltaire's historicization of the black African served, not only to temporalize, but also to contextualize and to explain the liabilities of the *race nègre*. From his point of view, an ordered, civilized polity could only come into being after many centuries of self-organization: "In order for a nation to unite into a body of people, for the nation to be powerful, hardened, learned, it is clear that an enormous amount of time is needed."[112] According to this schema, the African (both the *nègre* and the Hottentot) had barely advanced past the most basic era of human development. The lives of the Africans Voltaire came to know in travelogues reminded him of a hypothetical past, an era in which beast-like groups of potentially speechless humans spent their lives either hunting animals or eating roots.

John Locke famously wrote in his *Two Treatises on Government* (1690) that "in the beginning, all the world was America."[113] For Voltaire, however, in the beginning, all the world was also like Lapland, Caffreria, or West Africa. The peoples of these latter two lands in particular seemed not only primeval to Voltaire, but also oblivious to their own limited aptitudes and misery. As Voltaire put it succinctly: "The majority of *Nègres*, all the Caffres, are submerged in . . . stupidity, and will remain huddling there for a long time."[114] Among the cultural practices that were most indicative of this infantile level of development, nothing was more significant for the philosophe than African religion. Although he did not react to African animism like a missionary—with outrage—he nonetheless ridiculed all African religious practices, reducing them to the outlandish rudiments of metaphysics from whence evolved superior monotheistic traditions:[115]

> At some point, all peoples were for centuries what the inhabitants of the southern coasts of Africa, many islanders, and half of [Native] Americans are today. These peoples have no concept of a single God who created all, is omnipresent, and existing everywhere and eternally. However, one must not call them atheists as one might ordinarily, because they do not at all deny the existence of a Supreme Being; they do not know Him; they have no concept of Him. The Caffres take an insect as their protector, the *Nègres* a serpent.[116]

Voltaire's historical assessment of *nègre* and Hottentot (or *caffre*) religions reflects the African's overall dubious status within the philosophe's typology of the world's peoples. Anatomically suspect, intellectually inferior, and morally repellent, these bug- and snake-worshipping Africans clearly stood on one side of an abyss separating them from reason, advanced thought, and civilization.[117]

Such were the liabilities that led Voltaire to assert a causal link between the *essence* of Africanness and the destiny of this same group in the colonies. While he in no way sanctioned slavery, it is important to note that Voltaire's disquisitions on race provided his era with a compelling historico-materialist account of the existence of African chattel slavery. Indeed, Voltaire implicated Nature herself in the *nègre*'s human bondage. As he put it, "there is in each species of man, as in plants, a fundamental characteristic that distinguishes them from others. Nature has subordinated the different degrees of genius and national character to this principle, which is why the latter [national character] seldom changes. This also explains why *nègres* have become the slaves of other men, and are purchased on the coast of Africa like animals."[118]

Voltaire's pessimistic views regarding Africans and African chattel slavery have rarely been cited by historians of science, in large part because they do not fit in neatly with a narrative of science's "progress." In his famous *Les sciences de la vie*, for example, Jacques Roger cites Voltaire's ideas on race only episodically within the seemingly more important debates involving what he called *franc-tireurs* or daring materialist thinkers.[119] While it is true that Voltaire's positions on race seemed out of step with the more "serious" and "scientific" developments taking place in the eighteenth-century life sciences, the teleological *avant-gardisme* that asserts that, as Gustave Lanson writes, Voltaire "does not count" in this domain neglects an important reality: the often-republished and often-cited Voltaire was in the forefront of an intellectual movement that, throughout the eighteenth century, increasingly systematized the perceived differences of the African type into a zoological construct, or "race," a distinct category of inferiority where biology overlapped with aptitudes and vocation.[120] Indeed, it can be argued that Voltaire contributed significantly to one of the major transformations that the African underwent in European thought during the eighteenth century: from a barbaric heathen (a moral and religious category) who could be redeemed through slavery, to a subhuman (racial category) for whom human bondage seemed the logical but regrettable extension of the race's many shortcomings. Although Voltaire was, and is, often portrayed as lagging behind the rest of the philosophes with respect to the life sciences, Voltaire's conceptualization of the black African actually anticipated the direction that race theory would take by the end of the century.

Processing Africa and Africans in the *Encyclopédie*

Although Voltaire and Montesquieu wrote at a time when more properly scientific thinkers had established the basic parameters for the interpretation of African difference, each processed the "incontestable" physiological and psychological liabilities of the *nègre* in his own way. Voltaire, the sometimes outspoken anti-slavery thinker, declared that human difference was specific, revealing, and measurable, and that Africans were a separate species. As for Montesquieu, he saw difference from a less trenchant point of view, yet his understanding of the African's dubious mind and altered body nonetheless led to one of the most defining truths of *De l'esprit des lois*: Enlightenment-era political theory had no universal application and needed to be qualified in light of the incontrovertible reality of the African's many liabilities.

While some eighteenth-century scholars have suggested that the discourse on the African was a monolithic block of information that foreordained particular answers, the processing and incorporation of notions of difference into wider worldviews such as Voltaire's or Montesquieu's belies such an assertion. This fact becomes even clearer when we examine how the African's "difference" was subjected to the multiple perspectives and methodologies that constituted the *Encyclopédie, ou Dictionnaire raisonné des sciences des arts et des métiers*. Far from unilogical, the "African" that took shape in the *Encyclopédie*'s seventeen volumes of text is a complex hybrid of pro-planter rhetoric, anti-slavery diatribes, and philosophical and anti-clerical digressions.

As co-editor of the project, Diderot had envisioned anything but a complicated or fractured meditation on the African. In fact, in the short article "Afrique," he stated flatly that the *Encyclopédie*'s focus on Africa would be primarily commercial. Borrowing this orientation from Jacques Savary des Brûlons's *Dictionnaire universel de commerce* (1723), Diderot initially evoked an Africa whose significance was limited to its European strongholds, factories, and spheres of influence.[121] To find out more about Africa and European commercial interests, he suggested, the curious reader would only have to refer to more geographically specific headwords.

Not unexpectedly, the *Encyclopédie*'s presentation of Africa spilled out over these geo-mercantilist limits. Mirroring the mass deportation of Africans themselves from places such as Senegal, Ouidah, and Angola, the topic of "Africa" and "Africans" found its way into a variety of other interpretative realms, many of which were related to the *Encyclopédie*'s assessments of plantation life, human bondage, and sugar-making in the New World. This disciplinary and geographical partition of the "subject" of Africa was further compounded by the range of perspectives brought to bear on the topic in the *Encyclopédie*. As was often the case for multi-disciplinary questions, the examination of Africa and Africans was undertaken by a host of contributors of differing ideological and methodological orientations.

During the first two years of *Encyclopédie* production, it was Diderot himself who furnished the vast majority of articles on the subject of Africa. In addition to the dozens of short entries on the continent (including "Abex," "Abissinie," "Abramboé," "Acambou," "Angola," "Assiento," "Ambuella," "Ancober," "Arder," "Asbisi," "Asses," etc.), Diderot provided lengthier examinations of Africa that reflected his interest in subjects extending beyond the continent's trade and geography. It was in such articles that he used African ethnographical subjects to transmit a very European message to a European public. In the article on "Ansico," for example, Diderot pondered the sensationalistic accounts of cannibalism associated with this central African region, a land where the inhabitants purportedly "eat their fathers, mothers, brothers, and sisters as soon as they are dead" and human body parts were sold in numerous public butcher shops.[122]

Reacting to these reports with neither condemnation nor Montaigne-inspired relativism, Diderot opened an intriguing digression on both the supposed existence of cannibalism in Ansico and the veracity of such "facts." The more extraordinary such reports were, he argued, the "more witnesses were needed to make such [stories] credible."[123] While one might expect these reservations to preclude further commentary, the article does not end on a note of skepticism. Rather, Diderot's questioning of his era's travelogues paradoxically gives way to a common philosophe refrain, namely, an implicit indictment of fanaticism and fanatical practices in any form. Admitting, for the sake of argument, that cannibalism may in fact exist in Ansico, Diderot suggested that when such practices occur, they were surely the natural extension of extremism: "If, however, this country actually engaged in such horrible acts of cannibalism, and [we had determined that] the nation felt it was actually a great honor to be eaten by one's leader, [then] we would have found facts sup-

Anthropophagic Jaga. Centuries of travelogues portrayed the Jaga, or Giagues, as ruthless cannibals. From Olfert Dapper, *Description de l'Afrique* (1686). Courtesy Bibliothèque d'Orléans.

porting the prejudices, and extraordinary enough to give a certain plausibility to the cannibalistic practice alluded to here."[124] Cannibalism, as Diderot made clear, was really just an example of one more fanatical belief put into practice, no better or worse than the case of Hindu widows throwing themselves onto funeral pyres for husbands whom they had detested. More than a simple recycling of the account of the flesh-eating inhabitants of Ansico provided by Jean-Baptiste Ladvocat, this treatment of African ethnography is an example of ideological geography.[125]

Similar philosophe preoccupations come to the fore in Diderot's contemporaneous article "Bénin." Drawn from Olfert Dapper's lengthy account of this West African kingdom, Diderot's "Bénin" provides a range of information including details on human sacrifice, the all-powerful King of Benin, and the courage, loyalty, and generosity of the Beninians.[126]

It was presumably Dapper's portrayal of Beninian natural religion that inspired Diderot to contribute a substantial article on the kingdom. This becomes evident when one compares Diderot's account of Beninian religion to its original version in Dapper's *Description de l'Afrique*: if the Dutchman portrayed the Beninians as believing in a natural God, Diderot made the inhabitants of this part of Guinea, despite their idolatry, into something more akin to Voltairian deists: "these peoples do not celebrate any religious services for God; they claim that this being, whose nature is completely good, needs no prayers or sacrifices."[127]

Diderot's early articles on Africa certainly make one wonder what the ultimate form of the *Encyclopédie* discourse on Africa might have been had he been able or inclined to continue in this vein. With the notable exception of the article "Humaine espèce" (1765), Diderot effectively ceased to produce articles on African geography or peoples after the second volume. Like so many other subjects, those entries having to do with African geography (e.g., "Congo," "Nubie") as well as those related to "ethnographical" subjects (e.g., "Hottentots," "Mandingues") became the almost exclusive responsibility of the prolific Chevalier Louis de Jaucourt. This change in personnel had a profound impact on the representation of Africa: in lieu of articles such as Diderot's "Ansico," Jaucourt produced a less critical and less reflective series of entries on the continent, which were generally lifted word-for-word from the era's geographical dictionaries, travelogues, or natural history treatises. Obeying easily identifiable conventions, Jaucourt's entries habitually begin with a

The king of Benin. From Olfert Dapper, *Description de l'Afrique* (1686). Courtesy of Bibliothèque d'Orléans.

brief mention of topography and trade in a given region, and draw to a close with an aside on the regional "vices" of the inhabitants.[128]

Parallel to Jaucourt's anecdotal pigeonholing of Africans were other attempts to classify the African according to natural history categories.[129] Two basic rubrics for the black African were used in the *Encyclopédie*, both based on differences in pigmentation. The first of these two geo-ethnic appellative markers, *caffre*, had often been used mistakenly by naturalists like Buffon to designate the lighter-skinned Africans living near the Cape of Good Hope, in particular, the Hottentots, or Khoikhoi.[130] Appropriated by Portuguese explorers from the Arabic *kāfir*, meaning infidel, the designation *caffre* had, by the seventeenth century, found its way into common parlance in a number of European languages. Under the pen of mid-century thinkers like Buffon or Diderot, however, the term evoked more than just culture, morphology or pigmentation. Indeed, the term *caffre* had become part of European natural history; to be a *caffre* was to be a member of a specific subdivision within the larger black race.[131] *Caffre* thus implied a different genealogy or branch within the species, and existed in contradistinction to the second and much more common term, *nègre*.

Like many complex ideas in the *Encyclopédie*, the word *nègre* was defined as a function of context. This phenomenon is evident in the *Encyclopédie*'s four-part, multi-author definition of the term. In keeping with the way in which such proto-ethnographic information was organized at the time, the first of the four articles devoted to the subject begins with a coordinative natural history definition. Compiled by Johann Heinrich [Samuel] Formey, this assessment predictably associates physical traits and torrid zones: "From the Tropic of Cancer to the Tropic of Capricorn, Africa has only black inhabitants. Not only does their color set them apart [from other peoples], but they differ from other men by the features of their faces, their wide, flat noses, their fat lips, and the wool they have in place of hair."[132] As he did with many elements of the article, Formey poached this definition from a work containing one of the era's best-known scientific discussions of black Africans, Maupertuis's 1745 *Vénus physique*. While Formey nicely summarized parts of *Vénus physique*, he also misquoted Maupertuis in his discussion of some of the reigning pseudo-scientific explanations for dark pigmentation.[133] Incredibly enough, especially considering Maupertuis's well-known support of epigenesis, Formey cited *Vénus physique* in order to advance a preformationist explanation for the appearance of black humans within the species.[134] Making use of another paragraph from

Maupertuis in which the famous naturalist sought to refute preformationism, Formey gave credence to the idea that black Africans may have come about when a preexisting line of white ova gave way to black ova, much in the way that, "in a deep mine, when the seam of white marble has run out, one only finds stones of different colors."[135] This preformationist take on the African race is an important concept within the *Encyclopédie*'s representation of the black African; although such an idea does not, in this particular instance, give rise to the view that there are different races of men (polygenesis), Formey's view of the African nonetheless posits a fundamental difference between races on the most basic level: that of the human ovum.[136]

Formey's muddled and incomplete interpretation of Maupertuis's views on the origin of the *nègre* is followed by yet another natural history article: a meditation on the *nègre blanc*. Characterizing the albino as "of a livid white like that of a corpse" and essentially blind except "in the light of the moon, like owls," the anonymous author of this entry portrayed the *nègre blanc* as a physical anomaly or monster.[137] Just as Maupertuis had done in his well-known treatment of the question, this writer gave pride of place to the possible etiologies for such an occurrence. After citing leprosy, bestiality, miscegenation, or an improperly stimulated female imagination during gestation as the possible causes of this phenomenon, the writer ultimately admitted that the origin of the *nègre blanc* remained in doubt.[138] Such deliberate circumspection gave way, however, to an entirely speculative assertion that recalls the Plinian axiom that Africa could always produce something new.[139] According to the author of this entry, these freaks of nature might somehow be symptomatic of the continent as a whole: "it is possible that the interior of Africa, so little known to Europeans, contains numerous peoples of a species entirely unknown to us."[140] After all, as the author reminds his readers, "we do not know all the varieties and peculiarities of nature."[141]

In addition to this short natural history article on the albino, the *Encyclopédie* contains two more entries with *nègre* in the headword, both of which examine the African's role in the New World colonies. While this transition from Africa to the Caribbean seems elliptical, this shift simply echoes the physical movement of the African from his place of origin to his later location in the New World. More tellingly, such a change in perspective also mirrors the semantic transformation that the term *nègre* had undergone in European thought during the seventeenth century, from a general marker meaning "black inhabitant of Africa," to a signifier for chattel slave. Significantly, the final two articles with *nègre* in the headword are virtually the only *Encyclopédie* entries to relate the slave's deportation from Africa to the harsh life on various types of plantations.[142] The anonymous compiler of the first of the two articles, entitled "Nègres (commerce)," drew verbatim from the *Dictionnaire universel de commerce* in order to justify this mass movement of Africans with a familiar conflation of metaphysical and economic reasoning. Beginning with the commonly held theological view that slaves "find salvation for their souls in their loss of liberty," this article then affirms, more forthrightly perhaps, that Europeans have a "vital need" for black slaves in order to facilitate "cultivation of sugar, tobacco, indigo, etc."[143] Often cited by critics, this entry is followed by a neglected and more complicated article, "Nègres, considérés comme esclaves dans les colonies de l'Amérique," which was signed by the *Encyclopédie*'s expert on the French colonies and the sugar industry, Jean-Baptiste Le Romain.[144] Unlike those mid-century writers who suffered from what Roger Mercier has characterized as the era's "guilty conscience," Le Romain blithely accepted the economic realities driving the exploitation of African men and women in the colonies.[145] In contrast to the author of "Nègres (commerce)," this former *Ingénieur en chef de l'île de la Grenade* argued that the slave trade was not a necessary evil, but a necessary good: not only were *nègres* perfectly suited to the torrid climate of the sugar islands, but their constitution—"they are naturally vigorous and are accustomed to a coarse diet"—benefitted from the relative "comforts [of the Americas] that render animal life there much better than in [Africa]."[146]

To further this pro-slavery argument, Le Romain built on the belief that the colonies' *nègres* were a particular type of men whose overall character obeyed certain demonstrable laws. On their overall disposition, this colonial writer asserted that they were inclined to debauchery, to revenge, to rape, and to deception. Their stubbornness was so enduring that they "never admitted their faults, no matter what punishment they are made to endure."[147] Le Romain also asserted similar notions in the article "Sucrerie" ("Sugar mill"), which he contributed to volume 15, writing that the *nègre* is "a species of extremely vicious man, very tricky, and of a lazy nature [that], in order to get out of doing work, will feign hidden indispositions, headaches, stomach diseases, etc."[148] And yet, despite the fact that Le Roman effectively defined this *espèce d'hommes* with a series of negative traits, he also found it imperative to distinguish among various ethnicities. This was, of course, done for practical reasons; by recycling the type of ethnic typology that had appeared in Charlevoix and Buffon, Le Romain was able to dispense practical advice on choosing and managing Africans in a plantation setting.[149]

> The [slaves] of Cap Verd [modern Cap-Vert in Senegal], or Senegalese, are regarded as the most attractive in all of Africa . . . we use them in the home to take care of horses and other animals, to tend the gardens and to work around the house. The Aradas, the Fonds, the Fuédas, and all the *nègres* of the coast of Ouidah are idolatrous. . . . these *nègres* are believed to be the best for housework. The Mine *nègres* are vigorous and quite good at learning new skills. . . . The coast of Angola, the kingdoms of Loango and the Congo produce an abundance of attractive *nègres*. . . . Their penchant for pleasure makes them fairly unfit for hard labor, since they are generally lazy, cowardly, and very fond of gluttony. The least esteemed of all the *nègres* are the Bambaras; their uncleanliness, as well as the large scars that they give themselves across their cheeks from the nose to the ears, make them hideous. They are lazy, drunken, gluttonous, and apt to steal.[150]

Such typologies reflect the complexity of the construction of the term *nègre* at mid-century. If the word continued to denote those African peoples with a certain level of skin pigmentation and particular physical traits ("woolly" hair, wide noses, thick lips, etc.), it was also continually qualified against a series of esthetic and labor-related measurements. In this helpful *Encyclopédie* breakdown of the race—an analysis designed to be useful to the forward-thinking *armateur* (shipowner) or planter—the *nègre* exceeded his status as monological entity; defined by the light of the New World economy, he was divided by ethnicity into a typology of practical domination.

The Preternatural History of Black African Difference

In addition to the brutal assessments of the African produced by colonial writers, other *Encyclopédie* contributors, most notably Diderot, also addressed the conceptual status of the *nègre*. Although the *Encyclopédie* appeared before Johann Friedrich Blumenbach's breakdown of the human species in his *De generis humani varietate nativa* (1775), naturalists who wrote for the project nonetheless found themselves obliged to react to the era's systematic arrangements of animals and humans. Most prominent among these classificatory efforts was Linnaeus's *Systema naturae* (in its multiple editions). Unlike earlier taxonomical works, Linnaeus's *Systema* led philosophers and naturalists alike to grapple with humankind's place within nature as never before, by inserting man into a formal categorization alongside animals and plants.

In the 1735 edition of this work, humans were placed among the Anthropomorpha alongside monkeys and sloths and, significantly, were further divided into four varieties—European, American, Asian, and African.[151] By 1759, Linnaeus also attributed different qualities to these four varieties. The African, in this new edition, was described as "black, phlegmatic relaxed. *Hair* black, frizzled. *Skin* silky. *Nose* flat. *Lips* tumid. *Women* without shame. *Mammae* lactate profusely. *Crafty*, indolent, negligent. *Anoints* himself with grease. *Governed* by caprice."[152]

While the *Encyclopédie* is typically seen as one of the most emblematic texts of what Foucault called the "Age of the Catalogue," its contributors almost universally followed Buffon in disregarding the immutable categories for human "varieties" found in Linnaeus.[153] Equally significant, naturalists writing for this *Dictionnaire raisonné* rarely employed the word "race" to signify a group of people with common biological traits.[154] This disinclination to use classificatory terms or structures had nothing to do with a desire to produce a benevolent discourse on humankind's "varieties" or "races," however. To the extent that one can attribute this phenomenon to a particular cause or ideology, *Encyclopédistes* were simply following the thinly-veiled Anti-Linnaean policy that Diderot had made explicit as early as the "Discours préliminaire": "We do not in any way want to resemble this crowd of naturalists . . . who, working ceaselessly to divide nature's productions into categories and species, have spent time that could have more efficiently been employed to study these natural phenomena themselves."[155]

Accordingly, in the body of the *Encyclopédie*, authors providing entries on natural history reacted quite strongly against the classificatory tendency in general and Linnaeus's *Systema naturae* in particular. The authors of "Méthode" and "Classe" protested, for example, that the use of categories in natural history was necessarily reductive, absurd, and arbitrary. Although we may never know to what extent Diderot as editor exerted his influence in this respect, he (like his friend Buffon) was certainly responsible for some of the criticism of what he deemed uncertain classification schemes—especially schemes like Linnaeus's that were designed to affirm a divine and eternal order.

Diderot himself entered the classificatory debate by contributing two articles implicitly concerned with the question of human difference, both of which were adapted from Buffon's *Histoire naturelle*. Like the naturalist, Diderot separated the history of humankind into two different narratives: the story of the individual (in the article "Homme: *Histoire naturelle*") then the story of the species ("Humaine espèce"). In the first of the two articles, "Homme," he recounted what one might call the universal story of humankind—a narrative charting the shared physical experience particular to man from birth to death. Adapted from 350 pages of the *Histoire naturelle*, this article tells the story of the human body: the body during childhood, the body as it matures, the body during the age of virility, the body as subject to humankind's passions, and the body at death.[156] Despite the fact that Diderot alludes to a series of *moral* incongruities within this natural history of man, the differences in humankind's varieties are, in this context, completely subsumed by the more compelling narrative charting the fundamental materiality of human existence. Above all, this examination of the physical processes of *homme* is designed to produce axiomatic proclamations about the universality of the human condition.

The *renvoi* in "Homme" refers the reader to the logical conclusion to this inquiry, the article "Humaine espèce." It is in this entry that Diderot confronts the notion of the human species, an issue implicitly raising the question of humankind's categorization. Condensed from the *Histoire naturelle*'s long chapter on the "Variétiés dans l'espèce humaine," this seven-page article provides a conjectural history of humankind, a coherent explanation—and systematization to a certain extent—of the diverse peoples included under the previously established rubric *homme*.

Significantly, this account does not begin with a theory of human difference, but rather with an inventory of the species that, while centered on difference, in a way tends toward sameness. Commencing at the North Pole with a long and derogatory assessment of the supposedly unsanitary and half-witted Laplanders, Diderot then offered one- or two-sentence descriptions of the rest of the world's peoples. The article's focus shifts quickly: from the North Pole to Asia, to the Middle East, to North Africa, to Europe, to sub-Saharan Africa, and finally to the Americas. Some of these short observations include remarks regarding the relative aptitudes of the people discussed ("The Moors are small, skinny, and have ugly faces, but are spirited and perceptive"); the vast majority, however, concentrate on morphology and especially color ("those from Mexico are well-proportioned, energetic, have brown hair, and are olive-skinned").[157] While this index of subtle differences in degrees of pigmentation initially seems like a static list, Diderot's ultimate intention was to suggest a dynamic chronicle of human transmutation over time, from one group into many. Quoting the *Histoire naturelle* directly to emphasize this point, Diderot asserted that the varieties of the human species have a common source: "There was . . . originally only a single race of men, which, having multiplied and spread across the surface of the globe, in time produced all of man's varieties."[158]

In stark contrast to schemes classing human varieties as immutable entities, this monogenetic history of the human species asserts that racial differentia were not essential and arose, quite simply, as a function of diverse (and potentially adverse) environmental conditions.[159] As chapter 2 demonstrated, aspects of this dynamic view of humankind can be regarded, in historical context, as relatively open-minded. After all, this system construes the concept of "variety" as a nonessentialist appellation corresponding to a series of morphological tendencies that would in all likelihood change over time anyway.

Ann Thomson has argued this point quite effectively. According to Thomson, Diderot's commitment to the great chain of being, his rejection of any form of classification, and his belief in a fundamentally dynamic nature combine to produce a view of the human species that—despite certain racializing tendencies—more closely resembles a preternatural history than it does natural history.[160] This is particularly true given the fact that Diderot compressed Buffon's lengthy ethnographic portraits into seven pages of text. In contrast to Buffon's relatively slow exposition of humankind's varieties, Diderot's article "Humaine espèce" accelerated the dynamic processes that underlay the great naturalist's presentation of the "Variétés dans l'espèce humaine." In many ways, this continent-jumping inventory of remote peoples emphasized nature's mutability in a way that Buffon's article did not.

Despite some of the more optimistic elements of Buffon's and Diderot's explanations of racial differentia, the theory of a common root as it was developed in the *Encyclopédie* also grafted Europeans' self-perceived moral, physical, and esthetic superiority onto new epistemological structures.[161] In sum, if people living in the most temperate climate were "the whitest, the most attractive, and the best proportioned on earth" and if there had been initially but one common rootstock, it followed that "the white man seems to be of the natural, original color, that the climate, food, and customs have altered, creating the yellow and the brown, as well as the black."[162] This theory, which is not taxonomical per se, nonetheless suggested both the genealogical and the qualitative primacy of the white race. In the view of both Buffon and Diderot, the continuum of human pigmentation and morphology was nothing more than the degrees of degeneration from a white archetype, from the most beautiful Georgian to "the final shade of the tanned people."[163] Even more ominously, this morphological and color-based breakdown of the human species overlapped with assessments of cognitive potentials. Although "Humaine espèce" and even the much longer "Variétés dans l'espèce humaine" do not contain the vilifications of the *nègre* found in pro-slavery texts (indeed both articles lament the slave's plight), Buffon and Diderot accepted and diffused the idea that *nègres* generally lacked intelligence compared to Europeans, presumably as a result of certain environmental forces over time.[164] The way in which this conjectural history of the *nègre* (along with the belief in his limited acumen) functions within the whole of the *Encyclopédie* may have been quite unanticipated by Diderot. In fact, while Diderot's overall views regarding the plight of the African put him in direct opposition to pro-slavery writers such as Le Romain, his dynamic and regressive view of the history of the *nègre* operates (within the confines of the *Encyclopédie*) as part of a larger nexus of ideas rendering intelligible, if not justifiable, the *nègre*'s destiny in the colonies.

While the *Encyclopédie* was undoubtedly a radical departure from other, earlier compendiums of knowledge, this *Dictionnaire raisonné* brought forth no new methodologies in its interpretation of the African. Rather, the *Encyclopédie*'s African is the product of generally vulgarized or plagiarized ideas; very few of the articles that mention or discuss the *nègre* actually contributed something original to the debate on the "African variety." And yet, the *Encyclopédie*'s recycled and conflicted mosaic of *africanismes* reflects two important realities of the era, perhaps more so than most texts at the time. The first was that the hermetic, *renvoi*-lacking entries that examine the African reveal a general unwillingness or inability to think through the inconsistent portrait of the African as it took shape within differing contexts, be they natural law or human anatomy. The second was that the organization of African ethnographic "data" into stand-alone, alphabetical entries (e.g. "Hottentot," "Jagas") tended to isolate African ethnicities from the more optimistic endorsement of the unity of the human species that Diderot developed in "Humaine espèce." Broken down into a series of ethnographic subcategories, this constellation of disparaging ideas was far removed from the comforting context of Buffon's and Diderot's monogenetic worldviews.[165]

Teaching Degeneration: Valmont de Bomare's *Dictionnaire d'histoire naturelle*

During the eighteenth century, the *Encyclopédie* was far from the only reference work to compartmentalize the *nègre*'s problematic *difference*. The *Grand dictionnaire historique* (1674–1759), Abbé Laurent le François's *Méthode abrégée et facile pour apprendre la géographie* (1729), the *Géographie historique, ecclésiastique et civile* (1755), Jean-Baptiste Ladvocat's *Dictionnaire historique portatif* (1758), Andrew Brice's *A Universal Geographical Dictionary* (1759), and Daniel Fenning and Joseph Collyer's *A New System of Geography* (1764) all contributed mightily—if not necessarily thoughtfully—to the ongoing recycling of African stereotypes. This was also the case for the *Encyclopédie*'s immediate successor, the *Encyclopédie méthodique* (1782). Twenty years after *Encyclopédie* contributors including Jaucourt, Formey, and Le Romain synthesized their era's beliefs about the African, Didier Robert de Vaugondy provided his own highly recycled account of the *nègre* for the geography section of this massive project. Mirroring, yet again, the long-standing belief that the African race was incapable of thinking about the future, the acclaimed geographer affirmed that, in general, the people of Africa "only have ideas [corresponding to] one day. Their laws have no other principals other than those of an aborted morality, and no other consistency other than that produced by indolent and blind habits."[166]

Such comments hardly come as a surprise. Reference works such as the *Encyclopédie méthodique* inevitably drew from a shallow pool of shared sources and ideas. While eighteenth-century philosophes—Diderot included—often bemoaned the poor quality of such ethnographic information, these works nonetheless contributed significantly to the codification of a series of beliefs regarding Africans' liabilities.[167] Paradoxically, it was the very weakness of encyclopedias and dictionaries—the fact that these fragmented reference works drew from the same reductive and universally recycled ethnography—that made them so authoritative. Indeed, the seemingly universal set of alphabetized ethnographic truisms that these works put forward packaged the African into accessible categories of ethnic distinctiveness.

Looking back at the significant number of alphabetically organized reference works containing summaries or syntheses of African "ethnography," it might seem that beliefs about the *nègre* at this time were universally static, timeless, and unoriginal; it might also appear that all such reference works were unable or unwilling to engage with the increasingly subtle anatomical view of the *nègre* that was taking shape after mid-century. And yet, there is one critically important—albeit essentially forgotten—eighteenth-century reference work that runs counter to this trend. This is Jacques Christophe Valmont de Bomare's *Dictionnaire raisonné d'histoire naturelle.*

Like other authors who put together reference works containing thousands of articles, Bomare was a compiler of many of his era's ideas. And yet, in contrast to most authors who synthesized ideas on subjects such as the Black African, he was actively engaged with developments in natural history and the life sciences. Beginning his career as an anatomy student of none other than Claude-Nicolas Le Cat at the Hôtel-Dieu in Rouen, Valmont de Bomare also met or exchanged correspondence with Linnaeus, Diderot, d'Holbach, Réamur, Nollet, Daubenton, and Buffon, among others. While his contribution to natural history pales in comparison to that of Buffon, the accessibility of his *Dictionnaire,* coupled with the fact that he had become well known for his popular annual lecture series on natural history in his Paris *cabinet,* made this work an instant best-seller. Indeed, Valmont de Bomare's *Dictionnaire,* which he himself described as "useful, instructive, and interesting," ultimately became the "third most popular science book after Buffon's *Histoire naturelle* and [Antoine] Pluche's *Spectacle de la nature.*"[168]

Had Valmont de Bomare only produced one edition of the *Dictionnaire,* his work would be much less valuable for contemporary scholars. Valmont de Bomare was, however, the type of writer and thinker who constantly updated his *oeuvre.* The first edition of the dictionary contained five volumes and was published in 1764; the second appeared in 1767–68 with six volumes; the third edition came out in 1775 with nine volumes; and the final and fourth edition was published in 1791 with fifteen volumes. Much like the overall work itself, the naturalist reworked the article "Nègre" significantly in 1768, 1775, and 1791. In addition to the fact that he was obliged to process an increasingly large amount of anatomical "data" on the black African after 1764, the naturalist also thought more and more critically about the relationship of the black African to the problem of chattel slavery. Surveying these increasingly complicated treatments of the *nègre* sheds light on the way in which Valmont de Bomare—and natural-history-minded philosophes in general—attempted to negotiate the evolving information available to them regarding black Africans.

In the first edition of the *Dictionnaire,* Valmont de Bomare's entry for the term *nègre* comes to less than a page. Having already discussed the *nègre* variety more extensively in the article "Homme" (which he had essentially lifted from Buffon's "Variétés dans l'espèce humaine"), Valmont de Bomare described this degenerated human variety almost wholly through the lens of the slave trade. Repeating a series of untruths, the naturalist projected the blame for this institution squarely on "cruel" Africans, stating that "the *nègres* sell to the Europeans not only the black slaves that they have taken during times of war, but also their own children. Often a *négresse* mother sells her daughter to a stranger for a few cowries."[169]

In the second iteration of the article (1768), Bomare expanded on his description of the deplorable circumstances of the *nègre*, providing a more detailed description and explanation of the race's specific moral failings. According to Valmont de Bomare, black Africans were "more vicious than [men] in other parts of the world"; he also asserted that their "treachery, cruelty, impudence, heresy, uncleanliness and drunkenness seem to have suppressed in them the principals of natural Law and the pangs of conscience."[170] More significantly, the naturalist also acknowledged quite forthrightly that the animalistic characteristics of the *nègre* had led reasonable people to think that this race was in fact a different species: "Their customs are so extravagant and so unreasonable that their conduct combined with their color have long made many wonder if they were truly men descended from the same original man as we did."[171] Although Valmont de Bomare remained a staunch monogenist throughout his career, and explicitly rejected the notion that the black African was another *species* of man, he nonetheless followed up on the above doubt by qualifying the black African's alterity with three different adjectives: "One can envision the races of *nègres* as barbarous and degenerate or debased nations."[172] In addition to asserting that the African was *barbaric* (a classical category), Valmont de Bomare also maintained that the *nègre* was a *degenerate* creature (a Buffonian category). What was more, however, the naturalist also emphasized that *nègres* were *avilies,* or somehow lower than the larger category to which they belonged. It is this last adjective that distinguishes Valmont de Bomare from a thinker like Buffon. In a modification of Buffon's *horizontal* understanding of the human species, Valmont de Bomare envisioned the *nègre*'s degeneration in terms of a *vertical* (and inferior) relationship to the rest of humankind.

Valmont de Bomare also added new anatomical data to this overall negative portrait of the black African in 1768. In addition to citing Le Cat's assertion that the bluish brain of the African produced a dark *ethiops* that coursed through and stained the elemental structures of the black body, he also detailed the humoral differences between blacks and whites: "In a European or a white person, the lymph is white, except when it is mixed with bile, which gives a yellow tint to the skin. But in a *nègre*, whose lymph and bile are black, the skin, according to several reports, must, through similar reasoning, be black."[173]

Seven years later, in the third (1775) version of the *Dictionnaire*, Bomare supplemented this vision of an interior and material blackness with a layer of comparative anatomy.

> The *nègres* vary among themselves according to shades of color, but they also differ from other men according to all the features of their face: the round cheeks, the elevated cheekbones, the somewhat bulging forehead, the short, large, flattened, or crushed nose, the fat lips, and small ear lobes; the ugliness and irregularity of the face characterize their exterior. *Négresses* have a flattened lower back and monstrous buttocks, which gives their back the look of a horse's saddle.[174]

While reminiscent of the esthetic disparagement of the *nègre*'s body that had existed for centuries, the naturalist's description also reflected a new way of evaluating the black African. Written during an era where there was a growing interest in race-based studies of skulls and facial features, Valmont de Bomare's account mirrored the transition from purely esthetic judgments of African features—as *ugly*—to mathematical or geometrical assessments such as *irregular*, a term that Bomare employs in the above quotation. By 1791, and despite the fact that Valmont de Bomare fully endorsed the phenomenon of racial reversibility, the naturalist had nonetheless concluded that the black African was an anti-teleological being, a race whose pathological origins stood in stark contrast with the compelling logic of a primary whiteness. As he put it: "[The] black color in the human species is just as accidental as the brown, the red, the yellow, the olive, and the tan. We must see the Whites as the stem of all men."[175]

In making his final modifications for the final 1791 version of his article on the *nègre*, however, Valmont de Bomare also demonstrated that he had been influenced by the era's increasingly compelling anti-slavery politics. In the first place, he softened some of his more trenchant vilifications of the black African by adding qualifying clauses, for example, "One can, *to a certain point*, see the races of *nègres* as barbarous and degenerated nations."[176] More important, he announced that he had evolved from a practical *reformer* of the institution of slavery to an abolitionist, not only accusing planters of transforming the *nègre* into a "kind of livestock," but adding pathetic visions of African suffering and praising the creation of an anti-slavery movement in London.[177]

While this final version of the article "Nègre" prompted pro-slavery thinkers to castigate Valmont de Bomare as a thoughtless and irrational *négrophile*, the naturalist had hardly modified his unambiguously negative portrait of the black African in 1791.[178] Despite the fact that he expressed his solidarity with the abolitionist movement, he nonetheless continued to portray the *nègre* as a highly degenerate being whose corporeal and moral liabilities were, alas, both considerable and undeniable. Indeed, regardless of his politics, these "truths" were part of the record that Valmont de Bomare felt compelled to communicate to his students and readers. This was often the case with many thinkers associated with the High Enlightenment: from their point of view, there simply was no contradiction between lecturing in compassion and informing one's readers about the brutal reality of race.

CHAPTER FOUR

The Natural History of Slavery, 1770–1802

Disgraceful race! Contemptible *nègres*! Those who pity you do not know you.

FRANÇOIS VALENTIN DE CULLION, *L'EXAMEN DE L'ESCLAVAGE EN GÉNÉRAL* (1802)

[Negrophiles] have put the devil in the slaves' heads. Nothing is more diabolical than the strategy of the negrophiles. These brazen men give slaves liberty so that these *nègres* can become the executioners of their masters; negrophiles . . . change black into white, and white into black.

LOUIS-NARCISSE BAUDRY DES LOZIÈRES, *LES ÉGAREMENTS DU NIGROPHILISME* (1802)

The black race has always seemed mentally rather inferior to other races and, regardless of the care that may be taken to educate young *nègres*, they have never produced a man of great genius. . . . they will always be in servitude by their weakness, and in barbarism by their inability to think.

JOSEPH-JULIEN VIREY, "VISAGE ET PHYSIONOMIE," IN *NOUVEAU DICTIONNAIRE D'HISTOIRE NATURELLE APPLIQUÉE AUX ARTS* (1804)

Eighteenth-century natural history encompassed an extensive range of subjects: insects, plants, animals, minerals, and the "stars and meteors" that lit up the night sky.[1] Most memorably, however, natural history provided the forum for a new and wide-ranging exploration of the human species. While this particular area of inquiry was theoretically circumscribed by the primary function of the discipline—providing an accurate description of nature—natural history's practitioners invariably went beyond these strictures. Theologically or providentially oriented "naturalists" such as Abbé Antoine Pluche examined humans as part of a divine plan;[2] politically minded thinkers including Montesquieu drew from natural history in order to speculate on the effects of climate on the inhabitants of a given region. But of all the questions that a new generation of naturalists raised about the human species, none was more momentous than the subject of African chattel slavery. That natural historians wrote about slavery (or that pro-slavery thinkers engaged with and contributed to the field of natural history) is far from surprising. After all, during the eighteenth century the *nègre*'s enslavement was increasingly being understood from a zoological point of view.

Not unexpectedly, the natural history of the *nègre* has long drawn the attention of the most famous scholars working in the history of slavery. In discussing the intellectual underpinnings of human bondage in his *The Problem of Slavery in Western Culture*, David Brion Davis cited the "complex history of biological theory" within the overall refashioning of the image of the black African during the era of slavery.[3] Similarly, in his equally well known *The White Man's Burden*, Winthrop D. Jordan affirmed that natural history overlapped with and substantiated economic justifications of chattel slavery.[4] While numerous scholars including James Walvin have questioned the actual importance of *blackness* in the initial construction of the Atlantic slave system, it is nonetheless now generally conceded that the black African's presumed natural alterity played many roles during the eighteenth century: it allowed Europeans to produce new definitions of whiteness; it provided a coherent concept around which the first "scientifically based" human classification schemes were organized; and, most infamously, it replaced theological and even economic justifications as the most compelling rationale for African chattel slavery.[5]

Scrutinizing the political implications of the "Africanist" branch of natural history is a necessary step in understanding the slave-based societies that became the norm in the Caribbean. In addition to the fact that such a study fleshes out our understanding of one of the great crimes of human history, this type of inquiry also provides a salutary reflection on the potentially pernicious effects of Western science. And yet, the way in which scholars have examined natural history in the past—by studying it as a subplot within the larger and all-powerful history of slavery—actually confuses the relationship that existed between this eighteenth-century academic discipline and the "larger" matter of human bondage.[6] While natural historians who wrote on the black African undoubtedly derived the majority of their information from the context of slavery, their texts rarely operated in lockstep with pro-slavery discourse. In fact, natural historians and anatomists' relationship to the institution of slavery ranged from denunciation to collaboration. Bearing this in mind, in this chapter I reverse the traditional historiographical relationship between the "history of natural history" and the question of African slavery. Instead of simply identifying negative scientific views of *le nègre* as they show up in the era's pro-slavery thought, I chart the wider dialogue that existed between natural history and the discourse on human bondage. Indeed, by examining the multifaceted connections between science and slavery, I demonstrate how the anatomy of the black African ultimately became a tool wielded by both sides of the slave trade debate.

The Hardening of Climate Theory and the Birth of New Racial Categories circa 1770–1785

Twenty years after Buffon published the first volumes of his *Histoire naturelle* (1749), a significant number of eighteenth-century thinkers began to distance themselves from the seemingly over-optimistic elements of his degeneration theory. In the English world, Edward Long famously "rewrote" Buffon's environmentalist survey of the African in his 1774 *History of Jamaica*. Pilfering freely from the French naturalist's overall description of the "Negro," Long nonetheless contradicted Buffon when he concluded that blacks constituted a separate category of human that one should distinguish from "the rest of men, not in *kind*, but in *species*."[7] Several years later, the Italian economist and polygraph Abbé Ferdinando Galiani also refuted Buffon's theory of degeneration in the influential (but elite) *Correspondance littéraire* edited by Friedrich Melchoir Grimm and Denis Diderot. This was a three-step process. First, he flatly rejected the possibility of racial reversibility, stating unequivocally that "perfectibility is not a gift given to man in general, but only to the white and bearded race."[8] Secondly, he scorned climate theory: "Everything that is said of climates is nonsense, a *no causâ*, the most common error in our logic."[9] And thirdly, and well ahead of his time, he affirmed that the aptitude and destiny of humankind were not the products of environment: they were the results of a biology-based logic. In a declaration that is strangely prescient of Arthur de Gobineau's belief that race explained the destiny of the human species, Galiani asserted: "Everything is connected to race."[10]

Diogenes, who is said to have gone around ancient Athens with a lantern in broad daylight claiming to be seeking an honest man, illuminates Buffon, the discoverer of "humankind," while a putto places a laurel wreath on a bust of the late naturalist. From *Les Beautés de Buffon sous le rapport du style,* edited by Adélaïde-Gillette Billet Dufrénoy (Paris: Emery, 1823). Author's collection.

Along with Voltaire, Long and Galiani number among the most skeptical thinkers regarding monogenesis and climate theory during the eighteenth century. And yet, while one can point to increasing doubts regarding a single-origin theory of humankind by the 1770s, most French thinkers did not abandon Buffon's presumption; they simply retrofitted his understanding of the human species with a growing list of discernible and measurable anatomical differences separating blacks from whites. This more trenchant theory of degeneration, which was most clearly developed by Cornelius de Pauw in 1768, had a significant effect on the conceptual status of the black African. Such is the case in François Para Du Phanjas's 1772 *Théorie des êtres sensibles, ou, Cours complet de physique*. As a Jesuit philosopher and reader of both Buffon and de Pauw, Para Du Phanjas began his assessment of humankind's varieties by asserting that the *noir* and the *blanc* shared a "primitive origin."[11] The Jesuit also echoed standard degeneration theory when he explained that nonwhites had come into existence because of inhospitable climates and poor food. But Para Du Phanjas also painted a significantly more pathological view of blackness as well, asserting that the *nègre* had suffered from a tropical climate and diseases that not only denatured these "unfortunate victims" but also imprinted vices on the race "from father to son."[12] According to Para Du Phanjas, this pathological heredity expressed itself in both moral (e.g., penchants, taste, and judgment) and physical (e.g., color, traits) realms, and it could not be effaced in a different climate. Even if a group of Africans were to spend "an extended time of one or more centuries in the temperate zone," he wrote, their "original color" and thus their other shortcomings would not change "noticeably."[13] And here we approach a more properly *racialized* understanding of black Africans; according to Para Du Phanjas, the *nègre* was a stable category defined by inherited features that were the natural extension of an established and inferior root stock.

This belief in the fixed and hereditary liabilities of the *nègre* overlapped with the birth of more delimited *racial* categories. As numerous critics have pointed out, these classification schemes were first produced among German naturalists, all of whom seemed to have been careful students of Meckel, Le Cat, and, especially, de Pauw.[14] Among the German theorists of race, Johann Friedrich Blumenbach was perhaps the most attentive reader of the era's anatomical discoveries in general and Cornelius de Pauw's *Recherches philosophiques* in particular. In his *De generis humani varietate nativa* (*On the Natural Varieties of Mankind*) (1775), Blumenbach absorbed both the general foundation of

Buffon's monogenetic and environmentalist explanation of humankind and the increasingly fixed physiological understanding of the human that thinkers including de Pauw had championed.[15] This dual perspective generated a somewhat contradictory presentation of the human. On the one hand, and even more explicitly than Buffon, Blumenbach sought to strike a blow against the idea that there were separate *species* of men. An overt critic of the polygenists La Peyrère, Voltaire, Griffith Hughes, Henry Home Kames, and Simon Tyssot de Patot, Blumenbach maintained that the belief in the "plurality" of the human species was a loathsome position that, far from based on fact, stemmed from "ill-feeling, negligence, and the love of novelty."[16] In fact, the fundamental purpose of Blumenbach's race-based understanding of humankind was to demonstrate that there was a shared origin of the human species, one that "buttress[ed] the great monogenetic truth of the Old Testament."[17]

To a certain extent, this monogenetic belief was actually substantiated by Blumenbach's "use" of the era's anatomical and physiological discoveries. In order to give credence to his conviction that the human species had a common root, Blumenbach cited a series of shared cross-racial human traits that stood out in stark relief against those found in the rest of the animal world. All humans, he wrote, stood erect, were bipedal and bimanous, had a certain bone structure and "affections" of the mind, and were altogether "naked and defenseless" compared to other animals. Moreover, Blumenbach pointed out that, unlike the female members of other species, human women had a hymen, which was granted perhaps for "moral purposes."[18] In sum, after sifting through his era's anatomical knowledge (including his own observations on skulls), Blumenbach could forcefully assert that the "constitution, the stature, and the color" of humankind's various races could be attributed to a climate-induced degeneration from an original root species.[19]

In contrast to this essential or original sameness, Blumenbach also cited important anatomical changes (in addition to blackness) that functioned as the defining characteristics of the African race. These bodily characteristics, which Blumenbach claimed were "beyond all doubt" and identified by an "industry of learned men,"[20] were essentially a combination of anatomical discoveries and notions that had been circulating since antiquity. According to Blumenbach, the Ethiopian had a "dry" and "heavy" body, sun-darkened bile, black blood, and, per Meckel, a "brain and spinal marrow" that was an "ashy color."[21] To his credit, Blumenbach drew no specific conclusions about African intelligence from these data. Indeed, in his third edition of the *Generis humani* (1795), he would cite the authors Gustavas Vassa (Equiano) and Ignatius Sancho (among other Africans) as living proof of the "perfectibility of the mental faculties and the talents of the Negro."[22]

Blumenbach's overall intellectual legacy, however, has little to do with his effusive late-century defense of the African. Rather, scholars generally remember the German naturalist as one of the founders of physical anthropology and, more infamously, as the person responsible for a schematic classification of the human species into four (1775) and then five (1781) races, including the Caucasian (1795). Among the variables that contributed to this new schematization of the human species, nothing was more important than Blumenbach's belief that the African had undergone a series of significant anatomical changes (in addition to simply turning black as earlier environmentalists had claimed). Such was the ambivalence of the African's anatomy in the *Generis humani*. On the one hand, comparative study of human bodies served to refute the possibility of essential differences between human groups.[23] On the other hand, the notion that the physical features of the African and other races were measurable constituted the basis for real categories. While Blumenbach clearly saw no contradiction in these positions, many thinkers whom Blumenbach influenced simply dropped the optimistic aspects of his monogenetic presentation of the human species, putting forward their own racial classification schemes based on an increasingly physiological conception of human categories.[24]

Between 1775 and 1777 alone, Immanuel Kant, Johann Christian Polykarp Erxleben, and Eberhardt August Wilhelm von Zimmermann all published schematic renderings of essential human categories based, in large part, on the new view of anatomy and a fixed degeneration.[25] French thinkers seized on this new way of constructing the human as well. In his influential "Essai sur la cause physique de la couleur des différents habitants de la Terre" (1781), Abbé Nauton presented an influential understanding of the measurement of race. This began with a remarkably corporeal vision of degeneration that conflated a hodgepodge of recent anatomical discoveries with classical humoral theory:

> Constantly acting upon generation after generation, the varying degrees of heat in specific climates pass down to [the] oily element common to the blood, the bile, the gall, the sperm, and the mucus, a complexion commensurate with the local temperature. To better illustrate my point, imagine that a perfectly healthy

> white man and white woman travel from our temperate zone to one of the hottest areas in the torrid zone; imagine further that this couple conforms to the customs and manners of the indigenous peoples, and do not fall ill with any sort of fever; their skin will harden and turn brown; it will flake and peel off. In effect, the excessive heat must, then, produce a greater agitation in the blood and the other humors, and the animalistic ether of the mucous bodies will become more saturated with phlogiston.[26]

Like the majority of naturalists of his generation, Nauton believed both that "[t]he varieties and shades of the skin's complexion are noticeable in the principal fluids of the human body" and that the detrimental effect of heat had imprinted itself forevermore on the African.[27] Moreover, Nauton argued that after thirty generations, this originally "accidental tint" became something real and enduring, "a quality, an affect of nature, that is passed on from generation to generation, just as other deformities or diseases are passed from fathers and mothers to their children."[28] It was just such a pathological view of human varieties—which explained the very existence of the African in terms of *generations* of damage—that allowed for a trenchant new view of race. Clearly, much had shifted since Buffon had first introduced the notion of degeneration; his comparatively vague idea of a palette of human colors and, thus, *varieties,* no longer held sway. As Nauton himself put it, the European "naturalist-physician," like himself, was now making the logical leap from "the anatomical and physiological study of the individual" to a more complete study of "the species."[29] More important, he argued, European naturalists were now for the first time able to classify the human species in terms of three "striking" variables: "the difference between complexion and color, that between shape and size, [and] that of temperament, nature, and national genius."[30]

Numerous essentializing typologies began to appear in French thought in the early 1780s. In 1782, following the Germans, the *Encyclopédie méthodique* proposed a racial breakdown that included categories like the "Gothic-Germanic," "the Western European (Celtic) race," and "the *nègre* race."[31] In 1786, Para Du Phanjas maintained that there were perhaps three races: *Europééne, Nègre,* and possibly *Tartare.*[32] Other naturalists went even further in their breakdowns of humankind. In an article published in the *Mémoires de la société des sciences physiques de Lausanne* in 1784, J. P. Berthout van Verchem argued that there were ten distinct races. What was important in all these classification schemes was not the number of races, however. The significance of human taxonomy lay in the fact that an increasing number of the era's naturalists had claimed the right to classify the "measurable" effects of climatic degeneration, which, as van Verchem wrote, had produced the different races of men, or *general varieties*.[33]

In addition to having a notable effect on the overall representation and understanding of the black African in the era's thought, the synthesis of heredity, anatomy, and taxonomy inspired new methods of ethnographic data-gathering and experimentation. Perhaps the most striking example of this tendency can be found in the guidelines that the Paris Société de médecine prescribed to the explorer Jean-François de Galaup, comte de La Pérouse, in May 1785. Two months before his expedition set off on its ill-fated scientific voyage to the Pacific (the ship sank near the Solomon Islands), the Société not only encouraged La Pérouse and his crew to gather information on foreign customs and comparative anthropology (e.g., the shape of the skull), it also invited these explorers to collect data on "the color of the skin" and the color of indigenous "humors" and organs.[34] In particular, the *savants* aboard La Pérouse's ships were asked to determine whether or not: "the spermatic fluid of men who are more or less swarthy, their brain matter, and their blood correspond . . . to the complexion of their skin."[35] What exactly the Société hoped to achieve with comparative data on the color of indigenous sperm, brains, and blood is not directly spelled out in this query; nor, I should hasten to add, did the Paris-based men of science provide guidelines as how to collect such information. It is probable, however, that the people who dictated this command to La Pérouse hoped, first and foremost, to confirm the now-general belief that darker bodies had darker organs and humors. But in addition to this objective, the Société also presumably sought to identify a spectrum of sperm colors (and heredity) ranging from the whitest white to that of black Africans. While the ostensible goal of such a venture was isolating the precise source of ethnic or racial difference, the real objective, of course, was managing such concepts.

Among all the other clues regarding the underlying motives of this erudite group of armchair naturalists, none is more important than the directive to obtain sperm from human albinos (*nègres blancs* and *blafards*).[36] As we saw in chapter 2, the albino or *blafard* had been a subject of interest long before the voyage of La Pérouse. A critical juncture in the forging of the concept of race earlier in the century, the *blafard* had allowed thinkers like Maupertuis and Buffon to posit a shared origin between black and white human varieties; it had also permitted them to assert that the white variety was the prototype stock from which *nègres* had degenerated, and to which they occasionally reverted in the form of a *nègre blanc*.[37] By asking La Pérouse and his crew to determine if these albinos did in fact have darkened sperm, the naturalists were presumably trying to determine if the defining trait of Africanness (black sperm) was present in a white (albeit albino) human. The prospective finding of this outlandish experiment—that albinos might have the seed of blackness despite their pallid complexions—reveals the increasingly physiological view of race taking form in the last years of the ancien régime; it also underscores a new tendency among scientific institutions to gather physical data in order to establish both normal and degenerative categories of humans, categories that were based on interior physiology as well as on exterior color and morphology. This shift of the locus of race from the exterior to the interior, which had been taking place since the late 1700s, would not be lost on the more overtly racist thinkers of the nineteenth century. As the polygenist Julien Joseph Virey put it succinctly in the 1803 *Nouveau dictionnaire d'histoire naturelle*: "the *nègre* is . . . not only *nègre* on the outside, but in all parts of his body as well, down to the innermost of his parts."[38]

Toward a Human Biopolitics circa 1750–1770

Throughout the eighteenth century (and beyond), natural history overlapped with the question of slavery in distinct and complicated ways. By the time that the theologically and ethnographically based *Code noir* came into being in 1685, European thinkers were already looking to "natural history" as a means of explaining the politics of servitude.[39] And yet, while early eighteenth-century French authors including Charlevoix and Labat commonly cited certain physical justifications of slavery—in particular, the suitability of Africans to labor in extreme conditions—such *corporeal* reasons were still less relied upon than four other rationales: (1) Europeans were simply participating in a longstanding African cultural practice; (2) Europeans were taking slaves who were *legally* enslaved; (3) Europeans were substituting one form of slavery for a much worse fate in Africa; and (4) slavery came with the possible benefit of eternal salvation, something that Africans could not hope to achieve as heathens.

By 1750, the one "positive" rationale in this list—that enslaving (and baptizing) Africans was a work of charity—not only began to lose its luster, it became a target for violent ridicule by philosophes, who took particular pleasure in juxtaposing the tenets of Christianity and the horrors of the slave trade. While some writers including Buffon criticized the hypocrisy of pro-slavery Christians quite delicately, others, most notably Helvétius, derided religious justifications directly. In his 1758 *De l'esprit* (1758), Helvétius famously wrote that the power of "the church and the kings" to authorize and to justify the slave trade on the basis of metaphysical principles was an absurdity.[40] What was one to make of a religion, he wrote, that "curses in the name of God him who brings trouble and dissension into families, [but that] blesses the merchant who roams the Gold Coast or Senegal [in search of slaves]?"[41] While it would be an error to assert that metaphysical justifications of the slave trade disappeared from pro-slavery discourse after mid-century, an increasing number of thinkers began to view them as cynical arguments. As the physiocrat-oriented *Journal oeconomique* summed up in 1768, "the conversion of *nègres* is very rarely a cause that brings Europeans to engage in commerce on the African coast."[42]

Pro-slavery writers who came of age after mid-century responded to such (anti-clerical) criticism of human bondage by seeking out more fruitful territory for advancing their case. If religious justifications of slavery were increasingly falling on deaf or dubious ears, it was entirely logical that pro-slavery writers would draw on more secular justifications. In addition to underscoring the critical role that African chattel slavery played within the French and French colonial economies, pro-slavery thinkers also began increasingly conjuring up the desacralized scientific portrait of the *nègre* being put forward by naturalists and anatomists. This represented a new and much more complete naturalization and normalization of the *nègre*'s plight.

The vast majority of the scientifically oriented writers or naturalists who provided the grist for this pro-slavery portrait of the black African were seemingly oblivious to the political implications of their work. Once again, it was Buffon who established what would become something of a convention in this respect. Treating the colonial portraits of the black African with the same dispassionate language that he used to contemplate a specimen in the king's cabinet, Buffon concentrated on the "facts" before him and not the context within which these facts were derived. While it is true that he did in fact chide certain Caribbean planters for their biased portrayals of black African slaves in general, his tone effectively hid the full political implications of the new "science" of man. Other naturalists and anatomists were even less aware of the politics of natural history. Indeed, when Meckel "discovered" black brains, he did so without any explicit reference to the implications for slave owners. In Le Cat's book-length treatment of the source of African *ethiops*, he wrote at great length about the color of Africans after death, their color at birth, the different shades of black on their body, the color of drowned Africans, and the origins and implications of albinos; and yet he never mentioned the state of the Africans in the colonies. De Pauw, too, separated his hugely influential explanation of the African's extreme degeneration from the overall question of slavery, although he did proclaim at one point in his *Recherches* that slavery itself "horrified humanity."[43] In sum, although natural history in its various forms had overlapped with the politics of slavery for centuries, the thinkers who were most clearly responsible for supplying the physical data used to codify the notion of race often did so in a political vacuum.

Pro-slavery and anti-black thinkers, however, deployed these ostensibly apolitical assessments of the *nègre* with great verve. Among mid-century French writers to put forward a harsh new biopolitics of servitude, Jacques-Philibert Rousselot de Surgy stands out. De Surgy's 1765 *Mélanges intéressans*—published during the heyday of physiological speculation on the African—is a watershed in the French history of race. Some thirty-five years before the full potential of polygenist thought was harnessed by the generation of pro-slavery thinkers writing after the revolution on Saint-Domingue, de Surgy refuted any commonality between whites and blacks; indeed, he maintained that the African occupied a taxonomical space that was so close to that of *des bêtes*, that he was unable to decide whether these creatures were closer to the orangutan or to the white man. This posited taxonomical affinity with the animal kingdom was based not only on morphology and anatomical differences but also on performative criteria. As de Surgy wrote, the *nègre*'s cognitive abilities were so severely limited that *nègres* had "no ideas, no knowledge, that belong to men."[44] Little wonder, then, that de Surgy believed that slavery corresponded perfectly to Africans' animal-like aptitudes; in his opinion it was clear that they were not "made for a condition better than that to which they have been reduced."[45] Unlike most French writers in the 1760s, de Surgy put forward an unequivocal politics of racial superiority and dominance.

In addition to serving as fodder for pro-slavery writers, the natural history of the black African also played an important role in the political assessment of slavery among philosophes. As I argued in chapter 3, the earliest and perhaps most telling example of how natural history infiltrated the classical liberalism of the Enlightenment came in the 1748 *De l'Esprit des lois*. While Montesquieu famously censured the institution of slavery in this text as a violation of natural law, his capacious treatment of the question of human bondage in this text also led him to try to understand it as a practice whose origins, while illegitimate, could nonetheless be attributed to certain physical variables. Examining the enslavement of black Africans, Montesquieu ultimately admitted that the bondage of the African could perhaps be understood as a function, not only of the climate in which the *nègre* lived, but also of climate theory's corollaries: African mores and African physiology. It was these final variables that ultimately led Montesquieu to entertain the idea that the existence of slavery in certain climates and, by extension, among certain peoples, was a philosophically logical, albeit morally undesirable, phenomenon.

Similar epistemological dissonance would infuse much of the post-1750 examination of slavery by other "conceptual" thinkers. Eight years after Montesquieu published his reflections on slavery in *De l'esprit des lois*, François Victor Riquetti, marquis de Mirabeau, generated an equally ambiguous portrayal of human bondage in his 1757 runaway bestseller, *L'ami des hommes, ou Traité de la population*.[46] Drawing deeply, as had his friend Montesquieu, on the tropes of natural history for his so-called treatise on population, Mirabeau contended that the number of inhabitants in a given country (along with the success of its agricultural policy) was synonymous with either the wealth or poverty of this same nation. Within this thesis-driven text, Mirabeau dedicated a sizable chapter to the status of African slaves in European colonies. It was in this same section that he forcefully and famously vilified the institution of slavery as emblematic of the shortsighted and unhealthy state of the typical European colony.[47]

Much of the historical significance attributed to the *L'ami des hommes* obviously derives from its status as a forerunner of the positions of later antislavery thinking among the physiocrats, including that of Pierre Samuel du Pont de Nemours and Anne-Robert-Jacques Turgot.[48] To a large degree, this narrative is entirely accurate: like later thinkers, Mirabeau not only accused the colonial project of languishing in a "foolish infancy"; he criticized the colonization of the West Indian islands as a hopelessly backward endeavor based on greed, the overconsumption of a select few, and a financially illogical purchase and importation of slave labor.[49] And yet, Mirabeau was the first and perhaps the only French thinker to marshal economic arguments against slavery based, in part, on a profoundly negative assessment of the "natural history" of the African. This begins with Mirabeau's "definition" of the black Africans found in French colonies: "Our slaves in America are a race of men separate and distinct from our own species by the most indelible of traits, that is, color."[50] More reminiscent of Voltaire's very schematic breakdown of humankind into different *species* than of the later (more benevolent) assessments of Africans found in the physiocrat writings of the 1770s, Mirabeau's seemingly polygenist evaluation of the African actually allowed him to explain the unhealthy form of slavery practiced in the Caribbean. Comparing Caribbean slavery unfavorably to the type of human bondage practiced during antiquity, Mirabeau argued that the slaves in the Americas were cultural others, whom he described as either "brutish or endowed with an instinct that is foreign to us."[51] Little wonder, he suggested, that Europeans ended up treating them like animals.

For Mirabeau, much of what transpired in the scandalous colonial system could be seen in terms of an ethnological economy, or clash between cultures. This framework was carried over into his speculations on the future of sugar producing islands in the Caribbean as well. In Mirabeau's estimation, these colonies had one of two possible futures. If Europeans continued to oppress their African slaves, populations would necessarily plummet. Not only would African women increase their practice of infanticide; the male slaves would become progressively deceitful and corrupt. On the contrary, if European masters were lax with their slaves, Africans would inevitably insert themselves into European households. In this latter scenario, masters would engage in debauchery, and an increasingly large mixed-race caste would be produced. Ultimately, this moral and biological decadence would lead to a decline in the number of whites and a growth in the number of *nègres*. The result of this shift in demographics would be inevitable: the *nègres* would one day take over European colonies, because, as Mirabeau explained it, "even the dullest of men have always been enlightened enough to understand the benefits of liberty."[52]

Having evoked these two equally bleak futures, Mirabeau proposed a solution to the problem of slavery that, much like the other parts of his argument, incorporated elements of natural history. To address the problems of colonial slavery, he argued, one need not abolish the practice. Rather, colonial policy should encourage free trade and a more productive agricultural policy. This, in turn, would supposedly attract a new, industrious, and superior group of European settlers. These productive and intelligent "artisans" would, according to Mirabeau, produce goods and services that would be far superior to anything that the African craftspeople could manufacture. Indeed, it would be inevitable that "the artisans of Europe . . . overtake the industry of the *nègres,* which is nothing but an exception among this race of men."[53] Like many early anti-slavery texts, the precise status of *L'ami des hommes* is far from unambiguous. While Mirabeau was clearly among the first thinkers to marshal economic arguments as part of an anti-slavery argument, much of his reasoning relied on a brutal race-based assessment of European racial superiority to do so.

The Politics of Slavery in the *Encyclopédie*

While something of an outlier in terms of its orientation, the publication of *L'ami des hommes* in 1757 coincided with a number of other significant developments in the then-nascent anti-slavery movement. Already, Rousseau had famously declared in the *Discours sur l'origine et les fondements de l'inégalité* (1755) that human enslavement was an artificial and dishonest institution that only came into existence due to the desperate need of miserable people. In 1758, Helvétius also put forward a virulent denunciation of slavery in his *De l'esprit* in the section entitled "De l'ignorance." Condemning the "consumption of men" taking place in the Americas in a way that no philosophe had done before him—not Montesquieu, not Rousseau, not Mirabeau—Helvétius measured the colonial project in terms of the amount of suffering endured by Africans.[54] This included not only the wars fomented by Europeans in Africa, but also the horror of the Middle Passage and the meaningless deaths caused by "the cupidity and [the] arbitrary power of the masters."[55] It was just such a perspective that allowed him to proclaim that "no cask of sugar arrives in Europe that is not tainted by human blood."[56] By the 1760s, these relatively isolated outbursts gave way to a growing consensus among progressive thinkers that African chattel slavery was not only an anachronistic and brutal institution out of step with the liberal principles espoused by the philosophes; it was a cause that warranted attention, much like the "sin" of religious intolerance. How this "cause" was championed in light of the simultaneous rise of a whole new set of anatomical data, however, is a complicated story.

Among canonical Enlightenment texts, the multifarious and multi-author *Encyclopédie* best reflects the era's evolving and oftentimes incoherent treatment of human bondage. While it may appear inconceivable to us now, the "subject" of human slavery came up only briefly in the first volumes of this massive *dictionnaire.* Essentially absent from articles such as "Afrique" and "Colonie," the Atlantic trade did not really seem to draw the editors' attention until 1755. This was the year that d'Alembert paid tribute to the recently deceased Montesquieu in the *avertissement* that opened the project's fifth volume. In addition to providing Montesquieu's intellectual biography, d'Alembert's *éloge* also included a summary of *De l'esprit des lois,* featuring a disproportionately large section dedicated to Montesquieu's writings on slavery. This is just the first of several points at which volume 5 of the *Encyclopédie* implicitly questions the foundations of this long-standing institution. In the article "Egalité naturelle," for example, Jaucourt affirms that since "human nature is the same in all men; it is clear that according to natural law, everyone must value and treat his fellow men as beings who are naturally equal to himself."[57]

Jaucourt followed up on this spirited praise of liberty in the article "Esclavage," condemning the unnatural introduction of political and civil slavery into the state of nature.[58] Much like Montesquieu's view of slavery, from which Jaucourt gleaned many of his ideas, "Esclavage" accurately reflects the tension existing between natural law and natural history during this era. On the one hand, Jaucourt affirmed that human freedom was an inviolable principle that had often been sullied (as in the case of slavery) through an illegitimate use of physical force. On the other hand, Jaucourt (like Montesquieu) also made clear that one of the root causes of human bondage could be attributed to a climate-induced suitability to "servitude."[59] While Jaucourt cited this environmentalist explanation of slavery alongside other "physical determinants," for example, "force" and "violence," this "natural" account of slavery was distinctive. Unlike the "human" influences that perpetuated human bondage, climate would presumably continue to engender pathological and social conditions conducive to slavery no matter what happened to the other variables.

This vexing force of climate disappeared in Jaucourt's later and more forcefully abolitionist article "Traite des nègres" (1766). Appearing in the sixteenth volume of the *Encyclopédie,* this entry reveals a significant shift in conventions and tone, both of which can be attributed to Jaucourt's discovery (and plundering) of the neglected but extremely important writings of the Scottish jurist George Wallace.[60] As David Brion Davis has demonstrated in a side-by-side comparison of these two works, the violent "Traite des nègres" flows almost directly from Wallace's text: indeed, it is no exaggeration to state that it was actually Wallace and not Jaucourt who was responsible for one of the first real denunciations of the colonial project in French thought.[61]

Following Wallace, Jaucourt asserted that the business of slavery "violates religion, morality, natural law, and all the rights of human nature."[62] He also declared that those Africans who had been taken as slaves, regardless of the conditions of their enslavement, had the right to declare themselves free.[63] All the equivocation of the article "Esclavage" had disappeared in "Traite des nègres," and all justifications were now clearly overruled by humankind's defining principle: liberty. To emphasize this point, Jaucourt, per Wallace, courageously weighed the economic importance of France's colonies against the suffering of African slaves.

> One might say [that these colonies] will be soon ruined . . . if they abolish the slavery of *nègres*. But were we to accept this premise, should we also conclude that humankind must be horribly wronged to enrich us or furnish our luxury? It is true that the purses of highwaymen will be empty if theft is completely abolished: but have men the right to enrich themselves through cruel and criminal channels? What right has a brigand to rob the passersby? Who is allowed to become opulent in bringing misfortune to his fellow man? Can it be legitimate to strip the human species of its most sacred of rights, only to satisfy one's own avarice, vanity, or individual passions? No . . . [I would prefer] that European colonies be destroyed, rather than do so much wrong![64]

Jaucourt clearly intended this anti-colonial stance to be the climactic moment of "Traite des nègres." And yet perhaps the most remarkable moment in this article comes later, when Jaucourt moves from his condemnation of slavery to a quixotic and utopian post-slavery vision of what the Americas could be if the colonial powers were to free their African slaves: "Liberate the *nègres*, and in but a few generations this vast and fertile country will have innumerable inhabitants. The arts and other talents will flourish; and in place of a country almost fully populated by savages and ferocious beasts, there will soon be only industrious men."[65] Reflecting the "liberty"-equals-prosperity thesis that was becoming increasingly popular in physiocrat circles during the 1760s (as much as in Wallace's text), Jaucourt's article sets forth a vision of a new colonial system that would differ markedly from the Caribbean factories of human misery.

An amalgam of ideas advanced by Montesquieu, Mirabeau, and Wallace, Jaucourt's "Traite" has been identified as the kernel of later, even more incisive, condemnations of the slave trade. Jean Ehrard, for example, has maintained that Jaucourt is "the first to make the step from anti-slavery to abolitionism."[66]

While this assessment of Jaucourt's text is not inaccurate, the full measure of his legacy can only be understood by recontextualizing his anti-slavery stance within the contradictions and paradoxes of his era and his own Africanist writings in the *Encyclopédie*.

One of the most startling aspects of Jaucourt's "Traite des *nègres*" is the fact that it was an anomaly within the *Encyclopédie*'s seventeen volumes of text; indeed, this memorable anti-slavery diatribe stands out in stark relief against a much more obvious collection of "neutral," non-moral disquisitions on various aspects of slavery. This can be readily seen when we consider Jaucourt's "Traite" alongside Jean-Baptiste Pierre Le Romain's article "Sucrerie," which appeared almost simultaneously. Much more indicative of the overall assessment of slavery during this era, Le Romain's dissertation on the sugar production in the Caribbean describes not only the manufacture of this colonial commodity but also the racialized servitude of the "extremely vicious and cunning species of man" who toiled on Caribbean plantations.[67]

As we look back at the pro-slavery and anti-slavery articles penned by thinkers such as Le Romain and Jaucourt, it is easy to envision the discussion of slavery in the *Encyclopédie* as something of an ideological binary, in which a pro-slavery faction using natural history to dehumanize the African is pitted against an increasingly vocal anti-slavery contingent emphasizing a softer view of the *nègre*. More accurately, however, the *Encyclopédie*'s treatment of slavery remains fragmented into different disciplinary treatments, unlinked by cross-reference. This overall ambivalence regarding the question of slavery overlaps with a revealing ellipsis. In a work that was ostensibly designed to provide an inventory of the technologies of eighteenth-century commerce, there is no mention of the specific European know-how and equipment—such as leg irons, stowage plans, or ship conversion—that made the Middle Passage possible. While the *Encyclopédie* plates dedicated to colonial agriculture feature a series of vignettes of African slaves working in the cotton and indigo industries, corresponding articles such as Diderot's meticulous and lengthy treatment of the production of cotton gloss over the slave labor at the core of the manufacturing process.

In addition to this these lacunae, *Encyclopédie* contributors were also seemingly unable to reconcile their knowledge of the natural history of the African with their belief in natural law. And here again, it is Jaucourt himself who provides us with the best example of this curious phenomenon. On the one hand, as we have just seen, it was Jaucourt who took the most courageous

An African slave caravan, illustrative of Portuguese "crimes," in G.-T. Raynal's *Histoire philosophique et politique des établissements et du commerce des Européens dans les deux Indes* (1780). Author's collection.

stand against slavery in the aforementioned articles "Esclavage" and "Traite des nègres." And yet, throughout the *Encyclopédie*, Jaucourt vulgarized numerous unfavorable assessments of Africans in his many geographical and ethnographic entries on West Africa. In addition to describing West Africans as a credulous, dirty lot with an animalistic sexuality who "know neither modesty nor restraint in the pleasures of love," Jaucourt painted *nègres* as universally corrupt and more than partly responsible for the continued existence of the slave trade.[68] While not nearly as palpable, a similar ambiguity can be found in Diderot's sole outburst against the slave trade in the natural history article, "Humaine espèce." In this short treatment of humankind's varieties, where he reduces Buffon's monogenetic narrative of different human varieties to several folio pages, Diderot restates what his friend the great naturalist had said about the African: the *nègre* may have little intelligence but is far from lacking in "emotions."[69] Having provided his "natural history" of the *nègre*, Diderot then goes on to criticize the horrific treatment of Africans, which makes a mockery not only of the spirit of Christianity but of humanity's pretense to reason: "We have reduced [the *nègres*] not only to the condition of slaves, but to that of beasts of burden; and we have reason! And we are Christians!"[70] This incongruous flare-up within what is ostensibly a naturalist account of blackness reflects the conflicted status of the *nègre* in this era. At times examined from the perspective of the naturalist, the African is forced into a speculative history of humankind that seeks to explain aberrant morphology, congenital vice, and innate obtuseness. At other times evoked in the context of universal human rights, the same black African (and his always latent status as a commodity within the colonial system) led Enlightenment thinkers to pity the *nègre* and to question the integrity of European humanism.

Examining the *Encyclopédie* from this perspective, it becomes clear that this massive *machine de guerre* did not lend itself to the type of comprehensive abolitionist discourse found in later anti-slavery works. What these abolitionists had realized—the Philadelphia Quaker Anthony Benezet as early as the 1760s—was that timeworn arguments in favor of the slave trade would have to be refuted with a comprehensive line of reasoning that combined the standard moral and philosophical objections to the trade with a repudiation of the economic and ethnological rationalizations that were increasingly popular among pro-slavery writers. The *Encyclopédie*, in which "ethnology" and human rights seemingly constitute independent disciplines with their own interpretative structures and their own sets of truths, could not produce such a consistent case.

Mercier and Saint-Lambert and the New Natural History

The first hints of French anti-slavery thought appeared among philosophes during the 1750s and 1760s. Some of these indictments of slavery, while they rarely, if ever, questioned natural history beliefs regarding the African, overlapped with and/or inspired fictional representations of the evil of slavery in the era's literature. As both Léon-François Hoffmann and William B. Cohen have amply demonstrated, many of these literary representations of the *nègre* conferred a new autonomy and psychology on this maligned group of humans.[71]

To the extent that black Africans had been present in earlier eighteenth-century French fiction, they had generally functioned as "décor" or part of an exotic backdrop. After mid-century, however, the French began tapping into the sizable mythology of blackness that had long been a part of the English literary canon. French authors referenced Shakespeare's *Othello* and were fascinated by Aphra Behn's *Oroonoko: or, The Royal Slave* (1688), the compelling story of a defiant black African prince whose intelligence and comportment often put Europeans to shame. As the abundant scholarship on this subject has made clear, Behn's novella inspired numerous adaptations both in England and France. Whether restaged in England by Thomas Southerne or in France by Pierre-Antoine de La Place or Pierre-Joseph Fiquet du Bocage, the character of Oroonoko allowed audiences to envision the suffering of black Africans on an understandable, individual level.[72] In contrast to the era's natural history or mercantile treatments of the *nègre*, dramaturges like La Place staged slavery in terms of a moral binary according to which the most sympathetic character, a rebellious but noble black African, struggled against a horrific and unfair system populated by cruel and unfeeling oppressors. In addition to the fact that this story had the undeniable advantage of projecting the guilt of slavery onto a specific class of Europeans—planters and colonists—it forced sensitive audiences to moralize about a subject that had long been seen as an unfortunate, but unavoidable reality.

To a large degree, the increasing number of literary works to treat the subject of African slavery reflected the beginning of a historical *prise de conscience*. As I stated in the Introduction, Voltaire provided the most notable example of individualized black suffering when he conjured up the dying *nègre de Surinam* in his 1759 *Candide*. While the anguished slave in this philosophical *conte* is, like the other characters, nothing more than a philosophical puppet, his celebrated accusation ("This is the cost of the sugar you eat in Europe") signaled the growing perception that the era was failing to take responsibility for the misery and agony of the black African.[73] Ten years later, a number of plays, poems, and novels would confer a more strident and accusatory voice to the suffering African. Within the present study of the natural history of slavery, two narrative works merit special attention: Jean-François de Saint-Lambert's "Ziméo" (1769) and Louis-Sébastien Mercier's *L'an 2440: Rêve s'il en fut jamais* (The Year 2440: A Dream If Ever There Was One) (1770).

While Saint-Lambert's view of the African clearly inspired Mercier's, the respective contributions of these two authors to the genealogy of anti-slavery thought are best understood in reverse chronological order. Mercier's *L'an 2440* is a utopian novel that takes place in the twenty-fifth century. The protagonist, who wakes up after a seven-hundred-year nap, explores and describes, in a series of episodic chapters, an idyllically clean and rational Paris inhabited by Enlightenment sages, the seeming antithesis of the corrupt and unhealthy capital of 1770. Among the discoveries that this ancient time traveler makes is a "singular monument" that depicts a series of European nations on their knees begging forgiveness for any number of early-modern sins, including sectarianism and intolerance.[74] The most noteworthy failings of the prostrate European nations, however, involve their brutal colonial practices, chief among them the enslavement of Black Africans. Mercier's protagonist is forced to confront this sin directly when, after turning away from the first set of statues, he encounters a second historical "installation" piece, a huge figure of an Oroonoko-type rebellious slave posed on a dais. In sharp contrast to the passive figure of the African that we find in works such as *Candide*, Mercier's African statue stands tall among the wreckage of colonial empire: "I was leaving this place, when, to my right I saw upon a magnificent dais a *nègre*, with a bare head, outstretched arm, proud eye, and a noble, imposing attitude. Around him were the broken pieces of twenty scepters. At his feet read the words: *To the avenger of the new world!*"[75]

Having sprung to life violently and organically from a nature with which he is seemingly synonymous, this black "genius" is, in the utopian world of 2400, universally praised for having undertaken the slaughter of the colonial whites who have oppressed his people. Strangely prescient of what would occur on Saint-Domingue some twenty years later, this allegorical victory of a vengeful African over a group of disgraced European nations is a significant moment in anti-slavery discourse. In addition to projecting the guilt of slavery onto entire nations (as opposed to only the planters), Mercier also suggested that this culpability would weigh upon future generations.[76]

Mercier seems to have drawn many ideas from Jean-François de Saint-Lambert's "Ziméo," a short story that, in contrast to Mercier's tale, is set in the contemporary reality of Jamaica. Here too, we meet an audacious slave (John-Ziméo) who, after burning plantations and killing unjust planters, meets a gentle, fair Quaker who, unlike those who die at Ziméo's hand, loves his slaves and only makes them work two days a week. But the most intriguing part of this short story is the epilogue that comes after Ziméo returns with his group of maroon slaves to the forests of Jamaica. It is in this paratextual digression that the narrator undertakes a re-evaluation of African stereotypes.

Argued from the perspective of the Quaker, who is familiar with both Africa and the Caribbean, Saint-Lambert's epilogue is a fascinating departure from other such works of fiction. Anticipating what we find in the *Histoire des deux Indes* some ten years later, Saint-Lambert's narrator asserts that slave traders and colonists involved in the slave trade cannot possibly accurately portray Africans: "My stay in the Antilles and my travels in Africa have confirmed for me an opinion that I have long had. . . . The merchants who trade the *nègres* and the colonists who bond them to slavery have too many wrongs against [these Africans] to be able to speak the truth."[77]

This is only the beginning of Saint-Lambert's challenge of African stereotypes. Evoking the ethnographic diversity that his era came to know through African travelogues and compilations, the narrator asserts that Africa, like Europe, has both good and bad inhabitants. In his words, "the first of our injustices is to give the Africans a general character."[78] This blanket statement is followed by an enumeration of the different types of African moral systems, religions, and governments:

> The various types of governments, goods, and religions in these immense regions have also determined the respective characters of their inhabitants. Here you will see Republicans who have sincerity, courage, and a spirit of justice encouraged by freedom. There, you will see independent *nègres* who live without leaders and without laws, as ferocious and as savage as the Iroquois. Enter the interior of certain countries, or limit yourself to the coastlines, and you will discover great Empires, the despotism of princes and that of priests, feudal governments, regulated monarchies, etc. One sees different laws, opinions, and notions of honor everywhere; and consequently, one finds humane *nègres* and barbaric *nègres*; warriors and cowards; pleasant customs and detestable customs; the man of the earth and the corrupted man, yet nowhere the perfect man.[79]

While this schematic breakdown of the category of the *nègre* may recall aspects of both Buffon and, to a lesser degree, Montesquieu, what distinguishes

Saint-Lambert's "ethnography" is the fact that it overtly rejects the physico-environmental determinism that generally underlay most other theories during this era. If Saint-Lambert's Quaker readily admits the present superiority of the European over the African, he also asserts that this discrepancy stems neither from climate nor anatomy; it is simply the result of measurable "circumstances," such as the form of government or contact with more advanced peoples. According to the Quaker narrator, time and education, not anatomy and climate, are the only variables holding back the African. Saint-Lambert's message is clear: Do not enslave these people; rather, "let us bring them our discoveries and our knowledge; in centuries to come, they may add some of their own, and humankind will be better for it."[80]

The Synchretism of the 1770s: Grappling with "Nature's Mistreatment" of the *Nègre*

The portrait of the African during the last quarter of the eighteenth century has left critics flummoxed. This is quite expected given the competing theories, epistemologies, and generic conventions that were brought to bear on the question of both Africans and their unfortunate fate in European colonies. One way that historians have attempted to understand this confusing set of "data" is by separating thinkers into two distinct categories: polygenist/pro-slavery and monogenist/anti-slavery (table 4.1). This breakdown of perspectives works rather well when the sample is limited to the most prominent thinkers who actively endorsed environmentalism (e.g., Montesquieu, Buffon, and Blumenbach) and their opponents. Indeed, according to this schema, those thinkers associated with either a latent or overt anti-slavery orientation are all monogenist, whereas the pro-slavery thinkers such as de Surgy are polygenist.

While there is an often undeniable relationship between a given thinker's stated position on the "origin" of Africans and this same person's stance on the legitimacy of slavery, this correlation dissolves in many eighteenth-century contexts.[81] In addition to Voltaire—who is known for his outbursts against slavery, but who also believed that Africans constituted an entirely separate species of man—there are many other writers for whom this association is tenuous at best (table 4.2). Among Christian proponents of slavery, such as the Dominican missionary Jean-Baptiste Labat, a (Bible-based) monogenesis was a religious given. Cornelius de Pauw also provides an interesting case. While

Table 4.1. A Pro-slavery Polygenist and Three Anti-slavery Monogenists

	Montesquieu, 1748	Buffon, 1749	J.-P. Rousselot de Surgy, 1765	J. F. Blumenbach, 1775
Views on human origins	monogenesis	monogenesis	polygenesis	monogenesis
Position on slavery	anti-slavery	ambiguous but disapproving	pro-slavery	anti-slavery (esp. later editions)

the Dutch naturalist singlehandedly injected a new and brutal anatomical and hereditary component into Buffon's monogenetic framework—arguing that the inferiority of the African type or race was more fixed than previously thought—his work also shows a sensitivity to nascent anti-slavery sentiment. Even one of the most brutal polygenists of the early nineteenth century, Julien-Joseph Virey, was reluctant to embrace an unambiguous pro-slavery stance.

With the exception of those pro-slavery thinkers who fused the conceptual alterity of the African with a new and brutal biopolitics, many of the late-century writers discussed here simply did not produce what we might think of as a consistent treatment of this subject. This is also the case in the *Encyclopédie*'s contradictory treatment of slavery, as we have seen. But the most telling example of the era's incongruous relationship between natural history and slavery is found in Abbé Guillaume-Thomas Raynal's *Histoire philosophique et politique du commerce et des établissements des Européens dans les deux Indes* (first ed., 1770).

Unlike the *Encyclopédie*, the *Histoire des deux Indes* was supposed to treat questions such as the African and the African slave trade from a comprehensive point of view, one that combined ethnography, ethics, and economics. Raynal's treatment of Africa, Africans, and the African slave trade comprises an entire section, book 11. Although technically not part of the *deux Indes*, Africa warranted its own section because it was considered something of an "extension" of the Antilles, or West Indies.[82] As he did for subjects including China, India, and Peru, Raynal began his assessment of Africa (in 1770) with a meditation on the history and geography of this continent, including a digression on the possibility of a just war waged against Barbary despots. The text then abruptly moves from this hypothetical call to arms to a section on the natural history of the black inhabitants of this continent. Opening this section with an indictment of a religious-tainted "science" of the human, Raynal explained

Table 4.2. Views on Slavery in Relation to Position on Human Origins

	J.-B. Labat, 1722	Voltaire, c. 1733	Cornelius de Pauw, 1768	J.-J. Virey, c. 1800
Orientation	Dominican priest	philosophe	naturalist	naturalist
Views on African	uncivilized, redeemable; suited for slavery	animalistic, another species	degenerated; emphasis on anatomical differences based on heredity	separate race; characterized by essential corporeal differences; destined for slavery
Views on human origins	monogenesis, biblical Genesis	polygenesis	monogenesis, grave degeneration	polygenesis
Position on slavery	pro-slavery	anti-slavery	indifferent/ anti-slavery	reluctantly accepting of institution; expressed vague hopes for improvement[1]

1. Based on Virey's article "Nègre" in the *Nouveau dictionnaire d'histoire naturelle* (1803).

that to understand the black African "type" one should abandon Scripture and replace it with a more naturalistic point of view. Working backwards from the universally accepted "fact" of African inferiority, Raynal declared that "the *nègres* are beings who are *mistreated* by nature, and not damned by [the] justice [of God]."[83] Much was implied in this statement. In addition to the fact that Raynal was clearly rejecting any metaphysical (or biblical) views on humankind, he was also conjuring up an allegorical conception of nature, which, he maintained, had physically abused the *nègre*. A significant portion of book 11 attempts to explain or diagnose this so-called mistreatment.

Following established patterns in natural history discourse, Raynal began his discussion of the African's physical liabilities with an examination of the physiology of African skin color. Where he departed from thinkers like Buffon, however, was in the way he explained the origin and significance of this defining African feature. Although Raynal allowed that "famous naturalists" (meaning Buffon and his followers) had declared that the *nègre*'s color sprang up from the "climate in which they live," he downplayed this theory and emphasized the importance of anatomy in the overall constitution of the African.[84]

> Whatever the original, radical cause for the varieties in color in the human species, one must concede that the color of the complexion and skin comes from a gelatinous substance that is found between the epidermis and the dermis. This substance is blackish in the *nègres*, brown in olive-skinned and dark-skinned people, white in the Europeans, and spotted with freckles in pale blond or red-haired people.[85]

Up until this point, Raynal's understanding of skin is unremarkable. Yet for Raynal, this gelatinous substance was much more than simply an effect of a particular environment. Borrowing from de Pauw's *Recherches philosophiques*, he affirmed that the *nègre*'s skin was part of a larger set of corporeal liabilities that explained everything, including the ill-fated race's suitability to slavery.

> Anatomical science has found *nègres* to have blackened brain matter, a nearly completely black pineal gland, and blood of a darker red than that of the whites. Their skin is always overheated, and their pulse brisk. As a result, fear and love are excessive in these people; and this renders them more effeminate, lazier, weaker, and unfortunately more suitable to become slaves. Furthermore, since their intellectual faculties have been nearly exhausted by their overindulgence in physical love, they have neither the memory nor the intelligence to compensate (by cleverness) for the force that they lack.[86]

Like many of the era's post-1755 treatments of the *nègre*, Raynal's assessment of the African conflates a hundred years of negative Caribbean-born stereotypes with "ground-breaking" scientific discoveries regarding the *nègre*'s anatomy. In particular, Raynal asserted, once again following de Pauw, that the ultimate cause of blackness (and all that it entailed) could be located on the most elemental level, in the African's dark sperm. This "factual" microphysiological understanding of the African as having a different *seed* ultimately led Raynal to depart from de Pauw, however. Whereas the Dutch naturalist had explained black sperm in terms of a deleterious degeneration from a prototype race, for Raynal this essential difference indicated a conceptual distinction between whites and blacks on the level of species and even origins: "Finally, anatomical science has found the origin of the *nègres*'s blackness in their seeds of generation. It takes no more than this, it seems, to prove that the *nègres* are a different species of men, for if anything differentiates species, or classes within a species, it is certainly the difference in sperm."[87]

This stark understanding of the African's physiology represents an attention-grabbing departure from the general trend of natural history around 1770.[88] If the vast majority of the era's anatomists and philosophes were examining the African from an increasingly deterministic and physiological point of view, these same thinkers also remained wed to a climate-based, single-origin theory of humankind. Raynal, on the contrary, announced that the climate theory at the heart of the monogenetic enterprise was entirely misguided. While he readily acknowledged that there were certainly "climates suited only for certain species," he also refuted Buffon and de Pauw (although not by name), asserting that "climatic differences" did not "change the same species from white to black" and that "the sun does nothing to alter and modify the seeds of reproduction."[89] It was entirely wrongheaded, he asserted, to "attribute the color of the *nègres* to climate."[90]

The relationship between Raynal's conceptualization of African physiology and his views on slavery has understandably puzzled critics for decades. The question raised by the 1770 edition of Raynal's opus is quite simple: how is it that he could put forward one of the most deterministic and pessimistic views of the African of his time, while also denouncing the institution of slavery? In *The French Encounter with Africans*, William B. Cohen attempted to resolve this problem by suggesting that Raynal's early polygenetic views on the African were an extreme position that he (actually Diderot) ultimately corrected in the third and final edition of the *Histoire* in 1780. Cohen suggests two specific reasons for this shift. The first, which attributes a great deal of power to Enlightenment ideology, is that the era's belief in "the unity of man [finally] asserted itself." The second is that Raynal/Diderot seem to have retracted "the polygenism because it better suited [the] antislavery argument."[91] This latter explanation, which is perhaps much closer to the truth, merits further investigation.

Anti-slavery Rhetoric in Raynal's *Histoire des deux Indes*

Like much of the rest of the 1770 edition of the *Histoire des deux Indes*, the mosaic-like construction of the section on slavery allowed multiple truths and opinions to exist simultaneously. Beginning this section with a bit of historical context, Raynal explained that the present form of colonial slavery had arisen when European merchants transformed a relatively benign African institution into an inhuman drain on Africa. Examining the phenomenon from a somewhat physiocratic point of view, Raynal asserted that the trade was taking a terrible toll on African populations. He emphasized this point a little later in this section apropos of the mortality rate of the *nègres* brought to the Americas: "Every year in America [i.e., including the West Indies] one-seventh of the *noirs* that are brought from Guinea die. The 1.4 million wretches that one sees today in the European colonies of the New World are the remainder of the nine million ill-fated slaves that they received."[92]

Raynal's reactions to the reality and the shocking mathematics of the slave trade vary widely.[93] Perhaps most ambiguously, Raynal advocated a more "enlightened form of slavery" with the potential to improve the behavior of European masters by persuading them to treat their slaves with more "tenderness and humanity."[94] In one of his more curious suggestions along these lines, Raynal proposed that slave owners take advantage of the *nègre*'s innate love of music, which would not only produce happier slaves, but a better harvest of slave babies.[95] And yet toward the end of this appeal Raynal actually broaches the idea of a progressive emancipation of the slaves, once again based on vague physiocratic principles reminiscent of Jaucourt's "Traite des nègres":

> In according freedom to these miserable people, but progressively, as a reward for their economy, for their conduct, for their work, take care to make them adhere to your laws and your customs, and to offer them your superfluous things. . . . Give them a homeland, a small [business] concern, something to produce, and things to consume that correspond to their taste; and your colonies will not lack for [willing laborers] who, relieved of their chains, will then be more active and more robust.[96]

This idealistic vision of a happy and productive workforce, which was often as much a planter fantasy as it was the dream of philosophes and physiocrats, is followed by the most commented-on portion of Raynal's *Histoire des deux Indes*: the unapologetic condemnation of the institution of slavery, much of which was presumably drawn from paragraphs provided by Jean de Pechméja.[97] This section opens by drawing a line in the sand separating those who believe in slavery and those who justify its continued existence through reasoned arguments: "Whoever justifies so odious a system merits a contemptuous silence from the philosophe, and a stab of a dagger from the *nègre*."[98] The ensuing anti-slavery paragraphs build on this violent image in various ways. Like Montesquieu before him, Raynal began this section by conjuring up a pro-slavery persona. And yet, unlike what we find in *De l'Esprit des lois*, the justifications of slavery in Raynal's text do not remain unanswered; they become the target of an imagined philosophe interlocutor speaking from the standpoint of a slave. In many cases, this prosecutorial argument takes the form of pointed rhetorical questions:

> Will someone [really] claim that the person who wants to enslave me is not at all culpable, that he is exercising his rights? What are these, these rights? [and] who has given these rights a character that is sacred enough to silence mine?
>
> Can't you see, unfortunate apologists of slavery, that you cover the earth with justified assassins?[99]

In the ensuing paragraph, Raynal not only indicted the institution of slavery, but any institution that supported it:

> if there were to exist a religion that authorized, that tolerated, if only by its silence, such horrors; if [this religion] were otherwise occupied by idle or seditious questions; if it did not thunder endlessly against the authors or instruments of this tyranny; if it were to make it a crime for the slave to break free of his own chains; if it were to welcome into its midst the sinful judge who condemns the fugitive to death; if this religion were to exist, it should smother its ministers under the debris of its own altars.[100]

After this anti-clerical outburst, the first edition of the *Histoire des deux Indes* moves on to another target: pro-slavery arguments stemming from the discipline of natural history. This portion of the text, which contrasts markedly with the *Histoire*'s earlier polygenist presentation of the *nègre*, clearly struck Diderot as an important part of a new anti-slavery discourse. In editing this section of the *Histoire* for both the 1774 and 1780 editions, Diderot inflected these paragraphs in interesting ways. This is best seen in a schematic form (see table 4.3).

Diderot's ultimate contribution to Raynal's text is best remembered for its violent images—the avenging African and the evocation of a *code blanc*—but his most significant intervention may actually have been in his subtle targeting of the scientific justification of slavery, a set of beliefs to which the earlier editions of the *Histoire* had seemingly given credence. Ann Thomson's important work on the evolution of this perspective from the 1770 to 1780

Table 4.3. Evolving Assessments of the Natural History of the *Nègre* in Raynal's *Histoire des deux Indes*

First edition, 1770	Later editions
"But the *nègres* are a species of men born for slavery. They are narrow-minded, deceitful, evil. They themselves agree on the superiority of our intelligence, and almost acknowledge the justice of our empire."[1]	Same paragraph in 1774 and 1780.
"The *nègres* are limited [in intelligence]; because slavery shatters the springs of the soul. They are wicked; not enough. They are deceitful; because one owes no truth to one's tyrants."[2]	Slight modification (in boldface) in 1774 and 1780: "They are wicked; not enough **with you** [i.e., colonists or planters]."[3]
"They recognize the superiority of our minds, because we have exploited their ignorance; [they recognize] the legitimacy of our power over them, because we have exploited their weakness."[4]	Slight modification (in boldface) in 1780: "They recognize the superiority of our minds, because we have **perpetuated** their ignorance."[5]
	Added sentences in 1780: "You almost managed to persuade them that they were a separate species, born for abjection and dependence, for work and punishment. You spared nothing to degrade these unfortunate people, and then you reproach them for being vile."[6]

1. "Mais les nègres sont une espèce d'hommes née pour l'esclavage. Ils sont bornés, fourbes, méchants. Ils conviennent eux-mêmes de la supériorité de notre intelligence, et reconnaissent presque la justice de notre empire" (*HDI* [1770], 4: 172).

2. "Les nègres sont bornés; parce que l'esclavage brise tous les ressorts de l'âme. Ils sont méchants; pas assez. Ils sont fourbes; parce qu'on ne doit pas la vérité à ses tyrans" (ibid., 172–73).

3. "Ils sont méchants, pas assez **avec vous**" (*HDI* [1780], 6: 207).

4. "Ils reconnaissent la supériorité de notre esprit, parce que nous avons abusé de leur ignorance; la justice de notre empire, parce que nous avons abusé de leur faiblesse" (ibid.).

5. "Ils reconnaissent la supériorité de notre esprit, parce que nous avons **perpétué** leur ignorance" (ibid.).

6. "Vous êtes presque parvenus à leur persuader qu'ils étaient une espèce singulière, née pour l'abjection et la dépendance, pour le travail et le châtiment. Vous n'avez rien négligé, pour dégrader ces malheureux, et vous leur reprochez ensuite d'être vils" (ibid.).

editions of the *Histoire*—she too has charted this evolution schematically—has demonstrated convincingly that many of these changes were not original to Diderot but were adapted from the physiocrat Abbé Pierre Roubaud's *Histoire générale de l'Asie, de l'Afrique et de l'Amérique* (1770–75). As Thomson puts it, "The work of Abbé Roubaud was used in 1780 to carry out a reversal in the discourse about Africans, to fight against the prejudice which was making them out to be a separate race, and to reduce the gap between them and the Europeans."[101]

In contrast to the few critics who have written on this portion of the *Histoire*, Thomson underscores the fact that Diderot was carefully refashioning both the scope and focus of anti-slavery discourse: unlike the earlier iterations of the *Histoire*, the 1780 edition responded directly and unequivocally to the politics of natural history. In addition to adding the final paragraph cited above—where he effectively diagnosed the "liabilities" of the African as stemming from the colonial situation itself—Diderot completely retooled Raynal's understanding of natural history. Once again borrowing from Roubaud to do so, Diderot flatly rejected the increasingly humoral-anatomical understanding of blackness that had come into vogue after 1770:

> By the eighth day after birth, the children [of black parents] begin to change color; the skin browns and finally becomes black. However, the flesh, the bones, the organs, all the internal parts, are the same color in the blacks as in the whites: the lymph is equally white and clear; the milk of their nurses is the same everywhere.[102]

While Diderot allowed the belief in darkened blood to stand—this was, after all, a "fact" that his friend Buffon had accepted—he nonetheless forcefully refuted the conviction that the African's sperm was black. This was a critical decision, since darkened sperm was held to signify hereditary transfer of blackness and thus a more vigorous and deterministic view of race. In a sense, Diderot was consciously spurning the vision of human categories put forward by thinkers like de Pauw and was returning to the softer understanding of human *varieties* that Buffon had endorsed twenty years earlier. This act was much more than a simple recasting of environmentalism, however. Diderot, the author of the contemporaneous *Eléments de physiologie*, was actively challenging some of his era's most cutting-edge anatomical discoveries in order to advance a new biopolitics of essential sameness. This revised natural history of the black African made Diderot into one of the era's first *négrophiles*.

Not only had he expanded the era's critique of the colonial space to include heretofore-unimpeachable scientific "data" regarding the African, he violently refuted the political implications of the era's understanding of the black race. To a certain extent, this final 1780 treatment of the *nègre* in the *Histoire des deux Indes* rectifies the compartmentalization, contradictions, and ellipses that characterized many eighteenth-century philosophical works with anti-slavery leanings, including those present in the *Encyclopédie* and earlier editions of the *Histoire* itself.

The Era of Negrophilia

In the traditional historiography of anti-slavery discourse in France, scholars have generally identified two relatively distinct genealogies leading up to the birth of the main French anti-slavery organization of the eighteenth century, the Société des amis des noirs. On the one hand, there was the secular philosophe and physiocrat branch that grew out of Montesquieu's anti-slavery writings on through to Raynal; on the other, an evangelical branch that arrived in France in the late 1780s via Anglicans such as William Wilberforce, Thomas Clarkson, and Granville Sharp. While this is a useful schematic breakdown, it may be more fruitful, at times, to envision the evolution of French anti-slavery discourse as a hybridization of secular and religious thought (much like "scientific" monogenesis in a sense).[103] This is certainly the case for the important epilogue that Saint-Lambert appended to "Ziméo." Although the philosophe Saint-Lambert is certainly to be credited for publishing the landmark anti-slavery arguments discussed above, the entire architecture of his epilogue to "Ziméo"—the first inklings of what would be called *négrophilie* later in the century—surely came from the "other" side of abolitionist thought, perhaps even from the Quaker Anthony Benezet's *A Caution and Warning to Great Britain and her Colonies in a Short Representation of the Calamitous State of the Enslaved Negroes in the British Dominions* (Philadelphia, 1766).[104]

Many of the arguments contained in Benezet's short (forty-page) Quaker indictment of the slave trade intersected with the then-nascent anti-slavery movement both in England and in France. Like Wallace and Jaucourt before him, Benezet attacked both the trade and the traders as barbaric, condemning the "temporal evils that attend this practice" on both moral and religious grounds. Where Benezet distinguished himself from his predecessors, however, was in the way that he made use of his extensive knowledge of Caribbean and African travelogues. Citing numerous Caribbean sources, he recounted the horrors, tortures, and wasteful deaths of Africans on both Barbados and Jamaica. Quoting from a series of African traders, he also related the brutality of African royalty tempted by European riches. But Benezet's most significant contribution was the first true reevaluation of the extant ethnography associated with the African:

> Some who have only seen Negroes in an abject state of slavery, broken-spirited and dejected, knowing nothing of their situation in their native country, may apprehend, that they are naturally insensible of the benefits of Liberty, being destitute and miserable in every respect. . . . Although it is highly probable that in a country which is more than three thousand miles in extent from north to south, and as much from east to west, there will be barren parts, and many inhabitants more uncivilized and barbarous than others; as is the case in all other countries: yet, from the most authentic accounts, the inhabitants of Guinea appear, generally speaking, to be an industrious, humane, sociable people, whose capacities are naturally as enlarged, and as open to improvement, as those of the European.[105]

In order to advance this much more positive view of the "human geography" of Africa, Benezet drew from a variety of French and English travelogues, including the naturalist Michel Adanson's *Histoire naturelle du Sénégal*, William Bosman's *New and Accurate Account of the Coast of Guinea*, William Smith's *New Voyage to Guinea*, and André de Brüe's memoirs.[106] Anticipating the argument that he would make much more famously in his *Some Historical Account of Guinea* (1771), Benezet set out to prove that Africans were not only far from brutal or stupid; they were, in fact, generally peace-loving and kind, until they were either incited to violence by Europeans or ripped from a country of immeasurable fertility.[107]

While it would be wholly inaccurate to assert that Benezet's counterdiscursive salvos had an immediate effect on the well-established and very negative conception of the *nègre*, his writings (as well as the more positive views of Africans beginning to circulate in French thought) provided a new model for the way that anti-slavery arguments might be constructed. In addition to simply making a case against the institution of human bondage, Benezet sought to paint its victims as sensitive and full members of humankind. It was precisely this humanization that was later featured in the image that appeared on the title page of the 1787 edition of Benezet's *Some Historical Account*. Steeped in a biblical monogenesis (the emblem is preceded by Acts 18), this small, often-

"Am I Not a Man and a Brother?" Wedgwood medallion reproduced in Anthony Benezet, *Some Historical Account of Guinea* (1788). Courtesy Watkinson Library, Trinity College, Hartford, Connecticut.

reproduced Wedgwood figure of the kneeling African who asks, "Am I not a man and a brother?" underscores both the religious and ethnography-based *fraternity* of humankind.[108]

The humanization and demystification of the African played an important role in the anti-slavery movement. In Bernardin de Saint Pierre's best-selling *Voyage à l'île de France* (1773), the future author of *Paul et Virginie* not only described first-hand the misery of the slave system on Mauritius; he took special care to depict the Africans (actually the Malagasy) as skillful, intelligent, loyal, interested in love, and good musicians. Similarly, in 1781, when Condorcet (writing under the pseudonym Joachim Schwartz) published his *Réflexions sur l'esclavage des nègres*, he commenced his "Epitre dédicatoire aux nègres esclaves" by breaking down the categories that had long separated whites and blacks: "Although I am not the same color as you, I have always regarded you as my brothers. Nature has made you to have the same mind, the same reason, the same virtues as the whites. I speak only of the whites in Europe, for I will not insult you by comparing you to the whites in the colonies."[109]

One of the things that Condorcet sought to do in this multifaceted denunciation of the slave trade (with suggestions on how to destroy slavery by degrees) was to address the myriad natural history justifications and explanations of slavery that had, by the 1780s, become general knowledge. Particularly appalling to Condorcet was the planter-generated belief that Africans were a stupid and lazy breed. To combat this categorical assertion, Condorcet attributed these traits—which he did not deny—to the brutal oppression endured by Africans. This was done in one spectacularly clear sentence that refuted the various "explanations" of the African "type" either implicitly or explicitly crafted by writers including Montesquieu, Voltaire, and de Pauw: "It is not to the climate, nor the terrain, nor the physical constitution, nor the national character that one must attribute the laziness of certain peoples; it is to the bad laws that govern them."[110] Echoing the now familiar thesis that this "sweet, industrious, [and] sensitive people" had been disfigured by the institution of slavery, Condorcet explained that if African slaves were indeed "lazy, stupid, and corrupt," such was "the fate of all slaves."[111] In sum, while he admitted in the text that Africans seemed to suffer from "great stupidity"; he asserted that they were not to blame for this: "it is not toward [black slaves] that we should direct our reproach, it is toward their masters."[112]

Abolitionists on both sides of the Channel were quick to incorporate this type of argument. In his 1784 *Essay on the Treatment and Conversion of African Slaves*, James Ramsey emphasized the "shared rational capacity of Europeans and slaves," contending that it was obscured in slaves by their enslaved condition.[113] More pointedly, in the *Essay on the Slavery and Commerce of the Human Species* (1786), Thomas Clarkson attacked the supposed link between inferior intelligence, anomalous physical characteristics, and the justification of slavery.

> Had the Africans been *made for slavery*, or to become the property of any society of men, it is clear . . . that they must have been created *devoid of reason*: but this is contrary to fact. It is clear also, that there must have been many and evident signs of the *inferiority of their nature*, and that this society of men must have had a *natural right* to their dominion: but this is equally false. No such signs of *inferiority* are to be found in the one, and the right to dominion in the other is *incidental*.[114]

In France, Benjamin-Sigismond Frossard echoed many of these views in his *La cause des esclaves nègres* (1789). Although less well known now than either of the previously mentioned English abolitionists, this Lyonnais graduate of Oxford and member of the Société des amis des noirs provided the most complete published "interdisciplinary" arguments in French against human bondage of his era. Like Benezet, Frossard developed ethnographically based valorizations of the African, dedicating an entire section to favorable assessments of the "most advanced" peoples along the *côte des esclaves* in order to counter the belief that "life in [Africa] is so miserable, that [Africans] are very happy that [we] take them out of there."[115] On the contrary, the inhabitants of Guinea were not only content in their native land but potentially on their way to civilization, Frossard asserted.

The rhetoric of humanization—of painting African men and women as eminently redeemable—was not only altruistic; it was a *political* decision that implicitly argued for a new future. In particular, the image of a benevolent and potentially civilized African was designed to counter the planters' belief that any transition from slavery to liberty would be characterized by groups of violent and vengeful slaves running riot on Caribbean islands. In 1787, for example, Jean [Jacques]-Pierre Brissot de Warville, one of the founders of the Société des amis des noirs, explicitly conjured up an optimistic physiocrat-type conversion of miserable African forced laborers into grateful and contented workers.[116] In stark contrast to either anti-slavery writers' fears (or the dire *Code blanc,* which Diderot evoked in his editing of Raynal's *Histoire des deux Indes,* for that matter), Brissot envisioned a new day when sensitive and appreciative Africans would make the best of manumission.

> I . . . am sure that tears of joy will pour from . . . the eyes of the *nègres* [after their liberation]. No, no, it is not revenge that they will dream of when they see their chains fall. Give them bread, finally allow them to rest, allow them the freedom finally to hold their children, to enjoy the sweet pleasures of domestic life with their wives, and they will be far from thinking of revenge. On the contrary, they will see us, love us as their liberators. The *nègre* is a loving husband, a good father; revenge does not inhabit a soul that embraces these sentiments, to whom one grants the right to enjoy them.[117]

In the same basic era, this optimistic portrayal of the African's promise was being articulated with the most efficacy by people of African descent.[118] When the Afro-British writer Olaudah Equiano related his childhood in Africa (supplementing his own memories by "incorporating whole sentences" from Benezet),[119] literary self-representation reached its apex: Equiano's text served as a de facto refutation of the era's formulaic presentation of Africans. By proving himself to be a literate author, devout Christian, businessman—in addition to being a freed African slave—Equiano cast himself as a living confutation of the existence of a degenerate or inferior race that could not better itself. In a sense, Equiano dared to "write himself into the human community," to imagine himself, on a certain level, not only as equal, but as morally superior to an oppressive white culture.[120]

By the late 1780s, the anti-slavery writings of Benezet, Saint-Lambert, Diderot-Raynal, Ramsey, Clarkson, Condorcet, and Frossard (and freed slaves including Equiano) had established a new vision of the African whose foundation was based in a new and more optimistic natural history. While none of these thinkers denied the "liabilities" of African slaves (i.e., their limited intelligence and seemingly dissolute comportment), they did not explain such failings by looking strictly to the African body itself. In fact, without exception, these activists practiced a new politicized environmentalism where they attributed the slavish African *type* to the deleterious effects of human bondage and the influence of cruel planters.

Epilogue: The Natural History of the *Noir* in an Age of Revolution

Founded in 1788, the Paris-based Société des amis des noirs was the de facto champion of a more progressive and positive view of the black African. This group, which, at its peak, theoretically had two hundred men and women on its membership rolls, drew its general ideology from French philosophes and physiocrats as well as from its spiritual and organizational forebearer, the London Society for the Abolition of the Slave Trade.[121] Ultimately imitating much of what their counterparts had undertaken in England, the Société's core members—Brissot de Warville, Condorcet, Lafayette, Mirabeau, Etienne Clavière, and Henri Grégoire—sought to bring the French colonies into line with a new commitment to human rights.

As numerous historians have made clear, the complicated story of the Société takes place against the backdrop of a series of changing revolutionary events. From the very inception of the organization, the principal members of the group—who were also revolutionaries—found themselves engaged in battle against a powerful and organized pro-slavery opposition that effectively integrated itself into each of the many revolutionary governments. After encountering stiff resistance to their ideas from pro-colony representatives in the Etats généraux in 1788, the Amis des noirs then faced off against an even more powerful group of pro-slavery politicians the following year in the Assemblée nationale. As Robin Blackburn has written, this constituency effectively controlled the colonial agenda by often shouting down anti-slavery speeches or enacting arbitrary procedural regulations; not only were "[c]olonial issues . . . referred to specialist committees, on the grounds of their delicacy, [but] [a]bolitionist advocates like the Abbé Grégoire had to content themselves with antislavery interjections in debates on other subjects."[122] Outside the legislative body, the Amis des noirs were also excoriated by the club Massiac, a cabal of planters, colonists, and lawyers working (and writing) together with more efficiency than did the Société. This comparatively large pro-slavery lobbying group, which coordinated with likeminded elements in the Assemblée nationale and its successor the Assemblée nationale constituante, put forward a collective, public, and national response to the activities of the Amis des noirs. The success of the pro-slavery lobby during the first years of the Revolution is easily measured: while the Amis des noirs were instrumental in passing legislation that gave citizenship to the *gens de couleur libres*, they were unable either to ban French participation in the slave trade or alter the way in which human bondage functioned in the French colonies.[123]

The effective end of the first iteration of the Société des amis des noirs came about in 1792. Accused of facilitating the slave uprisings that began to take center stage in colonial affairs in 1791—a point to which we shall return—the Amis were also overwhelmed by a series of political developments that had little to do with anti-slavery politics. As the revolution lurched toward a more violent phase in September of 1792—the monarchy had been overthrown in August—the co-founder of the Société, Brissot, broke with Robespierre and his more radical understanding of the process of national transformation. From this point forward, members of the National Convention were effectively forced to choose between these two polarizing figures. When Robespierre and the Jacobins ultimately seized control of the Comité de salut public in June of 1793, the *Brissotins* or Girondins (among them, some of the most prominent members of the Société) would soon become victims of this revolutionary body. Brissot was guillotined in October; Clavière (1793) and Condorcet (1794) committed suicide in jail. While French hagiography has often conflated the Société des amis des noirs with the fateful vote to free French slaves (February 1794), the reasons behind this decision extended well beyond the Société's extensive efforts on behalf of black Africans earlier in the decade. Indeed, as Marcel Dorigny has amply demonstrated, the resolution to do away with slavery on 16 Pluviôse had as much to do with geopolitical reasons—driving the English out of Saint-Domingue—as it did with the disinterested emancipatory politics of the disbanded Amis.[124]

Despite this fact, it is nonetheless undeniable that the Société des amis des noirs forever altered the landscape within which slavery and the status of the black African would be understood in France. While it may be a bit too triumphalist to claim, as Jean Ehrard has, that "the eighteenth century did not liberate the slaves, but it did liberate the era's thought [on their behalf]," the Société's projet nonetheless represented a particular expression of Enlightenment humanism and universalism that replaced vague assertions with political action.[125] Indeed, the anti-slavery program of the Société, which generated a sizeable corpus of pro-*noir* and anti-planter pamphlets, was not guillotined with Brissot; much of its rhetoric was passed on to the revolutionary-era actors who played an important role in the Caribbean (e.g., Léger-Félicité Sonthonax) as well as to later abolitionists (e.g., Abbé Grégoire).[126]

Among the most effective rhetorical strategies handed down by the Société des amis des noirs, two merit special attention. The first is the aforementioned politics of humanization. Coming into stark relief against the era's nascent raciology, the posited fraternity of the *noir* allowed anti-slavery writers to ask their readers to imagine themselves in the slave's unfortunate situation. The second and related tactic that the Société bequeathed to later activists was much more accusatory: blaming the planters for the perceived moral and intellectual deficiencies of African slaves, in essence, for creating the so-called *nègre* out of a *noir*.[127]

Accusing Caribbean planters (while simultaneously lauding the *perfectibility* of the *noir*) had been an effective argument in the era's anti-slavery writings since the 1770s. This dual strategy worked much less well for the Amis, however, after reports of huge uprisings on the northern portion of Saint-Domingue began arriving in France in late 1791.[128] Pro-slavery thinkers predictably seized upon this occasion to attack the Amis as never before. In addition to accusing the members of the Société of secretly conspiring with the slaves and the English in order to overthrow the French colonies, they also increasingly disparaged the Amis' mind-set directly, labeling them foolish *négrophiles*. While this term now evokes someone who participated in the "current of literature . . . in which black characters were depicted as possessing heroic qualities," this expression had an altogether different meaning in the mouth of a member of the club Massiac, for example. More than a literal synonym of *ami des noirs*, the word *négrophile* referred to those people who had an "unrealistic" appreciation of the natural history of the African as well as a lack of understanding of the realities and advantages of France's colonial empire.[129] Indeed, the term was commonly used to pathologize anti-slavery thinkers, associating them with the horror of blackness that they supposedly loved. As the anonymous author of *Le danger de la liberté des nègres* (1791) suggested, *négrophilie* was a disease of sorts stemming from a preposterous and quixotic excess of *humanité*. This warped worldview, he continued, generated the "fantasy" that, once liberated, grateful Africans would work more efficiently in a slavery-free society.[130]

Literally denigrating the Amis was but the first way of interpreting the terrifying events on Saint-Domingue; vilifying the *nègre* race as an unworthy animal-like group was the second.[131] While the pro-slavery lobby had often cited the black African's "undeniable" inferiority when justifying slavery—in 1789, for example, Pierre Malouet famously responded to the Amis' positive views of the *noir* by describing Caribbean slaves as "monkeys" who were "subject to all sorts of vices" that could only be controlled by absolute subjugation—this type of slander intensified as writers contemplated the loss of Saint-Domingue.[132] Consider, for example, the *fragment politique* that the former apologist of the black African Louis-Sébastien Mercier published on the *nègre* during the dark days of 1792. While Mercier claimed that this text, as well as his other *fragments*, had been languishing in his portfolio for some time, this particular diatribe certainly seems to have been chosen in order to provide an accusatory explanation of the so-called *nègre*'s recent actions in the Caribbean:

> The *nègres* are equally prone to all acts of treachery and villainy. . . . These peoples, who are too wicked to [create] a national government, suffer justly for its absence. They have lost the feeling of nature; they have retrogressed because they did not know how to advance toward civilization. Their lapses and corruptness have rendered them the playthings of foreign countries; and the false, evil, and perfidious mind of these people, shying away from any worthwhile edification, has reduced them to the basest and most credulous superstition, [a state] in which they cherish fetishes and soothsayers, and surround themselves with spells.[133]

A number of thinkers seconded more "general" ethnographic condemnations with horrifying accounts of the specifics of the slave rebellion on Saint-Domingue. This was even true of one of the greatest advocates of the *gens de couleur*, Julien Raimond. A wealthy planter of mixed origins who was present during the revolts in Cap Français, Raimond published a pamphlet in

1793 that not only conjured up the massacres in the northern part of the island, it tapped into the collective fear of what he called "anthropophagic" and "bloodthirsty" *nègres* who were purportedly relishing the suffering of their victims, many of whom were either burned alive or had their arms, legs, and genitals amputated and were left to die in horrific pain.[134]

After 1794, however, much of the (written) defamation of the black African provisionally faded. In addition to the fact that the Terror phase of the Revolution had curtailed virtually all publications, writers who had previously produced anti-black tracts as part of their pro-slavery argument were no longer engaged in a difficult battle with a vocal anti-slavery organization; as such, they clearly had less reason to lash out at *négrophiles* and, along with them, the *nègre* himself. Perhaps more important, however, by 1795, the Directoire (which instituted republican colonial policies) was increasingly putting its hope for salvaging Saint-Domingue in the hands of the brilliant military tactician, politician, and ex-slave Toussaint Louverture.[135] While it is difficult to know exactly what went on in the minds of anti-black thinkers and writers living in France at this time, the temporary drop in *published* anti-black sentiment, during an era when Toussaint had driven out both the Spanish and the British and had succeeded in rebuilding much of the island's export economy, seems more than a coincidence.[136]

This short-lived dip in anti-black publications was reversed by the end of the decade. In addition to the fact that the pro-slavery lobby had reorganized during the last years of the century, the 1799 coup d'état of the decidedly pro-colony and pro-slavery Napoleon Bonaparte ushered in a new era for the black African. While the full interplay between Napoleonic politics and the era's "natural history of the *nègre*" extends well beyond the purview of this study, suffice it to say here that France's colonial policies during the Consulate had a decisive impact on the representation (and treatment) of the *nègre*. In the first place, Napoleon himself had long thought that the self-determination, independence, and liberty that had been accorded to the black Africans of Guadeloupe and Saint-Domingue by the revolutionary government in 1794 were not only horrible mistakes; they were undeserved. According to an account of the Conseil d'état in 1802, Napoleon reportedly proclaimed that the Convention would have never voted for the *nègre*'s freedom in 1794 had it fully recognized the palpable inferiority of the *nègre*: "How could have anyone given liberty to Africans," he asked pointedly, "to men who lacked any idea of civilization, and who had no idea of what France was?"[137]

Bringing the Caribbean's *nègres* back under the yoke of a new French colonial empire was the logical corollary to Napoleon's view of black Africans. A "thorough-going restorationist" who was advised by longtime partisans of human bondage, among them Médéric Moreau de Saint-Méry, Napoleon began planning the progressive reestablishment of France's West Indian empire as soon as he arrived in power.[138] This policy was implemented to varying degrees of success on the major (formerly French) islands of the Caribbean. On Guadeloupe, the French were able to slowly reenslave the island's previously liberated population after a besieged rebel army led by Louis Delgrès intentionally blew itself up on the slopes of the Souffrière volcano outside of Basse-Terre in 1802. The reincorporation of Martinique into the French sphere of influence was not so bloody. Captured by the British in 1794, Martinique once again became part of the French empire in 1802, one of the many territories that changed hands as part of the huge realignment brought about by the Treaty of Amiens.

Four years after he arrived in power, Napoleon had succeeded not only in reestablishing the plantation system and chattel slavery on Guadeloupe and Martinique, but in reinstating the race-based lines of separation that were an integral element of the colonial world during the ancien régime. Napoleon had clearly harbored similar fantasies for Saint-Domingue, although the political and military situation on this huge island obliged the First Consul to move more carefully. In his first dealings with Toussaint Louverture, for example, Napoleon tacitly recognized the fact that this assertive and effective "general-in-chief" of Saint-Domingue needed to be treated as an equal.

This arrangement changed radically by 1801, however. As Laurent Dubois's compelling account of the last phase of revolution on Saint-Domingue has demonstrated, Toussaint's takeover of Spanish Santo Domingo and, even more outrageous from Napoleon's perspective, the announcement of a new constitution for the island in 1801, prompted the First Consul to dispatch his brother-in-law, General Charles Victor Emmanuel Leclerc, and an initial force of 22,000 soldiers to the island, ostensibly "to disarm the blacks," and "make them free cultivators."[139] Recognizing immediately that this armada was there for more than disarmament, Louverture and his officers (among them, Jean-Jacques Dessalines, Jacques Maurepas, Henri Christophe, and Jean-Baptiste Sans-Souci) initially waged a brutal and effective guerrilla war against French forces for several months. By April 1802, however, a number of French victories led to the surrender of Maurepas, Christophe, Dessalines, and, most important, Toussaint Louverture himself. On an island where racial and military alliances shifted quickly as a function of political strategies, Leclerc allowed Louverture to retain his rank and "retire" to his plantation; the insurgency's soldiers, however, were immediately incorporated into the French army and often fought quite effectively against their former comrades under Dessalines.[140]

The success of this fragile strategy was short-lived. Although Leclerc ultimately arrested Louverture and deported him to France, a series of events beginning in July of 1802 brought disaster to the French mission. This began with a yellow fever epidemic that sprang up during the late summer and killed thousands of French soldiers, including Leclerc himself. As devastating as this plague was, the most problematic development for the French was perhaps the promotion of Donatien-Marie-Joseph Rochambeau to leader of the French expeditionary force on the island. His ruthless and bloodthirsty campaign on the island, which involved public tortures of both civilians and insurgents, "alienated even the most steadfast supporters of the French" and contributed mightily to the ultimate defeat of the French army.[141] By December 1803, the revolutionary forces of Saint-Domingue, aided by a British blockade of the island, triumphed over the remnants of French forces during the battle of Vertières, ushering in the birth of Haiti.

The Caribbean policies and historical events of the Consulate era utterly transformed Europe's understanding of the *nègre*. While the Haitian revolution has more recently been seen as one of the most successful resistance movements in the history of the world's oppressed peoples, in the early nineteenth century, the crushing *final* loss of Saint-Domingue, not to mention the death of more than 45,000 soldiers, was often explained in terms of a clash between cultures. "Those who wish the destruction of our colonies [now] understand that it is no longer possible to plead the cause of this barbarian race with success," *Mercure de France* declared in 1805.[142]

Such images were increasingly read against the more widespread scientific racialization of the *nègre* that had been taking place since the end of the eighteenth century. This process took several forms. As T. Carlos Jacques has asserted, much of what was ultimately written about the black African at this time can be tied to a belief in a new cognitive determinism that was increasingly projected onto the *nègre*. In his 1797 *Etudes de l'homme physique et moral*, for example, J.-A. Perreau portrayed black minds of an entirely "degraded nature" and "doomed sometimes to stupidity, sometimes to the most extravagant

A British view of vengeful *nègres* executing Frenchmen during the Haitian revolution (1791–1804). From Marcus Rainsford, *An Historical Account of the Black Empire of Hayti* (1805). Courtesy Wesleyan University Library, Special Collections and Archives.

delirium of the imagination."[143] Along similar lines, in 1800, the influential philosopher Joseph-Marie de Gérando (Degérando) developed many of these ideas in *Des signes et de l'art de penser*, stating that the *nègre*'s intellectual faculties were characterized by a type of mental "void,"[144] an unfeeling nothingness that, implicitly, separated the black comprehension of the exterior world from that of whites. Generations of thinkers had alluded to the machinelike qualities of the African mind before de Gérando, but his reevaluation of this race positioned the *nègre* as a cognitive child who only saw "in nature what was

French massacre of black rebel troops on Saint-Domingue in 1803. The dogs in the surf are Cuban bloodhounds imported by Vicomte Donatien de Rochambeau, commander of French forces, to hunt down the insurgents. From Marcus Rainsford, *An Historical Account of the Black Empire of Hayti* (1805). Courtesy Wesleyan University Library, Special Collections and Archives.

directly related to the needs of his senses."[145] Superimposed on a whole host of other negative stereotypes, including the late-to-catch-on physiognomical notions proposed decades before by Johann Casper Lavater, such a view implicitly recognized Europeans as the de facto masters of this intellectually and physically inferior brood.[146] Perhaps as important, this new theory of mind also created a new barrier between white and black that refuted the belief in a shared sensibility among humans, one of the critical elements of anti-slavery discourse.[147]

In addition to diffusing a belief in a cognitive blackness, late eighteenth- and early nineteenth-century race theorists also increasingly emphasized the supposed differences between blacks and whites on a zoological level. Although some naturalists, including Georges Cuvier, carried on in the footsteps of Buffon by putting forward a (bleak) monogenist account of humankind, others, like Virey, asserted in works such as his *Histoire naturelle du genre humain* (1800) that the era's anatomical discoveries proved without a doubt that "the *nègre* differ[ed] very specifically from all other races of man."[148] As was often the case with polygenist thinking, Virey accompanied this assertion with a specific view of humankind's origins: "the gentle and fertile sky of Asia appears to have been the original cradle of humankind, as it was of religion; but the *nègre* species and the American races were undoubtedly born elsewhere."[149] Virey followed up on this belief in multiple human origins even more vehemently four years later in the article "Nègre" that he supplied for J.-F.-P. Deterville's *Nouveau dictionnaire d'histoire naturelle* (1804). In this substantial entry on the black African, presumably written after the loss of Saint-Domingue, Virey summarized the many areas in which the *nègre* was supposedly at variance with the European, emphasizing, in particular, the beastlike level of "cerebral function" that he attributed to a brain of inferior size.[150] Virey's conviction that the *nègre* was "radically different from the white species" was seconded by numerous thinkers during the same decade.[151] The prolific naturalist and philosopher Jean-Claude de La Métherie asserted in his 1806 *De la perfectibilité et de la dégénérescence*, for example, that there were clearly two primary aboriginal races at the beginning of humankind's history, the race *hindouse* (Europeans and Tartars) and *la race nègre*, the former of which, over time, "spread out over the surface of the earth."[152]

By the early nineteenth century, natural history was exiling the *nègre* from the family of man; race science had finally "slipped its biblical moorings and abandoned the scriptural genealogy of peoples set out in Genesis."[153] Not unexpectedly, these classificatory and conceptual understandings of the *nègre* intersected with a new wave of anti-black feeling among French pro-slavery thinkers, who, after years of frustration and perhaps restraint, could now fully express themselves with impunity. Among the numerous texts from this era that castigate the *nègre*, none is as telling as Valentin de Cullion's 1802 *Examen de l'esclavage en général*. As his predecessors had done, this lawyer and former owner of property on Saint-Domingue began his diatribe by looking back at the "mistakes" of the past, lashing out in particular at the now-disbanded Société des amis des noirs and accusing them of ushering in the loss of the world's most valuable colony. Following up on this idea, de Cullion also complained that foolish *négrophile* texts such as Henri Bernardin de Saint Pierre's *Etudes de la nature* (first ed. 1784) continued to have a pernicious influence, despite their lies. Horrified, in light of recent events, that Bernardin's (recently republished) book could assert that certain African peoples "outshone [the Europeans] in moral qualities,"[154] de Cullion retorted: "[Bernardin de Saint Pierre] attributes to the *nègres* a superiority over us in qualities of the heart. . . . [I]t is absurd to assume that what one mistakenly calls goodness in a stupid being is genuine goodness when, in reality, it is no more than a negation of evil."[155]

This was only the beginning of de Cullion's brutal ethnological and cognitive vilification of the *nègre*. Seeking to naturalize slavery, de Cullion posited that since "there is nothing to be expected from such an inferior a race of men," it was entirely expected that *les nègres* were "good for essentially nothing but slavery" and that "slavery is therefore [their] natural state."[156] Indeed, the *nègre*'s destiny was an imperative according to de Cullion: "These people are the dregs, the scum of the human race; therefore let them occupy the lowest rank. Let them serve. Nature has decided their fate."[157]

By the early nineteenth century, natural history had come to the fore of the slavery question. This was true not only in pro-slavery texts such as de Cullion's, but also in the way that an anti-slavery thinker such as Abbé Henri Grégoire was forced to recast his own arguments during this era. In his seminal 1808 *De la littérature des Nègres*—the first entirely positive assessment of Africans to appear in French since Frossard's *La cause des esclaves nègres* (1789)—Grégoire provided a comprehensive moral, spiritual, and intellectual apologia of the capacities of the black African; to a large extent, the publication of *De la littérature* was the culmination of decades of *négrophile* politics.[158] But what had changed between the era of the first iteration of the Amis des noirs and the reestablishment of slavery in the Caribbean was the way in which natural history functioned within the overall debate on slavery. While the discipline of natural history had always been linked to the overall justification of slavery, the *sciences de l'homme* were now totally imbricated with questions of the African's human bondage. As such, Grégoire was unable to begin his text by trotting out the now well-seasoned *négrophile* conviction that the shortcomings of African "faculties" could be explained by the institution of slavery itself; rather, he was obliged to construct a careful refutation of what had become an authoritative nexus of biopolitical notions projected onto the *nègre*. In addition to contesting the increasingly influential polygenist understanding of humankind, Grégoire refuted a number of essentialist anatomical views of Africans advanced by Meckel, Christoph Meiners, Edward Long, and Franz Joseph Gall, the latter who had put forward the sinister new science of *cranioscopy,* later renamed phrenology. While much of the overall debate on slavery remained unchanged by 1808—economic (and national) interests continued to diverge from universal human values—the *nègre*'s body had become an increasingly critical battlefield for pro-slavery and anti-slavery thinkers.[159] This had not always been the case. When the first generation of Enlightenment-era writers had argued about slavery, there had always been one thing about which there was little debate: the undeniable anatomical and cognitive inferiority of the *nègre*.[160]

Coda
Black Africans and the Enlightenment Legacy

> The history and the legacy of the Enlightenment [are] worth understanding and arguing about.
>
> PAUL GILROY, *THE BLACK ATLANTIC*

In one of the more lyrical moments of the *Encyclopédie*'s "Discours préliminaire," Jean le Rond d'Alembert reflected on the discipline of History. In his estimation, this branch of learning stemmed from an instinctive tendency to enter into dialogue with past and future inhabitants of the earth:

> It is not enough to live with and understand our contemporaries. Excited by curiosity and pride, and seeking through a natural inclination to embrace the past, the present, and the future, we desire to exist both with those who will follow in our footsteps and those who have preceded us. From this [desire] stems the origin and the study of History. We are linked to past centuries through the spectacle of [humankind's] vices and virtues, as well as its knowledge and mistakes, just as our own are transmitted to future centuries.[1]

D'Alembert's conviction that historians should engage with past and present events from an ethical or philosophical perspective remains one of many legacies associated with the Enlightenment era. The irony, of course, is that the Enlightenment itself has been judged quite severely against this same standard.[2]

Among the many regretful "vices" attributed to the eighteenth century—child labor, forced marriages, and a dehumanizing rationality come to mind—no failing stands out as jarringly as the European-orchestrated trade and exploitation of African slaves. The legacy of this deportation, which, on French ships alone, grew to thirty thousand Africans per year by the late 1780s, continues to exact a huge toll on Africa as well as on the diasporic populations that slavery created. Both the magnitude and the multidimensional character of this five-hundred-year history of forced migration have incited a wide variety of historical treatments. Statistically minded historians, beginning with Philip Curtin, have pieced together the numerical history of slavery in a way that colonial administrations themselves could not have imagined.[3] Africanists such as Elikia M'Bokolo, Serge Daget, François Renault, Paul Lovejoy, P. E. H. Hair, John Thornton, and Robin Law have meticulously reconstructed the encounters between Africans and Europeans that took place in the slave-driven economies of cities like Ouidah in present-day Bénin.[4] Maritime historians including Stephanie Smallwood, Robert Harms, and Marcus Rediker have recounted the devastating realities of the Middle Passage.[5] Historians of the Caribbean, including Laurent Dubois, have meticulously reconstructed the complex history of African agency in the New World. Caribbeanists and Latin American specialists such as Gwendolyn Hall and Paul Gilroy have taken up the cultures and identities of Africa's diasporic peoples.[6] Legal historians such as Sue Peabody and Keila Grinberg have looked at the network of laws regulating the circulation of bodies and sexual mores of African slaves both in the Caribbean and in European countries.[7]

Contiguous to these material, cultural, and economic assessments of human bondage is the thorny question of how slavery could not only continue to exist but even flourish during an era of supposed Enlightenment. The first real studies to grapple with the problem of slavery *and* Enlightenment appeared, not surprisingly, in the 1960s and 1970s, during and just after the era of decolonization and the appearance of pioneering works such as Chinua Achebe's novel *Things Fall Apart* (1958) and Frantz Fanon's *Les damnés de la terre* (1961).[8] Critical works from this era include many now-classic studies cited in this book: among them works by Yves Benot, Michèle Duchet, Roger Mercier, Leon-François Hoffmann, and William B. Cohen, the latter an American who was castigated by the French press when his *The French Encounter with Africans* appeared in 1980.[9] While the respective methods of these scholars vary in significant ways, each conjured up a global eighteenth century involving the French Empire, French-organized slavery, and French Enlightenment ideas and structures. Duchet, in particular, famously demonstrated how Enlightenment thinkers including Raynal worked under the umbrella of a colonial administration that, particularly after the Seven Years' War, was keen on facilitating the ongoing slave trade. While it may be hard to imagine now, this was a shot across the bow of Enlightenment studies; in the era before scholarly works like Duchet's were published, eighteenth-century scholars had generally considered slavery and empire ancillary concerns at best. To the extent that such questions were even raised, this was done either to laud the anti-slavery movement among the philosophes or to quarantine these "problems" from the more important and progress-driven view of the high Enlightenment.

Four decades after Duchet's *Anthropologie et histoire au siècle des Lumières* first appeared, the intersection between the Enlightenment and the problem of slavery remains a beguiling area of inquiry. Akin to the way that eighteenth-century naturalists pored over descriptions of non-Europeans, scholars interested in resolving the problem of race and slavery have combed through the era's thought in search of some sort of answer to this nagging concern. Among the generation of scholars influenced by postcolonial approaches to such subjects—which seek, *grosso modo*, to question or to rewrite the Western narrative to varying degrees—the prevailing tendency has understandably been to critique the role of the Enlightenment in the overall representation and subjugation of the black African. In French circles, this effort has been led by Louis Sala-Molins.[10] In his two very widely read indictments of what he deemed to be the failures and hypocrisy of the long Enlightenment, Sala-Molins has argued that this intellectual movement not only failed to live up to its own rhetoric and emancipatory values; it produced a horrific synthesis of race and law that grew out of the 1685 *Code noir*.[11] While among the most polemical, Sala-Molins is far from the only critic to make this point. Echoing aspects of Theodor W. Adorno and Max Horkheimer's *Dialectic of Enlightenment*,[12] a number of scholars have actually identified Enlightenment-era philosophy itself as containing the very structures of oppression used to control Africans. In *Race, Writing, and Difference*, Henry Louis Gates Jr. highlights the paradox of the Enlightenment's use of *reason*, writing that "while the [era] is characterized by its foundation on man's ability to reason, it simultaneously used the absence and presence of [this faculty] to delimit and circumscribe the very humanity of the cultures and people of color which Europeans had been 'discovering' since the Renaissance."[13] In a similar vein, David Theo Goldberg has argued in his *Racist Culture* that "subjection perhaps properly defines the order of the Enlightenment: subjection by human intellect, colonial control through physical and cultural domination, and economic superiority through mastery of the laws of the market."[14] Most provocatively, the philosopher Charles Mills maintained in *The Racial Contract* that the very basis of the Enlightenment—an era that corresponded to the golden age of contract theory—can be characterized by a racially based covenant of dominance between Europeans and non-Europeans, especially people of African descent.

While many of these claims are sweeping, and neglect to define just what "Enlightenment" was, it is nonetheless incontestable that one of the recurrent and constituent elements of high-Enlightenment thought, present not only in Montesquieu's *De l'esprit des Lois,* Voltaire's *Essai sur les moeurs,* and in Raynal's *Histoire des deux Indes,* is the fact that the imperatives of natural history often crisscrossed and negated the theoretically inviolable tenets of natural law.

An illuminating example of how the eighteenth century began teaching and institutionalizing belief in the inherited inferiority of the *nègre* can be found, appropriately enough, in Rousseau's *Emile, ou, de l'Education* (1762). More explicitly than any other thinker of his generation, Rousseau demonstrated the "application" of ethnographic knowledge. Drawing from Buffon's climate theory, Montesquieu's geo-ethnic relativism, and some vague notions about African physiognomy, Rousseau asserted quite straightforwardly that a student's intellectual potential was undoubtedly climate- and geography-dependent: "The country is not an inconsequential factor in the education of men; humans only achieve their full potential in temperate climates."[15] This belief, that humankind's promise could only be realized in a moderate environment, is consonant with a Buffonian worldview; given the fact that humans supposedly degenerated in extreme climates, it followed that the moral and intellectual promise of a student would also be linked to the conditions in which he was raised. But Rousseau also went further than Buffon when he suggested in a much more deterministic fashion that humans living in taxing locations would make poor students because of their brains. Presumably conflating the era's belief (as best expressed by the German anatomist Johan Friedrich Meckel) that African brains were different in color with the notion that Africans and Laplanders were intellectually inferior, Rousseau declared: "It also appears that the organization of the brain is less perfect at the two poles. Neither the *nègres* nor the Laplanders have the intellect of Europeans."[16] This was a significant departure from the type of environmentalism found in

Buffon's *Histoire naturelle*. Unlike the famous naturalist's understanding of the *nègre*, Rousseau's take on this human variety emphasized the importance of human anatomy in the *nègre*'s educability. This was the result of the increasingly material study of the African's many "defects." While Buffon's environmentalist view of the *nègre* continued to hold sway with most thinkers (including Rousseau), the growing list of discernible, measurable, and potentially essential differences between blacks and whites was clearly undermining the more optimistic aspects of the "unity of the human species." This was also obvious eight years later, when Abbé Raynal published the initial (1770) edition of the *Histoire des deux Indes*. While the third (1780) version of this text would ultimately be castigated by pro-slavery thinkers as a despicable example of *négrophile* ideology, the first two editions present one of the most pessimistic, anatomically driven, and polygenist views of the *nègre* existing in eighteenth-century French thought.

The admittedly anachronistic question raised by such texts is one of consistency: how could the proponents of the Enlightenment's classical liberalism fall prey to a reprehensible, race-based view of the *nègre*, one that undoubtedly helped perpetuate a grave injustice against millions of black Africans from Guadeloupe to Mauritius? In this book, I have attempted to demonstrate that this distressing paradox is not necessarily the inevitable outcome of an intentional European hegemony per se, but often the result of the two givens in the era's thought: the convenient disciplinary compartmentalization of the "subject" of the *nègre* and the prevailing authority of natural history. Such is the case in Raynal's *Histoire des deux Indes*. A multi-authored work reflecting divergent perspectives (and disciplinary orientations), the *Histoire* reflects a general tendency to unbundle the moral status of enslavement from the physical status of those people being enslaved.

This curious disjunction between the *politics* and the *body* of blackness is to be found, not only in compendiums like the *Histoire des deux Indes* and the *Encyclopédie*, but in the worldviews of individual thinkers as well. Jaucourt's view of the African, as demonstrated in chapter 4, provides a telling example of this phenomenon. Best remembered for one of the earliest and most sustained refutations of African chattel slavery in Enlightenment thought ("Traite des nègres," pub. 1766), Jaucourt was simultaneously responsible for producing the veritable nexus of stereotypical assessments of African ethnicities that appear in the *Encyclopédie*. In these short (and generally pilfered) entries, Jaucourt provided concise sketches of supposedly backward African ethnicities that functioned like ethnographic snapshots. Far from establishing real differences among the groups of Africans he wrote about, articles like Jaucourt's "Mindingues" contributed to an overall ethnography of the black African that evaluated different types of *nègres* as more or less ugly, more or less prone to ridiculous customs, and more or less useful in various situations of labor.[17] While very much the anti-slavery thinker, Jaucourt nonetheless contributed to the uncritical and tautological association of the term *nègre* with *esclave*.

This general blindness to the biopolitics of representation touched all disciplines. Among more properly anatomical thinkers, a group of men for whom the influence of natural history was unimpeachable, "research" on the *nègre*'s *reticulum mucosum*, "black" blood, or "black" sperm, was initially understood as having *nothing* to do with the reality of African chattel slavery; such inquiries were simply envisioned as part of an impartial quest to define the parameters of the human species. This seeming myopia is also found among more conceptually minded naturalists, including Buffon. Like all thinkers of his generation, Buffon inherited a nexus of anecdotal stereotypes and utility-based assessments of African ethnicities that functioned as facts: facts that, in his view, needed to be processed and interpreted. This was, after all, the project of the era's nascent human science. In the same way that botanists and entomologists were systematizing plants and insects, Buffon sought to integrate the *nègre* and all human varieties into a more comprehensive and explanatory theory. His premise, monogenetic degeneration, was the first to incorporate morphological and pigmentation "data" from travelogues, anatomical information from thinkers like Pierre Barrère, and basic stereotypes of the world's peoples into a speculative chronology suggesting that a primitive white prototype had given rise to the world's different *variétés* over deep time.[18] While Buffon did acknowledge the unfairness of the ethnography flowing from the Caribbean, the overall treatment of the *nègre* in the *Histoire naturelle* nonetheless put forward an understanding of blackness that was both derived from and compatible with the context of slavery.

As noted in the Introduction, Buffon's 1749 "Variétés dans l'espèce humaine" has become a subject of great debate. Indeed, it is no exaggeration to say that this section of the *Histoire naturelle* is now something of a flashpoint, a text where the reputation of the entire Enlightenment as either racist or non-racist seems to be at stake. While it may seem curious for specialists of this era to accord so much importance to one work, Buffon's ethnography-based theory of human degeneration lends itself to multiple and coherent readings by both the apologists and censors of the Enlightenment. Among the apologists (both for Buffon and for the Enlightenment in general), it has been argued that this account of humankind's common origin stands in stark contrast to the vague nexus of polygenist thought circulating in France after the 1740s. After all, while anatomically minded writers were emphasizing fundamental and essential differences between whites and blacks, Buffon argued that such differences were nothing more than ancillary changes that had taken place over time as a result of the environment. Generally classing different ethnic groups with the botanical term "varieties" as opposed to the zoological term "races," it is argued, Buffon maintained that Africans, Danes, Eskimos, and Chinese were not members of essential categories; they were potentially shifting members of a dynamic species defined by an ability to interbreed and morph over time as it moved about the globe. To a large degree, this "enlightened" view of humankind is the one that is generally espoused today.

This is not the sole reading of Buffon. To his critics, the essential sameness that Buffon preached is seen as nothing more than a disingenuous sham that, as Sala-Molins famously argued, confused *unité* with *égalité*.[19] Stating that Buffon's theory of human sameness is perhaps more insidious than theories of essential difference, Sala-Molins has emphasized the fact that Buffon's theory far from accords the *nègre* a real spot in the family of man; in fact, he has argued that the supposed conceptual affinities between Africans and Europeans in Buffon amount to nothing more than a mirage whose corollary—human *perfectibilité*—actually overlaps with the basic disciplinary structure of the 1685 *Code noir*.[20] Never, according to Sala-Molins, did the naturalist maintain that the African was a full-fledged member of the human species on the same level with whites.

The ambiguities of Buffon's view of humankind echo those of the entire Enlightenment, of course. Seen from a more hopeful vantage point, Buffon's unified view of humankind was unquestionably a de rigeur concept within the writings of anti-slavery thinkers, including the Quaker Anthony Benezet, the French abolitionist Abbé Grégoire, and the German naturalist Johann Friedrich Blumenbach. All three not only studied and cited Buffon; they combined monogenism with stories of "high-performing" Africans in order to assert that blacks had the same potential as whites. Less optimistically, as I have made clear in this book, Buffon's monogenism also generated a new type of toxic speculation on the African. While Buffon himself actively attempted to prevent this, his degeneration theory—an explanatory paradigm emphasizing human change over time—prompted more anatomically oriented thinkers to retell the story of the *nègre*. Rejecting Buffon's fluid color map, *savants* including Le Cat and de Pauw overlaid environmentalism with a bleak anatomy- and heredity-based understanding of the black African. Their theories, which included scientifically derived data regarding darkened African blood, bile, brains, and sperm, poisoned the positive politics of Buffon's initial theory of monogenesis. If Buffon had explained the African as an accident of climate, the new anatomy positioned the African as radically pathological. By 1770 the consequences of this groundbreaking view of the African race were quite evident: a number of writers including Edward Long in England and J.-P. Rousselot de Surgy in France had enough anatomical data to reject monogenesis as part of their pro-slavery agendas. In a sense, Buffon's monogenesis had contributed to its own downfall; by inviting more speculation on how the process worked, it set the stage for a much more physiologically oriented understanding of the Black African.

Throughout this book I have shied away from assigning thinkers like Buffon to a particular space within a legacy-driven narrative. This is more a question of method than it is a lack of ideological orientation. In avoiding an account that seeks either to accuse or to exculpate, I have sought, above all, to come to grips with an overall representation of the black African that advanced the prospect of the African's *varietal perfectibility* and *equality*, on the one hand, and the supposed hopelessness of an immutable, violent, animalistic, and hypersexual *race nègre*, on the other. I have also attempted to show how this natural history breakdown did not skew along political lines; as I have often pointed out in this book, anti-slavery thinkers initially held the same beliefs regarding the African as their pro-slavery interlocutors.

This leads me to a final word about my own objectives. While I have attempted to be as fair-minded as possible, this study is far from morally neutral.[21] Even the most dispassionate or clinical passages of *The Anatomy of Blackness* reveal my own interests and preoccupations. Led to this subject via a fascination with the Enlightenment-era life sciences, I soon came to focus not only on the anatomy of the black African, but also on the nefarious shift in the ontological status of "black" and implicitly, "white" anatomy during the eighteenth century. If, during the era before "Enlightenment," blackness came into relief against a synthesis of biblical exegeses and vague physical explanations dating from antiquity, during the eighteenth century, the concept of blackness was increasingly dissected, handled, measured, weighed, and used as a demonstrable wedge between human categories. More than just a descriptor, blackness became a thing, defined less by its inverse relationship to light than by its supposed materiality. This eighteenth-century belief in the deep-rooted physicality of Africanness helps us to comprehend the power of "representation" and its connection to "reality" in the era's thought. The anatomization of blackness, in short, not only reflects the violence enacted on black bodies in the pursuit of knowledge; it mirrors the increasingly rationalized brutality to which real Africans were subjected during this time in the colonial world.

Notes

In the notes that follow below, the *Encyclopédie, ou, dictionnaire raisonné des sciences, des arts, et des métiers,* edited by Diderot and d'Alembert (Paris, 1751–72), is cited as *ENC.* The three editions of Raynal's *Histoire philosophique et politique des établissements et du commerce des Européens dans les deux Indes,* commonly abbreviated as *Histoire des deux Indes* (4 vols., Amsterdam, 1770; 8 vols., The Hague, 1774; 10 vols., Geneva, 1780) are cited as *HDI.* Buffon's *Histoire naturelle, générale et particulière* (Paris, 1749–88) is cited as *HN.*

Unless otherwise noted, all translations are my own. I have also modernized French and English in both text and notes, although actual titles have been given in their original forms.

Preface

1. Kennedy, *Nigger,* 4.
2. Moreover, in a text such as Buffon's *Histoire naturelle,* the capitalization (or not) of the term *nègre* is seemingly arbitrary, shifting from section to section. Capitalization may have had more to do with the typesetter than with Buffon himself, although there are places where it seems more deliberate, for example, when Buffon is citing a list of human categories.
3. Eze, *Race and the Enlightenment,* 2.
4. Gordon, "Introduction," in *Postmodernism and the Enlightenment,* ed. id., 2.

Introduction

1. See Klaus, "A History of the Science of Pigmentation," in *Pigmentary System,* ed. Nordlund, 5. See also Riolan, *Manuel anatomique et pathologique.*
2. Leeuwenhoek, *Collected Letters,* 4: 245.
3. Alexis Littré is sometimes referred to as Littre. The experiment involved speculation on the relationship between the color of the foreskin and the skin on other parts of the penis. See *Histoire de l'Académie royale des sciences de Paris* ([1702] 1703), 31–32.
4. For a more complete contextualization of Malfert's text, see Ehrard, *Lumières et esclavage,* 105. As Ehrard's account makes clear, Malfert's article, "Mémoire sur l'origine des nègres et des Américains," was criticized for engaging somewhat positively with the idea that the two races came from different seeds.
5. Monsieur de J***, "Explication," in *Mémoires pour l'Histoire des sciences et des beaux arts* (1738). This is the so-called *Journal de Trévoux.*
6. "Quelle est la cause physique de la couleur des nègres, de la qualité de leurs cheveux, et de la dégénération de l'un et de l'autre?" The question was published in the *Jour-*

nal des sçavans and in a variety of other periodicals in 1739. For the essays in response, see Bibliothèque municipale de Bordeaux ms. 825/65.

7. See the more developed discussion of these findings in chap. 4 above.

8. See Augstein, *Race*, ed. id., xx.

9. See Porter, ed., *Cambridge History of Science*, vol. 4: *Eighteenth-Century Science*, 50–51.

10. "L'anatomie a été portée presque au dernier degré de perfection [puisque] [l]es anatomistes et les physiologistes les plus célèbres ont senti qu'il était temps de diriger leurs recherches vers les causes des mouvements des animaux et les ressorts cachés de leurs sensations" (Millin, *Magasin encyclopédique*, 4: 155).

11. See Morenas, *Dictionnaire portatif*, 387, on "la forme qui convient le mieux avec leur façon de vivre [et la satisfaction de] leurs besoins." According to Morenas, anatomical knowledge was among the most valuable contributions that science had to offer, inasmuch as it was essential in the "traitement des maladies qui sont l'objet de la médecine et de la chirurgie [treatment of diseases that are the object of medicine and surgery]" (ibid., 4:155).

12. Laqueur, *Making Sex*, 26–35.

13. The era's basic understanding of the *nègre* was also heavily influenced by symbolic concepts related to darkness. As critics including Gustav Jahoda have asserted, the blackness of the *nègre* went far beyond the strictly physical realm. Denoting both the presence of obscurity and the absence of color, blackness seemingly functioned as the antithesis of Christ the Light; it also recalled the curse that had been cast on the son of Ham. See Jahoda, *Images of Savages*, 26.

14. Buffon cites Lade quite frequently. See, e.g., his description of the Hottentot in *HN*, 3: 476.

15. Swift, "On Poetry : A Rhapsody," in id., *Poems*, ed. Williams, 2: 645–46.

16. Lobo, *Voyage to Abyssinia*, trans. Johnson from Joachim Le Grand's French translation of the Portuguese text, which appeared in 1728.

17. Lenglet du Fresnoy, *Géographie des enfans*, 92: "D. Qu'entendez-vous par la Cafrerie? R. La Cafrerie ou le Pays des Cafres est un pays sur la mer, habité par les peuples les plus barbares et les plus stupides de l'Afrique. Les endroits les plus considérables sont le Cap de Bonne-Espérance aux Hollandais, et Sofala qui appartient aux Portugais."

18. "on peut dire qu'un nègre est tel par toutes les parties de son corps, si on en excepte les dents. Tous les organes portent plus ou moins l'empreinte de cette couleur, la substance médullaire du cerveau est noirâtre, cette couleur domine plus ou moins dans les diverses parties de cet organe, la liqueur spermatique, le sang, etc." (*Bibliothèque universelle des dames* [1787], 1: 210).

19. All these dictionaires had entries for *nègre*. For a discussion of *Encyclopédie* definitions of the *nègre*, see "Processing Africa and Africans in the *Encyclopédie*" in chap. 3 above.

20. See the 1732 *Dictionnaire universel françois et latin*, 4: 64. The same definition is used in later editions. *ENC*, 11: 140, defines "Nigritie," as a "large country of Africa, that stretches from the east to the west of the [river] Niger. It is bordered on the north by the Barbary deserts, on the east by Nubia and Abyssinia, on the south by Guinea, and on the west by the western ocean. This country comprises various small kingdoms, as many north of the Niger as those in the middle, and on both coasts of this great river [grand pays d'Afrique, qui s'étend de l'est à l'ouest des deux côtés du Niger. Il est borné N. par

les déserts de la Barbarie, E. par la Nubie et l'Abyssinie, S. par la Guinée, O. par l'Océan occidental. Ce pays comprend plusieurs petits royaumes, tant au nord du Niger qu'au midi, et des deux côtés de ce grand fleuve]."

21. What is less apparent, and perhaps more interesting, is the fact that the author of the *Dictionnaire de Trévoux* article refers to the *nègre* as a *peuple d'Afrique*. A more natural history–oriented *dictionnaire* might have called the *nègre* a *variété*, an originally botanical term used to describe what was deemed a subset of the larger category of the human.

22. The term *nègre* was also used to refer to the "dark-skinned" peoples living in places such as Madagascar and even New Guinea, who were, of course, of mixed African, Malayo-Indonesian, and Arab ancestry, among others. Buffon considered them *Caffres*, however. Buffon (or his typesetter) spells the word both *Cafre* and *Caffre*. I have chosen to use *Caffre*.

23. Berthelin, *Abrégé du Dictionnaire universel françois et latin*, 2: 19.

24. This also overlapped with the French language's qualification of each noun by gender.

25. While this curious gendering of the term *nègre* is certainly linked to that fact that black male slaves generally outnumbered their female and adolescent counterparts in French plantation colonies, it is also likely that such an equation protected many Europeans' consciences from the more sobering reality of female and child slave labor. In the eighteenth century, French ships' registries recorded that the slaves they transported were 47.4% men, 26% women, and 26.6% children (Geggus, "Sex Ratio, Age and Ethnicity").

26. This was a revised version of Maupertuis's 1744 *Dissertation physique à l'occasion du nègre blanc*. Strangely, the *nègre blanc* did not become an important element in his cosmology until he rewrote the *Dissertation* the following year. See chap. 2 above for fuller discussion of this.

27. *HN*, 3: 519. The term *Noir* is equally fraught, but generally did not have the political connotations of *nègre* as "slave." In general, Buffon did not use the term *Noir* as a taxonomic marker; in his text it functions as a general adjective that occasionally is used as a noun.

28. Ibid., 453. Today, "kaffir" is regarded as one of the most derogatory ethnic slurs in South Africa, and its public use is actionable there. During the sixteenth century, the Portuguese used this Arabic word to refer very generally to the indigenous peoples of southern and eastern Africa. The Dutch later applied it to non-Khoisan, Bantu peoples, notably the Xhosa. Buffon uses *caffre* or *cafre* as a general marker to designate the Khoisan, however, very much in contrast with its denotation in southern Africa. The precise meaning of the word in its various spellings (*kaffir, cafre, caffre*) was very slippery during the early-modern era, particularly among armchair naturalists in Europe.

29. These maligned peoples were very often referred to as Hottentots. Labeled as stutterers by the Dutch, the Khoisan peoples suffered terribly in no small part due to the perception that their supposedly outlandish behavior and customs made them subhuman. Unlike the black ethnicities of the African littoral, the hunter-gatherers and pastoral peoples now known as the Khoi and San were rarely enslaved (although sometimes enserfed) and even less often deported; their fate and their legacy is not one of forced diaspora; they died from European diseases, were run off their land, and, in the case of the San, were exterminated in genocidal campaigns by both Boers and Xhosa. Since this book seeks, among other things, to explore the treatment of blackness in the plantation

economies of the Caribbean, the Khoisan peoples are not a major focus here. For more information regarding the Hottentot and European representations of them, see Armstrong and Tennenhouse, eds., *Violence of Representation*; Barnard, *Hunters and Herders*; Elphick, *Khoikhoi*; Holmes, *African Queen*; Hudson, "'Hottentots' and the Evolution of European Racism"; "'Hottentot Venus,'"; Merians, *Envisioning the Worst*; Rawson, *God, Gulliver, and Genocide*; Sharpley-Whiting, *Black Venus*; and Fauvelle-Aymar, *L'invention du Hottentot*.

30. In the early 1750s, however, Buffon's ideas on the African were hardly seen as the most sensational or compelling aspect of the *Histoire naturelle*. Given the overall epistemological debates that were raging at the time, it was the potentially *materialist* implications of the *Histoire naturelle* that preoccupied most thinkers. See, e.g., Lelarge de Lignac's *Lettres à un Américain*, 5, where it is affirmed quite correctly that Buffon's universe has no need of a God.

31. *ENC*, 8: 347.

32. "Je crois que les nègres sont moins beaux pour les nègres mêmes, que les blancs pour les nègres ou pour les blancs" (Denis Diderot, *Salon de 1767*, in id., *Oeuvres complètes*, 16: 236).

33. "celle-ci jusqu'à la tête de l'homme, du nègre, à celle du singe" (Diderot, *Voyage de Hollande*, ibid., 17: 447).

34. "née pour l'abjection et la dépendance, pour le travail et le châtiment" (*HDI*, 6: 207).

35. See Hoffmann, *Nègre romantique*, 50.

36. "leurs yeux ronds, leur nez épaté, leurs lèvres toujours grosses, leurs oreilles différemment figurées, la laine de leur tête, la mesure même de leur intelligence, mettent entre eux et les autres espèces d'hommes des différences prodigieuses" (Voltaire, *Essai sur les moeurs et l'esprit des nations*, in id., *Oeuvres complètes*, 11: 6). Earlier titles and versions of the *Essai* include an unauthorized version, the 1745 *Abrégé de l'Histoire universelle*, and a version revised by Voltaire, the *Essai sur l'Histoire générale et sur les moeurs et l'esprit des nations* (1756). It received its final title used in 1769. See the introduction to Voltaire's *Oeuvres complètes*, i–vii, for the complete publication history.

37. "Quand nous travaillons aux sucreries, et que la meule nous attrape le doigt, on nous coupe la main ; quand nous voulons nous enfuir, on nous coupe la jambe: je me suis trouvé dans les deux cas. C'est à ce prix que vous mangez du sucre en Europe" (Voltaire, *Candide*, in id., *Romans et contes*, 189). For a brief assessment of blood-soaked sugar, see Hoffmann, *Nègre romantique*, 113–14.

38. Festa, *Sentimental Figures of Empire*, 2, writes: "In an era in which imperial reach increasingly outstripped imaginative grasp, sentimental fiction created the tropes that enabled readers to reel the world home in their minds. By designating certain kinds of figures as worthy of emotional expenditure and structuring the circulation of affect between subjects and objects of feeling, the sentimental mode allowed readers to identify with and feel for the plight of other people while upholding distinctive cultural and personal identities. . . . Sentimental depictions of colonial encounters refashioned conquest into commerce and converted scenes of violence and exploitation into occasions for benevolence and pity. In the process, sentimentality, not epic, became the literary mode of empire in the eighteenth century."

39. Substantive information regarding the fourth area of contact, the Middle Passage, was generally elided from the era's written sources until late in the century.

40. On slavery in the French Caribbean, see Kadish, ed., *Slavery in the Caribbean Francophone World*; Moitt, *Women and Slavery in the French Antilles*; Munford, *Black Ordeal of Slavery*; and Garraway, *Libertine Colony*.

41. See Butel, *Histoire des Antilles françaises*, 143–44, for basic demographic figures throughout the century. St. Domingue provides an illustrative case of the increase in the ratio of slaves to whites in the years leading to revolution in what is present-day Haiti. In 1779 there were 249,098 slaves and 32,650 whites. In 1789, there were 470,000 slaves and 40,000 whites. By comparison, on Martinique, there were 73,416 slaves and 10,634 whites in 1789. On Guadeloupe, there were 86,100 slaves and 11,100 whites in 1790. It should be noted that these demographic breakdowns occlude the complexity of the actual social ordering in French islands. As McClellan, *Colonialism and Science*, 47, writes, the *racial* axis of "black versus white" must be interpreted against the axes of *rich* versus *poor* as well as *free* versus *slave*.

42. The most populous colony in Africa during the eighteenth century was the Cape Colony, which had over seven thousand white inhabitants in 1765 (Mostert et al., *Demography*, 26).

43. Cohen, *French Encounter*, 121.

44. Ibid., 121.

45. See Peabody, *"There Are No Slaves in France,"* 4. See also Erick, *Etre noir en France au XVIIIe siècle*.

46. Boulle, *Race et esclavage*, 24.

47. Ibid., 25. This was never enforced.

48. See Peabody, *"There Are No Slaves in France,"* 86–87.

49. Jahoda, *Images of Savages*, 1–12, and others have demonstrated that some of the important elements of the early-modern understanding of black Africa can be attributed to vestigial beliefs dating from antiquity and the medieval era, but eighteenth-century readers were much more familiar with later texts. See also Snowden, *Blacks in Antiquity* and *Before Color Prejudice*. Snowden writes that it "is important to emphasize that the overall, but especially more detailed Greco-Roman, view of Blacks was highly positive. . . . There was clear-cut respect among Mediterranean peoples for Ethiopians and their way of life. And, above all, the ancients did not stereotype all Blacks as primitives defective in religion and culture" (*Before Color Prejudice*, 59).

50. Hair, *Africa Encountered*, 11, 22, describes the Portuguese *descobrimentos* along the West African coast as an example of an "outthrust," a concerted and significant movement of one ethnic group in another world territory.

51. See chap. 2, pp. 107–13.

52. Based on arcane biblical exegesis, La Peyrère's *Praeadamitae* asserted that there had been two acts of divine creation: the first generated tribes of Gentiles, and the second produced Adam, father of the Jews. Before La Peyrère, in the late sixteenth century, Giordano Bruno had also postulated pre-Adamite races in order to explain the existence of Africans. Bernier's much more secular "Nouvelle division de la terre" is seen by many scholars as the first race-based text.

53. As chap. 2 above shows, the Linnaean system had seemingly contradictory effects on the French life sciences. On the one hand, Linnaeus's incorporation of humankind into an overall schema of the world's fauna implicitly invited naturalists (including French naturalists) to use the same methods to understand man as they did animals. On the other hand, his schematic view of both animals and especially humans was

seen as reductive and unrealistic and had the curious effect of inspiring more "empirical," environmental, and inductive approaches to explaining the relationship among humankind's different varieties. While the Linnaean system of binominal taxonomy would ultimately "win out" in France—Louis XV gave the order for the Linnaean system of nomenclature to be adopted at the Jardin du Roi in 1774—this was not the case at mid-century. Indeed, the rise of the life sciences in France—and the contiguous understanding of the human species—would take shape in contradistinction to such a priori categorization.

54. In the tenth edition of the *Systema* (1758), Linnaeus went further in this direction by adding humoral dispositions to his breakdown of the human species, e.g, the African was described as "Black, phlegmatic." Although perhaps rear-guard in terms of its humoral theory, Linnaeus's scheme uniquely accounted for Africans' "liabilities." Indeed, dark skin, lethargy, and a dominant humor were more than simply differentia: they were the basis for a real and evaluative taxonomical system. See Sloan, "Gaze of Natural History," in *Inventing Human Science,* ed. Porter et al., 123.

55. This is discussed at length in the "Coda," pp. 221–23.

56. Since "racist" implies a simple binary, whereas the issues involved are infinitely more complex, I use the term "racializing" here, which Hudson, "From 'Nation' to 'Race,'" 259, defines as "the subjection of populations to scientifically invalid forms of classification based on arbitrary selection of phenotypical or genetic differences."

57. "La Chronologie . . . place les hommes dans le temps" and "[la géographie] les distribue sur notre globe" (d'Alembert, "Discours préliminaire," in *ENC,* 1: xj).

58. Indeed, this interpretive framework provided the point of departure for a potentially *progressive* conjectural history of *Africanness,* within which the African, much like any other human, could be expected as a group to "change, improve, and perfect its nature." See Wokler, "Anthropology and Conjectural History in the Enlightenment," in *Inventing Human Science,* ed. Porter et al., 34.

59. See Hannaford's *Race.* This book provides a very teleological view of race history.

60. White, *Content of the Form,* ix.

61. See Curran, "Pourquoi étudier la représentation de l'Afrique dans la pensée du dix-huitième siècle?"

62. Miller, *French Atlantic Triangle,* 74.

63. See, among those who have reacted to Miller's *Blank Darkness,* Aravamudan, *Tropicopolitans,* 4, who writes that the colonial subject can be simultaneously "fictive construct," "actual resident," and even potentially an "agent of resistance."

64. This is particularly the case in Linda Merians's otherwise excellent study, *Envisioning the Worst.* Merians attributes British views of the Hottentot to feelings of inadequacy suffered by a national (British) imagination: "[The views of the Khoikhoi] suggest no matter [how great] Britain's imperial dominance around the world, the collective British *imagination* still had a need for demonizing 'Hottentots'" (237). Contrary to what is asserted here, the overall discourse on Africa is anything but the product of a national mentality or a Lovejoy-type "unit-idea" that cuts across a variety of disciplines and genres. For a discussion of how the "unit-idea" functions within the intellectual history of Arthur Lovejoy, see Macksey, "History of Ideas."

65. My emphasis. Féraud, *Dictionnaire critique,* 1: 108: "On dit [aussi] faire l'anatomie d'un discours, anatomiser un ouvrage." Féraud put together this definition based on

ideas that had existed since the 1691 edition of the *Dictionnaire de l'Académie françoise*. Furetière's *Dictionnaire universel* (1690) also cites this use of the word.

CHAPTER ONE: Paper Trails

1. Dom Henrique's father was King João I of Portugal.

2. Had this project been commissioned today, I suspect, this unapologetic rendering of Portuguese agency in the Renaissance era would certainly be tempered: rather than a thrusting ship, one might find—as one does now in Portugal—recognization of the contemporary cultural melting pot, not only in Brazil and the former Portuguese African colonies, but also in Portugal itself.

3. After Henry's death, responsibility for the exploration of Africa was ultimately delegated to a rich Lisbon merchant, Fernão Gomes, in 1469.

4. See Russell-Wood, *Portuguese Empire,* xiii–xxv, on how anglophone historiography has neglected Portuguese exploration.

5. See Hair, *Africa Encountered,* 1.

6. However, Pliny the Elder's cosmology, not to mention the fantastical medieval fables found in Reisch's *Margarita Philosophica* (1503), provided the framework onto which new discoveries were initially mapped. See Broc, *Géographie de la Renaissance,*18.

7. According to Hair, *Africa Encountered,* 112, many of these original sources were summarized and then destroyed because they were deemed "no longer useful" by archivists. Moreover, the main repositories of "Africanist knowledge" were also destroyed in the 1755 earthquake that leveled much of Lisbon.

8. *Chronique de Guinée,* ed. Paviot, 9. The English edition, based on a nineteenth-century edition, is *Chronicle of the Discovery and Conquest of Guinea* (Cambridge: Cambridge University Press, 2010).

9. Paviot points out (ibid., 17) that the goal of Eanes de Zurara's *Chronicle* was the glorification of Henry; it was not supposed to be an ethnographic report on African peoples.

10. Such was the market for African slaves during this era that by 1500, 7,000 to 8,000 "captives" purchased from African intermediaries were already "supplying Portuguese markets" each year. See Enders, *Histoire de l'Afrique lusophone,* 25.

11. See Quenum, *Eglises chrétiennes,* 72.

12. "d'assez bonnes gens [qui] se fient fort de nous." La Fosse, *Voyage,* 25.

13. See Diop, "Mise à l'épreuve," in *L'Afrique du siècle des Lumières,* ed. Gallouët, et al.

14. Ca' da Mosto wrote that his expedition was only the second to reach this part of Senegal and that he was, in fact, seeking out Budomel because he had heard reports that the first expedition had been well received by him four years previously. Although Ca' da Mosto called the chieftain Budomel, the proper title of the ruling king of the area known as Budomel was Damel. Budomel was actually the title of the ruler of the region.

15. Ca' da Mosto, *Voyages,* trans. and ed. Crone, 40.

16. Ibid., 38. He does, however, say that it reflects the lubricity of his hosts.

17. It should be pointed out that this may also have more to do with the syncretic orientation of popular medieval Christianity than Ca' da Mosto himself. On the relationship between popular medieval European Christianity, official doctrines of the Church, and magic, see Kieckhefer, *Magic.* For a slightly different reading of Ca' da

Mosto's voyage, see Medeiros, *Occident*, 263–65, who asserts a much more direct relationship between Portuguese expansion and the 1455 papal bull *Romanus Pontifex,* on which see also Mudimbe, "Romanus Pontifex," in *Race, Discourse, and the Origin of the Americas,* ed. Hyatt and Nettleford.

18. Ca' da Mosto, *Voyages*, 49.

19. While the Serers engaged in hostilities with the Wolofs, the languages were dialectic cognates.

20. Ca' da Mosto, *Voyages*, 54.

21. As Claude Lévi-Strauss might have said, for the Serers as for the Europeans, humanity ceased to exist at the borders of their own ethnicity. See Lévi-Strauss, *Race et histoire*, 21.

22. Russell-Wood, *Portuguese Empire*, 211.

23. The first version of this work was published in 1589.

24. Ramusio's mid-century text was appropriated by Hakluyt among others.

25. Zhiri, "Leo Africanus," in Blumenfeld-Kosinski et al., *Politics of Translation*, 161.

26. Ibid., 162.

27. Ibid., 163.

28. Broc, *Géographie de la Renaissance*, 147–49.

29. Leo Africanus, *History and Description*, trans. Pory, 3: 831. Pory, whose translation appeared in 1600, "prefixed" information from other sources.

30. Ibid., 823.

31. Ibid., 825.

32. Ibid.

33. Ibid., 824. And see Caillié, *Journal d'un voyage à Temboctou.*

34. The first edition was Filippo Pigafetta's Italian translation, *Relatione del Reame di Congo e della circonvicine contrade* (1591). The "Congo" in question was the kingdom of Kongo, comprising parts of present-day Angola, the Angolan enclave of Cabinda, the Republic of the Congo, and the Democratic Republic of the Congo. The English title cited here is that of Lopes, *Report on the Kingdom of the Congo,* trans. Hutchinson. The famous brothers De Bry of Frankfurt published a Latin edition in 1598 with new illustrations.

35. See Willy Bal's introduction to Lopes, *Royaume de Congo,* 11.

36. Ibid., 22.

37. Lopes, *Report on the Kingdom of the Congo,* 13.

38. Ibid., 27.

39. For a more complete assessment of the visual rendering of Africa, see Graille, "L'Afrique noire illustrée," in *L'Afrique du siècle des Lumières*, ed. Gallouët et al., 167–96.

40. Lopes, *Report on the Kingdom of the Congo,* 70.

41. Bal notes that the careful ethnographic, climatological, and other information gathered and compiled by Lopes is recounted in an indistinguishable fashion from tales of the Jaga's cannibalism, stories from slave traders, or information about Prester John's kingdom in Abyssinia. See Lopes, *Royaume de Congo*, 21.

42. Leo Africanus, *History and Description of Africa*, 1: 28.

43. Ibid., 1: 46. Francisco Alvares was a Portuguese missionary sent by Manuel I to the negus (king) of Abyssinia. After living there for more than half of a decade, he published his *Verdadeira informação das terras do preste João das Indias.*

44. Leo Africanus, *History and Description of Africa*, 1: 55.

45. Ibid., 76.

46. Ibid., 78.

47. Ibid., 82.

48. Dutch, English, and French ships had, however, occasionally rounded Cape Bojador.

49. Russell-Wood, *Portuguese Empire,* 24–25.

50. Early European ideas about black Africans derived from three sources: texts from antiquity "rediscovered" during the Renaissance, particularly Herodotus, Solinus, and Pliny the Elder; Islamic-sourced travelogues such as that of Leo Africanus; and the travel writings of Europeans who traveled to West Africa with the Portuguese. See Cohen, *French Encounter,* 3.

51. Hannaford, *Race,* 164–65.

52. See Russell-Wood, *Portuguese Empire,* 196–97.

53. Tyson, *Ourang-Outang* (1991), 6.

54. Mocquet, *Voyage à Mozambique et Goa,* 78–80.

55. Boogaart, "Books on Black Africa," in *European Sources for Sub-Saharan Africa,* ed. Heintze and Jones, 55.

56. See Jones, "Olfert Dapper," in *Objets interdits,* 78–79, for a list of his sources.

57. "Ces nègres passent pour être les mieux faits de toute la Guinée, mais ce sont des barbares et des perfides" (Dapper, *Description de l'Afrique,* in *Objets interdits,* 155). The English translation provided in the text is a translation from the French of 1686 because I am talking primarily about the French reception of Dapper at this point. In chap. 2, I cite John Ogilby's 1670 English translation.

58. See Jones, "Olfert Dapper," in *Objets interdits,* 81.

59. Ibid., 79.

60. Dapper, *Description de l'Afrique,* in *Objets interdits,* 258.

61. Ibid., 321.

62. In the seventeenth century, the Dutch were more active in industrial production, international trade, and finance capitalism than the rest of Europe combined, dominating trade with Africa, the Far East and the Caribbean and were in possession of "at least half of the world's total stock of shipping" (Rogoziński, *Brief History of the Caribbean,* 59). During the early seventeenth century "no European country seemed in a better position to play a major role in establishing and profiting from the Atlantic slave economy" (Emmer, "The Dutch and the Slave Americas," in *Slavery in the Development of the Americas,* ed. Eltis et al., 70).

63. Recife, which the Dutch West India Company had seized in 1630, was the capital of Pernambuco—one of Brazil's two most prominent sugar-producing regions—and had been a main port and a key asset. The Dutch established a Caribbean plantation colony in Surinam, captured from the English in 1667.

64. Emmer, "The Dutch and the Slave Americas," 73.

65. Ibid.

66. Walvin, *Making the Black Atlantic,* 21.

67. Ibid., 27.

68. Ibid., 20.

69. Rogoziński, *Brief History of the Caribbean,* 58.

70. David Brion Davis, *Slavery and Human Progress,* 65.

71. Eltis, "Volume of the Transatlantic Slave Trade," 45. It is estimated that from 1605 to 1800, Barbados alone imported 444,800 slaves (ibid., 186).

72. Morgan, *Slavery, Atlantic Trade and the British Economy*, 50.

73. David Brion Davis, *Slavery and Human Progress*, 45. See also Eltis, "Volume of the Transatlantic Slave Trade," 45.

74. Eltis, "Volume of the Transatlantic Slave Trade," 45. For further information regarding the influx of slaves and subsequent increase in sugar production, see Butel, *Histoire des Antilles françaises*, 85, 143; see also *Extending the Frontiers*, ed. Eltis and Richardson.

75. Although the pope issued various edicts against this forced labor, an accompanying loophole permitted Spanish soldiers to enslave those who did not submit to Spanish authority.

76. See Williams, *From Columbus to Castro*, 40.

77. Daget and Renault, *Traites négrières en Afrique*, 77–80, believe that the French were sponsoring voyages to Sierra Leone, Benin, and Elmina dating back to 1539. While the exact cargo of the ships is undocumented, Daget and Renault indicate it was almost certainly African slaves.

78. This rapid increase in slaver expeditions arose around 1672, when "the slave trade was thrown open to private merchants," according to Morgan, *Slavery, Atlantic Trade and the British Economy*, 9.

79. There were, of course, prominent exceptions. British and French sailors reached the coast of Guinea in the early sixteenth century, some Africans were taken to France in the sixteenth century, and in 1571, a "ship owner placed some Blacks on sale in Bordeaux, but they were ordered released on the grounds that slavery did not exist in France," according to Cohen, *French Encounter*, 5.

80. "Outre les marchandises ci-dessus mentionnées, dont on fait commerce dans toutes les isles Françaises d'Amérique, il s'y porte aussi des nègres que l'on va acheter en Afrique sur les côtes de Guinée: ce commerce est d'autant plus avantageux qu'on ne se peut passer de nègres dans lesdites isles pour travailler aux sucres, tabacs et autres ouvrages" (Savary, *Parfait negociant*, 1: 139).

81. "Ce commerce paraît inhumain à ceux qui ne savent pas que ces pauvres gens sont idolâtres ou Mahometans" (ibid., 139–40).

82. "les marchands Chrétiens en les achetant de leurs ennemis, les tirent d'un cruel esclavage, et leur font trouver dans les isles où ils les font porter, non seulement une servitude plus douce; mais même la connaissance du vrai Dieu et la voie du salut par les bonnes instructions que leur donnent des prêtres et religieux qui prennent le soin de les faire chrétiens" (ibid.).

83. P.-F. de Charlevoix, born on October 24, 1682, in San Quentin, France, came to Québec as a teacher in 1705. He returned to France in 1708 and was ordained as a Jesuit priest in 1713. In 1720, he embarked for Québec again, this time assigned with the task of determining the veracity of reports of a sea between the Americas and Asia. These investigations took him all over North America, traveling far down the Mississippi River to New Orleans in 1722. In September 1722, Charlevoix arrived in Saint-Domingue, and shortly thereafter, he left for Le Havre. His source material for Saint-Domingue came from an unpublished manuscript by J.-B. Le Pers, a Jesuit colleague. See "Charlevoix," in *Dictionary of Canadian Biography*, ed. La Terreur et al., 104–6. See also McClellan, 112.

84. For a summary of the kind of penal laws that existed for slaves in the Americas at the time, see Chowdhry and Beeman, "Situating Colonialism, Race, and Punishment," in *Race, Gender, and Punishment*, ed. Bosworth and Flavin, 21–22.

85. "Effectivement, outre qu'un nègre fait autant de besogne, que six Indiens, il s'accoutume bien plus tôt à l'esclavage, pour lequel il paraît né, ne se chagrine pas si aisément, se contente de peu de choses pour vivre, et ne laisse pas, en se nourrissant mal, d'être fort et robuste. Il a bien naturellement un peu de fierté; mais il ne faut pour le dompter, que lui en montrer encore davantage, et lui faire sentir à coups de foüet qu'il a des Maîtres. Ce qu'il y a d'étonnant, c'est que le châtiment, quoique poussé quelquefois jusqu'à la cruauté, ne lui fait rien perdre de son embonpoint, et qu'il en conserve assez peu de ressentiment pour l'ordinaire" (Charlevoix, *Histoire de l'isle Espagnole,* 1: 288).

86. "tout le poids de la servitude" (ibid., 1: 422).

87. "Ces Sénégalais sont de tous les nègres les mieux faits, les plus aisés à discipliner, et les plus propres au service domestique. Les *Bambaras* sont les plus grands, mais voleurs; *les Aradas,* ceux qui entendent mieux la culture des terres, mais les plus fiers: *les Congos,* les plus petits, et les plus habiles pêcheurs, mais ils désertent aisément: *les Nagos,* les plus humains: *les Mondongos,* les plus cruels: *les Mines,* les plus résolus, les plus capricieux, et les plus sujets à se désespérer. Enfin les nègres *créoles,* de quelque nation qu'ils tirent leur origine, ne tiennent de leurs Pères que l'esprit de servitude, et la couleur. Ils ont pourtant un peu plus de d'amour pour la liberté, quoique nés dans l'esclavage; ils sont aussi plus spirituels, plus raisonnables, plus adroits, mais plus fainéants, plus fanfarons, plus libertins que les *Dandas,* c'est le nom commun de tous ceux qui sont venus d'Afrique" (ibid., 2: 498).

88. This was a common contradiction. Jacques Savary, too, qualified his stance that Africans were a knowable, unified group in order to describe the planters' general preference for slaves from the Gold Coast (*Parfait négociant,* 1: 139). English buyers of slaves provided similar assessments. "The negroes most in demand at Barbados are the gold coast or, as they call them, Cormantines, which [*sic*] will yield £3 or £4 more a head than the Whidaw or Papa [papaw, Popo] negroes," Captain Thomas Phillips wrote in 1694 ("Journal of a Voyage," in *Documents,* ed. Donnan, 1: 398).

89. Walrond, "Act," in *Slavery,* ed. Engerman et al., 105.

90. Ibid., 110.

91. Ibid., 106.

92. Ibid., 105.

93. "en nos Isles d'Amérique, la discipline de l'Eglise Catholique, Apostolique et Romaine, pour y régler ce qui concerne l'état et la qualité des esclaves dans nos dites îles." See also Kom and Ngoué, eds., *Code noir,* and Sala-Molins, *Code noir.*

94. The *Code noir* was issued in March; the Revocation was issued in October of 1685.

95. This was not always done of course.

96. Labat, *Voyage aux isles,* 152, reminds us that this was the policy from 1674 when the king brought the islands under his control.

97. Many scholars, however, assert that the *Code noir* was not enforced nearly as ardently as its script demanded. "The enactment of rigorous laws [of the *Code noir*], however, and the rigorous enforcement of those laws are two different things; at least they were under the notoriously lax French and Spanish regimes," according to Roach, *Cities of the Dead,* 252, which provides a thorough analysis of the *Code*'s regulations regarding slave religious and ritual practices. "For the next century [after the 1685 edition of the *Code noir* was issued] slave masters [in Saint-Domingue] brazenly, openly, and consistently broke almost every provision of the code. . . . [T]he Code Noir was always

'judged absurd' and its implementation 'never attempted,'" says Dubois, *Avengers of the New World*, 30.

98. Labat was born in 1663 in Paris, and joined a Dominican convent there at the age of 22. In 1694, he left Martinique for the Antilles, where he lived until 1705. Labat began writing of his travels in the Caribbean in 1712, although the *Nouveau voyage aux isles* was not published until 1722. He died in 1738. See Michel Le Bris, "Un sacré bonhomme," in Labat, *Voyage aux isles*, ed. Le Bris, 13–14. The legacy of Labat is contested. He was a "notorious, slave-trading, swashbuckling 'pirate' priest . . . who was also a prolific (and plagiarizing) writer of travel accounts," Miller, *French Atlantic Triangle*, 18, claims. Le Bris argues, however, that his writing is "of such a constant blissful expressivity" that one wonders why his book on the Caribbean "does not rank with the great classics of French literature" (Labat, *Voyage aux isles*, ed. Le Bris, 9).

99. "Il vint beaucoup de nègres à bord; ils n'avaient pour tout habillement qu'un simple caleçon de toile, quelques-uns un bonnet ou un méchant chapeau, beaucoup portaient sur leur dos les marques des coups de fouet qu'ils avaient reçus: cela excitait la compassion de ceux qui n'y étaient pas accoutumés, mais on s'y fait bientôt" (Labat, *Voyage aux isles*, ed. Le Bris, 38)

100. "Je fis attacher le sorcier et je lui fis distribuer environ trois cents coups de fouet, qui l'écorchèrent depuis les épaules jusqu'aux genoux. Il criait comme un désespéré et nos nègres demandaient grâce pour lui, mais je leur disais que les sorciers ne sentaient point de mal et que ses cris étaient pour se moquer de moi. . . . [Plus tard je] fis mettre le sorcier aux fers après l'avoir fait laver avec une pimentade, c'est-à-dire avec de la saumure dans laquelle on a écrasé du piment et des petits citrons. Cela cause une douleur horrible à qui le fouet a écorché, mais c'est un remède assuré contre la gangrène, qui ne manquerait pas de venir aux plaies" (ibid., 118).

101. "moyen infaillable . . . pour inspirer le culte du vrai Dieu aux Africains, les retirer de l'idolâtrie, et les faire persévérer jusqu'à la mort dans la religion chrétienne qu'on leur ferait embrasser" (ibid., 221).

102. Ibid., 225.

103. "Tous les nègres ont un grand respect pour les vieillards" (ibid., 229).

104. "[Les] habits qu'on leur donne et à quelque autre douceur qu'on leur témoigne, les rend affectionnés et leur fait oublier leurs pays, et l'état malheureux où la servitude les réduit" (ibid., 228).

105. "Pour peu qu'on leur fasse du bien, et qu'on le fasse de bonne grâce, ils aiment infiniment leur maître, et ne reconnaissent aucun péril quand il s'agit de lui sauver la vie, aux dépens même de la leur" (ibid., 229).

106. Ibid., 288.

107. Labat, *Nouvelle relation*, 1: iij.

108. It should be pointed out that "de Brüe's memoires" actually came from an unpublished manuscript of Michel Jajolet de La Courbe's. This was first noted by Pierre Cultru in his introduction to La Courbe's *Premier voyage*.

109. Labat, *Nouvelle relation*, 1: 213.

110. Ibid., 230–35.

111. Ibid., 4: 85.

112. Catherine Gallagher, "Introduction: Cultural and Historical Background," in Behn, *Oroonoko*, ed. id., 9.

113. Green and Astley, *New General Collection of Voyages and Travels.*

114. Although the 1760s saw a veritable explosion in the publication of travel literature, Prévost clearly provided the foundation for the increasing interest in and speculation about African ethnography in mid-century France.

115. See Eche, "L'image ethnographique africaine," in *L'Afrique du siècle des Lumières*, ed. Gallouët et al., 207–22.

116. It should be noted that Prévost finally headed up the project in its entirety after the eighth tome, after the Astley edition was "interrupted." See Duchet, *Anthropologie*, 76–77.

117. Prévost, *Histoire générale des voyages*, 1: v.

118. Ibid., 3: iii.

119. Ibid., 2: 285.

120. "Moore paraît plus exact que Jobson dans ses observations sur les Foulis" (ibid., 3: 152).

121. "destinés d'être leurs serviteurs"; "des milliers d'esclaves en Guinée" (ibid., 2: 344).

122. "Tous nos Voyageurs s'accordent à donner la qualité anthropophage aux Jagas. Lopes assure qu'ils se nourrissent de chair humaine. Battel dit qu'il la préfèrent au boeuf et au chevreau, quoiqu'ils aient l'un et l'autre en abondance. Merolla répète souvent qu'ils mangent les hommes; et . . . il ne balance point à les regarder comme la plus barbare nature de l'univers" (Prévost, *Histoire générale des voyages*, 5: 100).

123. Various other examples abound. At one point, Prévost describes the lascivious dancing of the *négresses* (ibid., 3: 178). He also laments the *nègre*'s fondness of alcohol (ibid., 165). Finally, when describing the kingdom of Congo, he writes that they have no inkling of what constitutes a science (ibid., 4: 633).

124. "Le nombre des pêcheurs est fort grand à Rufisco, et dans d'autres lieux, sur les Côtes du Sénégal. Ils se mettent ordinairement trois dans une almadie ou un canot, avec deux petits mâts qui ont chacun deux voiles" (ibid. 3: 179).

125. "L'enfant reçoit son nom un mois après qu'il est né, avec la cérémonie de lui raser la tête et de la frotter d'huile, dans la présence de cinq ou six témoins. Les noms les plus communs sont pris des Mahométans. Ainsi les garçons s'appellent-*Omar, Guiah, Dimbi, Maliel*, etc., et les filles, *Fatima, Alimata, Komba, Komegain, Warsel, Hengay*" (ibid., 171).

126. "Depuis trois ou quatre cents ans que les habitants de l'Europe inondent les autres parties du monde, et publient sans cesse de nouveaux recueils de voyages et de relations, je suis persuadé que nous ne connaissons d'hommes que les seuls Européens" (Rousseau, *Discours*, in *Oeuvres complètes*, 2: 257).

127. The members of the African Association, formed in 1788, "were all men of wealth, influence, and social standing. . . . The association's goal was to improve the knowledge of African geography and markets, and its chosen method was to sponsor individual explorers who would be willing to take the risk of traveling on their own, or alone but for such native guides as they fell in with along the route," Kate Ferguson Marsters writes in her "Introduction" to Park, *Travels*, 9.

128. "Le joaillier Chardin, qui a voyagé comme Platon, n'a rien laissé dire sur la Perse. La Chine paraît avoir été bien observée par les jésuites. Kempfer donne une idée passable du peu qu'il a vu dans le Japon. L'Afrique entière et ses nombreux habitants,

aussi singuliers par leur caractère que par leur couleur, sont encore à examiner. Toute la terre est couverte de nations dont nous ne connaissons que les noms: et nous nous mêlons de juger le genre humain" (Rousseau, *Discours*, in *Oeuvres complètes*, 2: 257).

129. "Nous verrions nous-mêmes sortir un monde nouveau de dessous leur plume, et nous apprendrions ainsi à connaître le nôtre" (ibid., 213).

CHAPTER TWO: Sameness and Science, 1730–1750

1. The description of the royal natural history cabinet was actually the rationale for the whole project (see *HN*, 3: 524–25).

2. "Cette tête a été préparée comme les précédentes, on ne la distinguerait pas de la tête d'un blanc, si on ne savait d'ailleurs qu'elle vient d'un nègre" (ibid., 150).

3. "Cette pièce est à peu près dans le même état que la précédente, mais on y connait bien mieux les traits de la physionomie des nègres" (ibid.).

4. "des individus d'une même espèce les uns après des autres" (ibid., 4).

5. Kidd, *Forging of the Races*, 21.

6. Ibid. Several translations omit of "one blood." Less important than the correct translation, however, is the fact that it was translated as such.

7. Gen. 9:21 (KJV).

8. Ibid.

9. Ibid., 9:25.

10. Some pro-slavery thinkers commonly evoked the curse of Ham as a justification for slavery, more commonly in the eighteenth and nineteenth centuries. For a survey of the evolution of this myth from fifteenth to the nineteenth centuries, see Quenum, *Eglises chrétiennes*, 25–32.

11. St. Augustine's *Civitas Dei* initiated and shaped this argument, basing much of it in fact on an analysis of Noah's three sons. Augustine treated "all men as rational and moral" (Hannaford, *Race*, 95).

12. See Jacques Le Goff, "Préface" in Medeiros, *L'Occident et l'Afrique*, 9–10.

13. For more on eighteenth-century theories concerning human and animal origins, see Jahoda, *Images of Savages*, 25.

14. Braude, "Sons of Noah"; Evans, "From the Land of Canaan."

15. "toutes les nations . . . si différentes entr'elles de corsage, de stature, d'inclinations et d'esprit, quoiqu'elles descendent d'un même père?" (Dubos, *Réflexions*, 2: 15, 251).

16. Ibid., 15, 253.

17. "la conformation de [leurs] organes"; "aux qualités de [leur] sang" (ibid., 14, 239).

18. "deux hommes qui auront le sang d'une qualité assez différente pour être dissemblables à l'extérieur, seront encore plus dissemblables par l'esprit. Ils seront encore plus différents d'inclination que de teint et de corsage" (ibid., 15, 255).

19. Ibid., 15, 256.

20. For a useful summary of climate theory, see Reeser, *Moderating Masculinity*, 223–24.

21. "Il est peu de cerveaux qui soient assez mal conformés pour ne pas faire un homme d'esprit ou du moins un homme d'imagination sous un certain ciel; c'est le contraire sous un autre climat" (Dubos, *Réflexions*, 2: 15, 257).

22. *Pour et contre*, 12: 176.

23. "une autre espèce que nous" (ibid.).

24. "[S]i l'on considère aussi le portrait que les historiens nous font de certains siècles du monde, et l'état où les peuples qui passent aujourd'hui pour les plus éclairés ont langui longtemps avant que de parvenir aux lumières dont ils font leur gloire, on revient à penser malgré soi que l'esprit et la politesse étant sujets aux mêmes révolutions que tout le reste des êtres . . . il n'est pas plus surprenant que les nègres soient tombés dans le dernier degré d'abrutissement" (ibid.).

25. "Quelle est la cause physique de la couleur des nègres, de la qualité de leurs cheveux, et de la dégénération de l'un et de l'autre?" As mentioned in the Introduction, a note left with the manuscript indicates that some of these essays arrived after the deadline and were thus not considered; they were nevertheless included at the back of the collection. The essays themselves are grouped together and are numbered 1–16, a convention that I have upheld here. I have not cited the different titles on the manuscripts, which generally mirror the question posed by the Academy. The reference for the text is Bibliothèque municipale de Bordeaux ms. 825/65.

26. Although I have no way of proving this, the fact that the prize was never given may be due to the fact that one of the essayists, the soon-to-be-well-known Pierre Barrère, published his essay submission the same year (#5 here), perhaps because his essay was classed among those that arrived too late. The title of his published essay, *Dissertation sur la cause physique de la couleur des nègres, de la qualité de leur cheveux, et de la dégénération de l'un et de l'autre*, came directly from the contest question, and Barrère indicated in an introductory note to his book that he would have never published it had he not been encouraged by "illustrious Academicians."

27. See Wheeler, *Complexion of Race*, 3.

28. See, e.g., Kidd's discussion of Buffon in his otherwise excellent *Forging of the Races*, 84. Put simply, polygenists believed in separate origins for different species of men; monogenists believed in a common origin from which all human varieties developed.

29. "les nègres ne sont pas descendus d'Adam, mais d'un autre homme qui était noir, et que par conséquent ils sont une espèce d'hommes toute différente de la nôtre" (Bibliothèque municipale de Bordeaux ms. 825/65, #4, p. 3).

30. Incredibly enough, this conviction comes after this same author adopts François Bernier's early and very schematic seventeenth-century classification of human types (adding one more species, the dog-eaters of Ile des Chiens), a schema generally associated with the birth of scientifically-based racial categories (ibid., p. 6).

31. There was also a good deal of syncretism among theological thinkers as well, for example, in the four-element theory found in essay #8.

32. "Si les blancs et les nègres sont différents par la couleur, ils se ressemblent beaucoup par l'amour propre. Alors considéré par cette face, c'est bien la même espèce" (Bibliothèque municipale de Bordeaux ms. 825/65, #7, p. 4).

33. Maupertuis clearly consulted Barrère before writing on the *nègre* in his *Vénus physique* (1745).

34. "Il y a bien des gens qui s'imaginent que les nègres sont une espèce d'hommes dégénérés, ou comme l'on dit, abâtardis, et par conséquent, différents des autres, que ce qui vient de leur sang mêlé avec un autre sang différent, peut être comparé à peu près aux mules et aux mulets. Mais ce qui renverse ce sentiment, c'est que si cela était, les mulâtres, les *Saccatras*, c'est à dire, ceux qui tiennent du nègre et de l'indien et autres

hommes de différents teints issus tous originairement des nègres, devraient être, stériles et nullement propres pour la génération, ainsi que les mules et les mulets . . . , ce qui ne s'accorde pas avec ce qu'on voit tous les jours" (Bibliothèque municipale de Bordeaux ms. 825/65, #5, p. 11).

35. "On ne peut donc chercher, ni dans l'histoire sacrée, ni dans la profane, la cause de l'origine de cette noirceur qu'on ne peut découvrir, que par des réflexions sur l'état de la situation du monde connu comparées avec les soins, que la Providence prend journellement de tous les hommes" (ibid., #1, p. 2).

36. "plus en état [de] supporter la chaleur [de] l'Afrique" (ibid., p. 16).

37. "Les nègres transplantés dans les pays des vrais blancs perdent leur noirceur et les blancs transplantés dans le pays des noirs perdent de leur blancheur" (ibid., #5, p. 8).

38. "[F]aites cesser les causes [of blackness, viz., the sun and heat] et les effets cesseront. Transporter des hommes et les femmes noirs dans des pays tempérés, donnez-leur des nourritures saines et des habits convenables; les frictions n'auront plus lieu; les glandes se recontracteront; les filtrations se feront d'une manière différente que de coutume: la nature du sang et des humeurs se changera; les enfants qui naitront de ces gens-là ou du moins les enfants de leurs enfants, auront les membranes réticulaires d'une ou deux nuances moins noires que celles de ceux dont ils seront issus" (ibid., #4, p. 17).

39. "peut expliquer la lenteur de l'esprit et le flegme des passions des peuples [de l'Afrique]" (ibid., #2, p. 11).

40. According to this essayist, the only way to revert to a white was through *métissage* (ibid., p. 12).

41. "Il n'y a aucun doute que la première espèce, (c'est-à-dire celle dont nous sommes nous-mêmes) ne soit la primitive, et pour ainsi dire la légitime; et que toutes les autres n'ayant dégénéré" (ibid., #14, p. 15).

42. Locke, *Essay Concerning Human Understanding,* 104, italics and capitalization his.

43. "il [était] permis à chacun d'en faire tant qu'il voudra" (Bibliothèque municipale de Bordeaux ms. 825/65, #4, p. 1).

44. For a discussion of antiquity's understanding of albinism, see Little, Nègres blancs, 24–25.

45. Cortés, *Letters from Mexico*, ed. Pagden, 109. Although Cortés mentioned these marvels in his second letter to the Spanish court, the explorer remained unruffled by the strange white beings. Compared to other Renaissance-era prodigies, these pigmentless humans simply did not demand much attention. It was also true that all travellers believed the world outside of Christendom to be populated by all sorts of strange creatures. What is interesting about this discovery, however, was that no one identified a link between these two populations. The Leucoaethiopes, after all, were an African phenomenon by definition.

46. Battel's *Strange Adventures,* his account of his life among the itinerant anthropophages known as the Giague or Jaga, is a fascinating read.

47. In the late seventeenth century, other albino humans were found; the Swedish traveler Nils Matson Kjöping described the albinos of the Moluccas, or Maluku Islands (a micro-archipelago in Indonesia), calling them *homo sylvestris* and *homo nocturnus,* terms also used by Linnaeus. Francis Leguat found pigmentless men on Batavia and used the term *Chacrelats*. See Little, Nègres blancs, 30.

48. Ogilby, *Africa,* 508–9. Ogilby shamelessly translated Dapper word for word, with

almost no attribution. When I discuss the genealogy of French ideas, I cite the French translation of Dapper.

49. Van Stekelenburg, "Ex Africa semper aliquid novi," attributes the saying to Pliny the Elder's *Naturalis historia* (c. 70 CE).

50. "paraissait peu intelligent, et destiné à être imbécile." Fontenelle, "Observations," in *Histoire de l'Académie royale des sciences* (1734 [1736]), 15.

51. " . . . tous ces traits de nègre si marqués, cette laine au lieu de cheveux?" (ibid., 16).

52. "In his account of the 'very singular' phenomenon, Fontenelle dwelt on the question of the child's parentage, wondering whether the father of the albino might have been a white Negro like those observed by travelers in Africa" (Terrall, *Man Who Flattened the Earth,* 208).

53. Maupertuis simply "used the albino boy as a hook to attract the attention of those witnesses who had seen him and argued about his origins" (ibid.).

54. *Vénus physique* is cited here from the 1752 Dresden edition of Maupertuis's *Oeuvres*.

55. This clear-cut refutation is interesting because Maupertuis punctuates the beginning of *Vénus physique* with caveats about the conclusions possible given the subject.

56. Terrall, *Man Who Flattened the Earth*, 213.

57. Maupertuis often uses *espèce, variété,* and *race* interchangeably to indicate black Africans. To reflect what was presumably his intention in evoking this subtype, I use "variety" here (although he also uses *race*). The definition of *variété* comes from the botanical context. See *ENC,* 16: 847.

58. For a discussion of the evolution of Linnaeus's thought, see Sloan, "Gaze" in *Inventing Human Science*, ed. Fox et al., 122–26, which is particularly useful in that it actually provides charts showing how the Swedish thinker's classification changed during the many iterations of the *Systema natura*.

59. Although race-mixing was a reality of the French Caribbean, it was banned in the 1685 *Code noir.*

60. "[P]ourquoi ces sultans blasés dans des sérails qui ne referment que des femmes de toutes les espèces connues, ne se font-ils pas faire des espèces nouvelles?" (Maupertuis, *Vénus physique,* 134).

61. "Pleasure, intellectual understanding, and physical organization are thus inextricably linked by the very nature of life [in *Vénus physique*]" (Terrall, *Man Who Flattened the Earth*, 213).

62. François Bernier's 1684 "A New Division of the Earth" "was the first text in which the term 'race' is used in something like its modern sense to refer to discrete human groups organized on the basis of skin color and other physical attributes. Later writers, including Gottfried Wilhelm Leibniz and Johann Friedrich Blumenbach, acknowledged Bernier's contribution to the classification of humanity," according to Bernasconi and Lott, *Idea of Race*, 1. See also Hudson, "From 'Nation' to 'Race'," 247–64, for additional information on the history of this term.

63. "Ce n'est point au blanc et au noir que se réduisent les variétés du genre humain; on en trouve mille autres" (Maupertuis, *Vénus physique*, 133).

64. Although Maupertuis did not state it explicitly, he clearly believed that such classifications should be made on the basis of anatomical compatibility, in other words, an ability to produce fertile offspring.

65. "Si les premiers hommes blancs qui en virent des noirs, les avaient trouvés dans les forêts, peut-être ne leur auraient-ils accordé le nom d'hommes. Mais ceux qu'on trouva dans de grandes villes, qui étaient gouvernés par de sages Reines, qui faisaient fleurir les Arts et les Sciences, dans des temps où presque tous les autres peuples étaient des barbares; ces noirs-là, auraient bien pu ne pas vouloir regarder les blancs comme leurs frères" (Maupertuis, *Vénus physique*, 127).

66. "Depuis le Tropique du Cancer jusqu'au Tropique du Capricorne l'Afrique n'a que des habitants noirs. Non seulement leur couleur les distingue, mais ils différent des autres hommes par tous les traits de leur visage: des nez larges et plats, de grosses lèvres, et de la laine au lieu de cheveux, paraissent constituer une nouvelle espèce d'hommes. Si l'on s'éloigne de l'Equateur vers le Pole Antarctique, le Noir s'éclaircit, mais la laideur demeure: on trouve ce vilain peuple qui habite la pointe Méridionale de l'Afrique" (ibid., 128).

67. See Abbé Demanet's fascinating synchretic plagiarization of Maupertuis. Demanet republished pages of the *Vénus physique* but tacked on Christian conclusions to Maupertuis's more materialist monogenesis. See Demanet, *Nouvelle histoire de l'Afrique françoise*, 2: 211.

68. "Tous ces peuples que nous venons de parcourir, tant d'hommes divers, sont-ils sortis d'une même mère? Il ne nous est pas permis d'en douter" (Maupertuis, *Vénus physique*, 132). This monogenetic account of the origins of the human species heralded Buffon's assertion to the same effect.

69. "dans quelque ancêtre inconnu" (ibid., 134). Less pathological and morally charged than Buffon's theory of climate-degeneration, this heredity-based view of generation explains both the relatively optimistic view of humankind's different varieties, as well as Maupertuis's tendency to give credence to the existence of the monstrous races found in the era's travel writing.

70. "des individus fortuits" (ibid.).

71. As indicated in the Introduction, the universal condition known as albinism or hypopigmentation did not yet exist. The *nègre blanc* was associated with Africans.

72. See Magner, *History of the Life Sciences*, 154–56, on the development of epigenesis and preformation theory. Magner explains that "[t]he theory of epigenesis, which Aristotle favored, is based on the belief that each embryo or organism is gradually produced from an undifferentiated mass by a series of steps and stages during which new parts are added. Various explanations for the means by which development occurred were proposed, but all advocates of epigenesis were opposed to the concept of preformation or preexistence" (154). On Lucretius's regenerative view of nature, see Campbell, *Lucretius*, 10.

73. "Maupertuis repeatedly asserted the contingency of the current order, arguing that things could have been otherwise. If the present order were disrupted, by a global catastrophe for example, a whole new order could come about without altering the primordial chemistry" (Terrall, *Man Who Flattened the Earth*, 357).

74. See Katzew, *Casta Painting*, on Spanish trends and forerunners in this area, e.g., the work of the Jesuit José Gumilla, whose imaginationist view of albinism also included the notion that the "first blacks must have been born to white parents" and that "if albinos only married among themselves, in time they would create a new race—the same way that blacks had centuries ago"(ibid., 47–48).

75. "le blanc est la couleur primitive des hommes, et . . . le noir n'est qu'une variété

devenue héréditaire depuis plusieurs siècles, mais qui n'a point entièrement effacé la couleur blanche qui tend toujours à réapparaître" (Maupertuis, *Vénus physique*, 141–42).

76. In Maupertuis, the notion of species and the notion of monster seemingly overlap. Nonetheless, Maupertuis's own use of the term monster to describe this *nègre blanc* is misleading. Not only is the etiology of the albino entirely naturalized in Maupertuis's cosmology; it is explained away as a result of what we might call a genetic modification.

77. "l'espèce la plus nombreuse aurait relégué ces races difformes dans les climats de la terre les moins habitables" (Maupertuis, *Vénus physique*, 144).

78. From 1736 to 1737, Maupertuis directed a scientific expedition to Lapland organized by the Paris Académie royale des sciences and brought back two Laplanders.

79. This explanation overlaps with one of the most important shifts in eighteenth-century embryology: accidentalist explanations for human monstrosities. See Tort, *L'ordre et les monstres*, for an excellent summary of the accidentalist and providentalist explanations for human monstrosities debated at the Paris Académie royale des sciences.

80. "il y a dans cette île de Java une nation qu'on appelle Chacrelas, qui est toute différente, non seulement des autres habitants de cette île, mais même de tous les autres Indiens. Ces Chacrelas sont blancs et blonds, ils ont les yeux faibles, et ne peuvent supporter le grand jour; au contraire ils voient bien la nuit, le jour ils marchent les yeux baissés et presque fermés" (*HN*, 3: 339).

81. "Les habitants de Malaca, de Sumatra et des îles Nicobar semblent tirer leur origine des Indiens de la presqu'île de l'Inde; ceux de Java, des Chinois, à l'exception de ces hommes blancs et blonds qu'on appelle *Chacrelas*, qui doivent venir des Européens" (ibid., 410).

82. "Les habitants de Ceylan ressemblent assez à ceux de la côte Malabar . . . [mais] il y a. . . des espèces de sauvages qu'on appelle *Bedas*, ils demeurent dans la partie septentrionale de l'île, et n'occupent qu'un petit canton; ces *Bedas* semblent être une espèce d'hommes toute différente de celle de ces climats . . . ils sont blancs comme les Européens, il y en a même quelques-uns qui sont roux. . . . Il me paraît que ces *Bedas* de Ceylan, aussi bien que les *Chacrelas* de Java, pourraient bien être de race européenne, d'autant plus que ces hommes blancs et blonds sont en très petit nombre. Il est très possible que quelques hommes et quelques femmes européennes aient été abandonnés autrefois dans ces îles, ou qu'ils y aient abordé dans un naufrage" (ibid., 415–16).

83. "Ce qui me paraît appuyer dans beaucoup cette manière de penser" [i.e., that Europeans are not the source of these whites] is that "c'est que parmi les nègres il naît aussi des blancs de pères et mères noirs; on trouve la description de deux de ces nègres blancs dans l'histoire de l'Académie" (ibid., 501).

84. "[O]n trouve parmi eux des hommes tout différents, et quoiqu'ils soient en très petit nombre, ils méritent d'être remarqués: ces hommes sont blancs, mais ce blanc n'est pas celui des Européens, c'est plutôt un blanc de lait, qui approche beaucoup de la couleur du poil d'un cheval blanc . . . leurs sourcils sont d'un blanc de lait, aussi bien que leurs cheveux qui sont très beaux. . . . Au reste ces hommes ne forment pas une race particulière et distincte, mais il arrive quelquefois qu'un père et une mère qui sont tous deux couleur de cuivre jaune, ont un enfant tel que nous venons de décrire" (ibid., 500–501).

85. [If it is indeed the case that these children come from dark-skinned parents, then] "cette couleur et cette habitude singulière du corps de ces Indiens blancs, ne seraient qu'une espèce de maladie qu'ils tiendraient de leurs pères et leurs mères; mais en supposant que ce dernier fait ne fût pas bien avéré, c'est-à-dire qu'au lieu de venir des Indiens jaunes ils fissent une race à part, alors ils ressembleraient aux *Chacrelas* de Java, et aux *Bedas* de Ceylan dont nous avons parlé; ou si ce fait est bien vrai, et que ces blancs naissent en effet de pères et mères couleur de cuivre, on pourra croire que les *Chacrelas* et les *Bedas* viennent aussi de pères et mères basanés, et que tous ces hommes blancs qu'on trouve à de si grandes distances les uns des autres, sont des individus qui ont dégénéré de leur race par quelque cause accidentelle. J'avoue que cette dernière opinion me paraît la plus vraisemblable" (ibid., 501).

86. "Ce que j'en ai vu, indépendamment de ce qu'en disent les voyageurs, ne laisse aucun doute sur leur origine; ces nègres blancs sont des nègres dégénérés de leur race, ce ne sont pas une espèce d'homme régulière et constante, ce sont des individus singuliers qui ne sont qu'une variété accidentelle" (ibid., 502).

87. Buffon had posited the existence of an "interior mold" or *moule intérieure*, a basic theory of hereditary transfer. See Ibrahim, "La notion de moule intérieur."

88. As Buffon himself summarized: "one is tempted to believe that men of all races and all colors sometimes give rise to *blafards* [on serait donc porté à croire que les hommes de toute race et de toute couleur, produisent quelquefois des individus blafards]" (*HN, Suppl.* [1777], 4: 556).

89. Ibid.

90. Ibid., 560.

91. See Douthwaite, *Wild Girl,* 209.

92. "Remuée par la honte . . . de se faire voir nue," Geneviève is a "pucelle," whose breasts are "grosses, rondes, très-fermes, et bien placées" (*HN, Suppl.*, 4: 561). She has "dents bien rangées et du plus bel émail, l'haleine pure, point de mauvaise odeur comme les négresses noires," and [her] "cuisses et les fesses présentent une peau bien ferme et assez bien tendue" (ibid., 564).

93. An even more exhaustive list of Geneviève's features is appended in a footnote.

94. As Buffon put it himself, to "se propager, se multiplier et conserver à perpétuité par la génération, tous les caractères qui pourraient la distinguer des autres races" (ibid., 565).

95. "mais ce ne sera qu'en multipliant les observations qu'on pourra reconnaître les nuances et les limites de ces différentes variétés" (ibid.).

96. It should also be added that the engraving of the nude Geneviève (like the other images reproduced in this essay) is quoting deeply ingrained conventions in Western European art that derive from the Renaissance paintings of nude classical goddesses by Raphael, Botticelli, and Giogione, among others. These conventions were clearly "grafted" onto medical, ethnographic, and other naturalist images for publication during the eighteenth century before giving way to a more a clinical and dehumanized mode of scientific representation during the nineteenth century.

97. "Une animalité en pièces" (ibid., 20).

98. "[I]ls se rapprochent de la couleur primitive de laquelle ils ont dégénéré" (*HN, Suppl.* [1777], 4: 557).

99. "[L]es Dondos [*nègres blancs*] produisent avec les nègres des enfants noirs" and "les Albinos de l'Amérique avec les Européens produisent des mulâtres" (ibid.).

100. On Schreber, Buffon wrote that he doubted "all these facts" (ibid.).

101. Ibid., 564.

102. "elle ne produirait rien, parce qu'en général les mâles nègres blancs ne sont pas prolifiques" (ibid.).

103. Ibid., 4: 566. Although the engraving of this mottled girl occupies the same ontological space as the plate dedicated to Geneviève (several pages earlier), the way in which this young *négresse pie* is represented differs markedly from the (perhaps) more emotionally realistic and somewhat cheerless portrait of the *négresse blanche*. Unlike the distraught and sexualized Geneviève, this small girl appears to be a happy, integral, and organic part of the exotic—albeit presumably degeneration-inducing—colonial climate. See Katzew, *Casta Painting,* 109, on the varied compositional implications of colonial subjects.

104. "on serait moins mal fondé à rapporter l'origine de cet enfant à des nègres dans lesquels il y a des individus blancs ou blafards" (*HN, Suppl.* [1777], 4: 567).

105. "disgraciés de la nature" (ibid., 570).

106. "de la laine, les autres des cheveux, et . . . d'autres n'ont ni laine ni cheveux, mais un simple duvet; quelques-uns ont l'iris de yeux rouge, et d'autres d'un bleu faible" (ibid., 569).

107. "le blanc paraît être la couleur primitive de la nature . . . qui reparaît dans de certaines circonstances, mais avec une si grande altération, qu'il ne ressemble point au blanc primitif, qui en effet a été dénaturé par les causes que nous venons d'indiquer" (*HN,* 3: 502).

108. In many ways, the creation of the *blafard* seems to be yet another example of what has come to be called Foucauldian biopolitics. Hacking, *Historical Ontology,* 104, refers to the conceptual categorization of such human phenomena as "making up people," using "medico-forensic-political" language" with the end of "individual and social control." But the "making" of the albino by thinkers such as Buffon actually had an effect that raised as many troubling questions as it answered. In other words, if the *telos* of this construction was supposedly societal control, the theory of albinism actually gave rise more to chance than to order.

109. "de misérables chaumières . . . tandis qu'il ne tiendrait qu'à eux d'habiter de belles vallées" (*HN,* 3: 462).

110. "Si le nègre et le blanc ne pouvaient produire ensemble, si même leur production demeurait inféconde, si le Mulâtre était un vrai mulet, il y aurait alors deux espèces bien distinctes; le nègre serait à l'homme ce que l'âne est au cheval; ou plutôt, si le blanc était homme, le nègre ne serait plus un homme, ce serait un animal à part comme le singe, et nous serions en droit de penser que le blanc et le nègre n'auraient point eu une origine commune; mais cette supposition même est démentie par le fait, et puisque tous les hommes peuvent communiquer et produire ensemble, tous les hommes viennent de la même souche et sont de la même famille" (ibid., 4: 388–89).

111. The full version of this quotation is as follows: "Dès que l'homme a commencé à changer de ciel et qu'il s'est répandu de climats en climats, sa nature a subi des altérations; elles ont été légères dans les contrées tempérées, que nous supposons voisines du lieu de son origine; mais elles ont augmenté à mesure qu'il s'en est éloigné, et, lorsqu'après des siècles écoulés, des continents traversés et des générations déjà dégénérées par l'influence des différentes terres, il a voulu s'habituer dans les climats extrêmes et peupler les sables du midi et les glaces du nord, les changements sont dev-

enus si grands et si sensibles qu'il y aurait lieu de croire que le nègre, le lapon et le blanc forment des espèces différentes, si d'un côté on n'était assuré qu'il n'y a qu'un seul homme de créé, et de l'autre que ce blanc, ce lapon et ce nègre, si dissemblants entre eux, peuvent cependant s'unir ensemble et propager en commun la grande et unique famille de notre genre humain; ainsi leurs taches ne sont point originelles; leurs dissemblances n'étant qu'extérieures, ces altérations de nature ne sont que superficielles, et il est certain que tous ne sont que le même homme qui s'est verni de noir sous la zone torride, et qui s'est tanné, rapetissé par le froid glacial du pôle de la sphère" (ibid., 14: 311).

112. "Tout le travail de Buffon est de traiter les sous-espèces linnéennes comme les accidents de l'histoire" (Hoquet, *Buffon/Linné,* 97).

113. "Il paraît . . . qu'il y a autant de variétés dans la race des noirs que dans celle des blancs; les noirs ont, comme les blancs, leurs Tartares et leurs Circassiens" (*HN,* 3: 453).

114. "Il est . . . nécessaire de diviser les noirs en différentes races, et il me semble qu'on peut les réduire à deux principales, celle des Nègres et celle des Caffres" (ibid.). Buffon capitalized *nègre* here, it seems, because he was juxtaposing the black African against other categories that were generally capitalized. As I mentioned earlier, capitalization of the term *nègre* is very inconsistent throughout the eighteenth century, the *Histoire naturelle* being no exception in this regard.

115. "Dans la première [la race des noirs] je comprends les noirs de Nubie, du Sénégal, du Cap-verd, de Gambie, de Sierra-liona, de la côte des Dents, de la côte d'Or, de celle de Juda, de Bénin, de Gabon, de Lowango, de Congo, d'Angola et de Benguela jusqu'au Cap-nègre; dans la seconde [la race des Caffres] je mets les peuples qui sont au delà du Cap-nègre jusqu'à la pointe de l'Afrique, où ils prennent le nom de Hottentots, et aussi tous les peuples de la côte orientale de l'Afrique, comme ceux de la terre de Natal, de Sofala, du Monomotapa, de Mozambique, de Mélinde; les noirs de Madagascar et des isles voisines seront aussi des Caffres et non pas des Nègres. Ces deux espèces d'hommes noirs se ressemblent plus par la couleur que par les traits du visage, leurs cheveux, leur peau, l'odeur de leur corps, leurs moeurs et leur naturel sont aussi très différents" (ibid.).

116. "de la plus affreuse mal-propreté"; "indépendants et très-jaloux de leur liberté" (ibid., 471).

117. "espèce d'excroissance ou de peau dure et large qui leur croît au dessus de l'os pubis" (ibid., 473).

118. In addition to the numerous articles and books on the Hottentots and Hottentot ceremonies cited on p. 27, see also Lanni's collection of contemporary accounts, *Fureur et barbarie.*

119. "Par tous ces témoignages il est aisé de voir que les Hottentots ne sont pas de vrais nègres, mais des hommes qui dans la race des noirs commencent à se rapprocher du blanc. . . . Les Hottentots qui n'ont pû tirer leur origine que de nations noires, sont cependant les plus blancs de tous ces peuples de l'Afrique, parce qu'en effet ils sont dans le climat le plus froid de cette partie du monde" (*HN,* 3: 473, 81).

120. Cf. "Nomenclature des singes" in the *Histoire naturelle,* where Buffon discusses and rejects the notion of a possible lineage between monkeys and man. Here Buffon nonetheless paints a horrifying portrait of the Hottentot and even evokes, briefly, the possibility of cross-species coupling between *négresses* and monkeys that may have allowed hybrids to be incorporated into each of the two "species" (ibid., 14: 30–32).

121. "Les Ethiopiens . . . les Abyssins, et même ceux de Mélinde, qui tirent leur origine des blancs, puisqu'ils ont la même religion et le même usage que les Arabes, et

qu'ils leur ressemblent par la couleur, sont à la vérité encore plus basanés que les Arabes méridionaux, mais cela même prouve que dans une même race d'hommes le plus ou moins de noir dépend de la plus ou moins grande ardeur du climat; il faut peut-être plusieurs siècles et une succession d'un grand nombre de générations pour qu'une race blanche prenne par nuances la couleur brune et devienne enfin tout-à fait noire, mais il y a apparence qu'avec le temps un peuple blanc transporté du nord à l'équateur pourrait devenir brun et même tout-à-fait noir, surtout si ce même peuple changeait de moeurs et ne se servait pour nourriture que des productions du pays chaud dans lequel il aurait été transporté" (ibid., 3: 482–83).

122. "Les hommes diffèrent du blanc au noir par la couleur, . . . la hauteur de la taille, la grosseur, la légèreté, la force, et . . . l'esprit; *mais cette dernière qualité n'appartenant point à la matière, ne doit point être ici considérée*" (ibid., 386–87; emphasis added).

123. "grands, gros, bien faits, mais niais et sans génie" (ibid., 456).

124. Ibid., 450.

125. "il y a parmi eux d'aussi belles femmes, à la couleur près, que dans aucun autre pays du monde, elles sont ordinairement très-bien faites, très-gaies, très-vives et très-portées à l'amour, elles ont du goût pour tous les hommes, et particulièrement pour les blancs qu'elles cherchent avec empressement, tant pour se satisfaire, que pour en obtenir quelque présent" (ibid., 457–58).

126. Ibid., 461–62.

127. "Les négresses sont fort fécondes et accouchent avec beaucoup de facilité et sans aucun secours, les suites de leurs couches ne sont point fâcheuses, et il ne leur faut qu'un jour ou deux de repos pour se rétablir" (ibid., 459–60).

128. I have supplied the full version of this quotation here. "Quoique les nègres de Guinée soient d'une santé ferme et très bonne, rarement arrivent-ils cependant à une certaine vieillesse, un nègre de cinquante ans est dans son pays un homme fort vieux, ils paraissent l'être dès l'âge de quarante; l'usage prématuré des femmes est peut-être la cause de la brièveté de leur vie; les enfants sont si débauchés et si peu contraints par les pères et mères que dès leur plus tendre jeunesse ils se livrent à tout ce que la Nature leur suggère; rien n'est si rare que de trouver dans ce peuple quelque fille qui puisse se souvenir du temps auquel elle a cessé d'être vierge" (ibid., 463). For a discussion of the complement to this idea, see also Bilé, *Légende*.

129. "s'accoûtument aisément au joug de la servitude" (*HN*, 3: 471).

130. A similar typology is cited in chap. 1, p. 54.

131. "On préfère dans nos isles les nègres d'Angola à ceux du Cap-verd pour la force du corps, mais ils sentent si mauvais lorsqu'ils sont échauffés, que l'air des endroits par où ils ont passé en est infecté pendant plus d'un quart d'heure; ceux du Cap-verd n'ont pas une odeur si mauvaise à beaucoup près que ceux d'Angola, et ils ont aussi la peau plus belle et plus noire, le corps mieux fait, les traits du visage moins durs, le naturel plus doux et la taille plus avantageuse. Ceux de Guinée sont aussi très-bons pour le travail de la terre et pour les autres gros ouvrages, ceux du Sénégal ne sont pas si forts, mais ils sont plus propres pour le service domestique, et plus capables d'apprendre des métiers. Le P. Charlevoix dit que les Sénégalais sont de tous les nègres les mieux faits, les plus aisés à discipliner et les plus propres au service domestique; que les Bambaras sont les plus grands, mais qu'ils sont frippons; que les Aradas sont ceux qui entendent le mieux la culture des terres; que les Congos sont les plus petits, qu'ils sont fort habiles pêcheurs, mais qu'ils désertent aisément; que les Nagos sont les plus humains, les Mondongos les

plus cruels, les Mines les plus résolus, les plus capricieux et les plus sujets à se désespérer" (*HN*, 3: 467).

132. "[Charlevoix] ajoûte que tous les nègres de Guinée ont l'esprit extrêmement borné, qu'il y en a même plusieurs qui paraissent être tout-à-fait stupides, qu'on en voit qui ne peuvent jamais compter au delà de trois, que d'eux-mêmes ils ne pensent à rien, qu'ils n'ont point de mémoire, que le passé leur est aussi inconnu que l'avenir" (ibid., 468).

133. "Quoique les nègres aient peu d'esprit, ils ne laissent pas d'avoir beaucoup de sentiment, ils sont gais ou mélancoliques, laborieux ou fainéants, amis ou ennemis, selon la manière dont on les traite; lorsqu'on les nourrit bien et qu'on ne les maltraite pas, ils sont contents, joyeux, prêts à tout faire, et la satisfaction de leur ame est peinte sur leur visage; mais quand on les traite mal, ils prennent le chagrin fort à coeur et périssent quelquefois de mélancolie: ils sont donc fort sensibles aux bienfaits et aux outrages, et ils portent une haine mortelle contre ceux qui les ont maltraités; lorsqu'au contraire ils s'affectionnent à un maître, il n'y a rien qu'ils ne fussent capables de faire pour lui marquer leur zèle et leur dévouement. Ils sont naturellement compatissants et même tendres pour leurs enfants, pour leurs amis, pour leurs compatriotes" (ibid., 468–69).

134. "je ne puis écrire leur histoire sans m'attendrir sur leur situation, ne sont-ils pas assez malheureux d'être réduits à la servitude, d'être obligés de toujours travailler sans pouvoir jamais rien acquérir ? faut-il encore les excéder; les frapper, et les traiter comme des animaux?" (ibid., 469).

135. "Comment des hommes à qui il reste quelque sentiment d'humanité peuvent-ils adopter ces maximes, en faire un préjugé, et chercher à légitimer par ces raisons les excès que la soif de l'or leur fait commettre?" (ibid., 470).

136. Ibid.

137. Ibid.

CHAPTER THREE: The Problem of Difference

1. See Holbach, *Système de la nature*, 1: 44–45: "Le raisonneur subtil qu'on nomme théologien dans les nations civilisées, et qui, en vertu de sa science inintelligible, se croit en droit de se moquer du sauvage, du Lapon, du nègre, de l'idolâtre, ne voit pas qu'il est lui-même à genoux devant un être qui n'existe que dans son propre cerveau." (That subtle philosopher we call a theologian in civilized nations, and who, by virtue of his unintelligible science, thinks he has the right to mock the savage, the Laplander, the *nègre*, [and] the idolater, does not see that he himself is on his knees before a being that exists only in his head.)

2. See Helvétius, *De l'homme*, 1: 218.

3. See Le Maire, *Voyages*, 139 : "Comme l'ambition est une passion inconnue à ces peuples, ils ne se sont pas mis en peine de bâtir des villes, des châteaux, et des maisons de plaisance." (As ambition is an unknown passion to these peoples, they have not bothered to build cities, palaces, or country homes.)

4. See chap. 1, p. 54.

5. "Pour ce qui est de l'esprit; tous les nègres de Guinée l'ont extrêmement borné; plusieurs mêmes paraissent stupides, et comme hebêtés. On en voit, qui ne peuvent jamais compter au-delà de trois, ni apprendre l'Oraison Dominicale. D'eux-mêmes ils ne pensent à rien, et le passé leur est aussi inconnu que l'avenir: ce sont des machines,

dont il faut remonter les ressorts à chaque fois, qu'on les veut mettre en mouvement" (Charlevoix, *Histoire de l'isle Espagnole*, 2: 499).

6. See chap. 2, p. 111.

7. "Ces hommes stupides . . . ne pensent guère à la culture de leurs terres. Contents de vivre au jour la journée sous un ciel qui donne peu de besoins, il ne cultivent que ce qu'il leur faut pour ne pas mourir de faim" (Poivre, *Voyage*, 5–6).

8. "L'Africain paraît être une machine qui se monte et se démonte par ressorts, semblable à une cire molle, à qui l'on fait prendre telle figure que l'on veut" (Demanet, *Nouvelle histoire de l'Afrique française*, 1: 1). Demanet's view is particularly significant, because this was an era in which Africa was beginning to be seen, not simply as a continent to be exploited, but as "the key to commerce" (see Røge, "'La clef de commerce'").

9. "Les Africains vivent pour la plupart au jour le jour. . . . Ils ne songent ni au passé ni à l'avenir; et comme leurs ancêtres ont été peu curieux de leur laisser des annales, ils ne sont pas plus disposés à en laisser à leur postérité" (Dubois-Fontanelle, *Anecdotes africaines*, 2–3).

10. Miller, *Blank Darkness*, 6.

11. Although similar criticisms existed much earlier, this expression—"une pieuse absurdité"—is actually found in Delisle de Sales's *Essai philosophique sur le corps humain*, 2: 143.

12. Paré, *Oeuvres complètes*, 210. For a useful discussion of humor theory, see Wheeler, *Complexion of Race*, 22.

13. See Klaus, "History," 5.

14. See Riolan, *Manuel anatomique*, 109: "La couleur [de la peau] dépend de l'humeur qui domine au corps." (The color of the skin depends on the dominant bodily humor.)

15. Riolan's methods generated a sizable number of imitators (and challenges) by the 1640s. Two Englishmen, in particular, replicated Riolan's experiment, but both explicitly refuted Riolan's climatic explanation of blackness, albeit in different fashions. The London anatomist Alexander Reed argued in 1642 that the black African's "skin color was derived from the body's inner humors." And the famous Anglo-Irish natural philosopher Robert Boyle likewise argued against "heat" theory in 1664, maintaining that the "principal cause of the blackness of *Negroes* is some peculiar seminal impression." See Klaus, "History," 6.

16. Marcello Malpighi (1628–94) was an Italian anatomist who is often considered the founder of histology. He is credited with the discovery of capillaries in the lungs, the branching tubes that carry air in insects' bodies, as well as the above-mentioned layer of skin. See Thomson, *Outline of Science*, 2: 286.

17. Roger, *Sciences de la vie*, 215. The three main layers of the skin are the outside layer, or scarf (also called the cuticle and epidermis); the Malpighian layer, or *rete mucosum*, where pigmentation is found; and the last cutaneous layer, or "true" skin (corium, cutis, or derma).

18. See Klaus, "History," 7, and Meijer, *Race and Aesthetics*, 70–71.

19. Alexis Littré is sometimes refered to as "Littre."

20. *Histoire de l'Académie royale des sciences*, 1720: 31.

21. Meijer, *Race and Aesthetics*, 72.

22. Jordanova, "Sex and Gender," in *Inventing Human Science*, ed. Fox et al., 162.

23. Barrère, *Dissertation*.

24. Klaus, "History," 7.

25. "si l'on prétend que c'est le sang ou la bile qui, par leur noirceur, donnent cette couleur à la peau, alors au lieu de demander pourquoi les nègres ont la peau noire, on demandera pourquoi ils ont la bile ou le sang noir; ce n'est donc qu'éloigner la question, au lieu de la résoudre" (*HN*, 3: 525).

26. Meckel's findings appeared as well in the 1768 *Mémoires de l'Académie royale des sciences de Paris*, 1: 414–38; 2: 288–95.

27. "[L]a couleur de la substance médullaire du cerveau . . . différait un peu des autres cerveaux; car cette couleur n'était pas blanche, comme on la trouve ordinairement dans des cerveaux aussi frais, mais elle était bleuâtre; et aussitôt qu'une partie détachée du cerveaux était exposée à l'air, elle devenait sur le champ tout à fait blanche" (Meckel, "Recherches anatomiques," 99). According to Gossiaux, "Anthropologie," Meckel actually put forward a polygenist thesis to the Berlin Academy in 1757.

28. "Elle n'était pas, comme on la trouve ordinairement, d'une couleur cendrée, mais d'un bleu noirâtre, et de sa base sortaient deux péduncules tout à fait blancs. . . . En les disséquant ils se trouvaient contenir de la substance médullaire disposée par raies entre la substance corticale, et qui était bleuâtre, ou noirâtre" (Meckel, "Recherches anatomiques," 100).

29. Formey, *Nouvelle bibliothèque germanique* 16 (January–March 1755): 252 : "une différence caractéristique entre [le cerveau des nègres] et celui des blancs."

30. Le Cat, *Traité de la couleur*.

31. Ibid., 57. See also Le Cat, *Traité des sens*, 62: "it is thus in the nervous system, and in its components, that we should look for the source of the color that tints the skin of animals, and in particular that of the *ethiops*, which gives the *nègre* his color [c'est donc dans le système nerveux, et dans ses appartenances qu'il faut chercher la fabrique des couleurs qui teignent la peau des animaux, et en particulier celle de l'oethiops, qui donne la couleur au nègre]." Le Cat was known for his assertion that communication between muscles and the brain was affected by a nervous liquid.

32. "Enfin voilà cette ancienne opinion de Strabon, que la couleur des hommes est dans la semence de leurs parents; la voilà, dis-je, établie par l'observation, car personne ne doute que le cerveau ne soit une partie spermatique, et comme l'amande féconde qui produit tout le reste de l'animal" (Le Cat, *Traité de la couleur*, 58). The *Encyclopédie* makes clear that the term *spermatique* exists in contrast to *sanguine* and that antiquity divided the body in either spermatic formations or sanguine formations. "The ancients divided the parts of the animal body into categories of spermatic and sanguine. The spermatic parts are those that, by their color, etc., resemble to some degree the semen, and that we imagine are composed of it; these include the nerves, the membranes, the bone, etc. The sanguine parts are those that we imagine to have been formed of blood after conception." (Les anciens divisaient en général les parties du corps animal en spermatiques et sanguines. Les parties spermatiques sont celles qui par leur couleur, etc. ont quelque ressemblance avec la semence, et qu'on supposait en être formées; tels sont les nerfs, les membranes, les os, etc., les parties sanguines qu'on supposait être formées du sang après la conception.) *ENC*, 15: 540.

33. E.g., the article "Sèche" in Henri-Gabriel Duchesne and Pierre-Joseph Macquer, *Manuel du naturaliste, ouvrage dédié à Monsieur de Buffon* (Paris: G. Desprez, 1771), 485–86. The philosopher and amateur naturalist J.-B. Delisle de Sales actually reversed Le Cat's theory, suggesting that an essential essence "tainted" the brain, changed the color of the blood and had a determining effect on the "spermatic liquid," all of which was presum-

ably linked to the African's inferior intelligence. See Delisles de Sales, *De la philosophie de la nature*, 4: 183–84.

34. De Pauw was doing considerably more than "looking back" to Buffon. See Garrett, "Human Nature," in *Cambridge History of Eighteenth-Century Philosophy*, ed. Haakonssen, 1: 189.

35. "l'entralas des nerfs optiques [était] brunâtre"; "le sang est d'un rouge beaucoup plus foncé que le nôtre" (De Pauw, *Recherches philosophiques sur les Américains*, 1: 179).

36. "colorée par le même principe qu'on trouve répandu dans leur membrane muqueuse" (ibid.). De Pauw was citing Meckel's *Recherches anatomiques*, and though he does not provide adequate citation information, it is probable he was using the edition published in *Histoire de l'Académie royale des sciences et des belles lettres de Berlin* for all references in his book.

37. See ibid., 1: 180. Buffon seems to have sensed the challenge that this represented to his system when he wrote about this supposed transmutation from black to white, and he chastized de Pauw for asserting this idea without "sans citer ses garants" (without citing his sources). *HN, Suppl.* (1777), 4: 504.

38. "If the savage life, if the lack of agriculture and literacy proved incontestably the newness of a people, the Laplanders and the *Nègres* would be the most modern of men [Si la vie sauvage, si le défaut d'agriculture et d'alphabet prouvaient incontestablement la nouveauté d'un peuple, les Lapons et les Nègres seraient les plus modernes des hommes]" (De Pauw, *Recherches philosophiques*, 1: 30); "s'ils demeurent continuellement sous la ligne, exposés à la plus grande chaleur qu'aucun point de la terre éprouve" (ibid., 99).

39. The significance of this stability was not lost on Buffon, who finally confronted the implications of the "new" physiology in the *Supplément* to his *Histoire naturelle*. Sensing, perhaps, that his climate theory was under siege by thinkers including de Pauw, Buffon began this section by defiantly stating that his initial hypothesis thirty years before remained entirely valid: "All that I have said about the color of *nègres* seems to me to be the greatest truth [Tout ce que j'ai dit sur la couleur des nègres me paraît de la plus grande vérité]" (Buffon, "Sur la couleur des nègres," *HN, Suppl.* [1777], 4: 502).

40. "Je ne doute pas nullement qu'il ne fallût aux nègres transmigrés dans les provinces de l'Europe septentrionale, un temps plus long pour perdre leur noirceur qu'il n'en faudrait à des Européens établis au coeur de l'Ethiopie, pour devenir nègres; parce que la liqueur spermatique et la substance moelleuse et glanduleuse des Africains, étant une fois colorée et impregnée de cette *matière âcre* qu'on nomme *AEthiopes animal*, conserverait très longtemps ce principe de père en fils et ne s'effacerait que par une suite très nombreuses de générations" (ibid., 1: 188).

41. See "Philosophical Enquiries concerning the Americans by M. de Pauw."

42. Forster, *Observations*, 256–84.

43. "la faiblesse de leur intelligence" (Delisle de Sales, *Essai philosophique sur le corps humain*, 2: 154).

44. "faisceau des fibres sensibles" (ibid., 156).

45. "inertie de l'esprit qui diffère . . . peu de la stupidité" (ibid.).

46. "la sécheresse et l'aridité naturelles des fibres" (Dazille, *Observations*, 4: 32).

47. "la plus malheureuse et la plus négligée, malgré [son] utilité" (ibid., iv).

48. "une nourriture insuffisante, le défaut de vêtements et un travail au-dessus de leurs forces." "[faisant] périr le produire annuel de la génération des nègres" (ibid., 22–23).

49. Cohen, *French Encounter*, 87.

50. J.-P. Rousselot de Surgy followed up on a particularly virulent [pro-slavery] and polygenist portrait of the African by musing: "One is tempted to believe . . . that the *nègres* constitute a race of creatures that reflect a certain stage in the development of nature's species, from Orangutan, to Gorilla, to man [On serait tenté de croire . . . que les nègres forment une race de créatures qui est la gradation par laquelle la nature semble monter, des Orang-Outangs, des Pongos, à l'homme]" (Rousselot de Surgy, *Mélanges intéressans*, 10: 166). Ten years later, Edward Long asserted similarly that the "humble" and "narrow" intelligence of the inhabitants of "Negro-land" could be attributed to the proximity of the "Guiney Negroe" to apes and the orangutan. Long, *History of Jamaica*, 2: 375.

51. "Si j'avais à soutenir le droit que nous avons eu de rendre les nègres esclaves, voici ce que je dirais : (1) Les peuples d'Europe ayant exterminé ceux de l'Amérique, ils ont dû mettre en esclavage ceux de l'Afrique, pour s'en servir à défricher tant de terres. (2) Le sucre serait trop cher, si l'on ne faisait travailler la plante qui le produit par des esclaves. (3) Ceux dont il s'agit sont noirs depuis les pieds jusqu'à la tête; et ils ont le nez si écrasé qu'il est presque impossible de les plaindre. (4) On ne peut se mettre dans l'esprit que Dieu, qui est un être très sage, ait mis une âme, surtout une âme bonne, dans un corps tout noir. (5) Il est si naturel de penser que c'est la couleur qui constitue l'essence de l'humanité, que les peuples d'Asie, qui font des eunuques, privent toujours les noirs du rapport qu'ils ont avec nous d'une façon plus marquée. (6) On peut juger de la couleur de la peau par celle des cheveux, qui, chez les Egyptiens, les meilleurs philosophes du monde, étaient d'une si grande conséquence, qu'ils faisaient mourir tous les hommes roux qui leur tombaient entre les mains. (7) Une preuve que les nègres n'ont pas le sens commun, c'est qu'ils font plus de cas d'un collier de verre que de l'or, qui, chez des nations policées, est d'une si grande conséquence. (8) Il est impossible que nous supposions que ces gens-là soient des hommes; parce que, si nous les supposions des hommes, on commencerait à croire que nous ne sommes pas nous-mêmes chrétiens. (9) De petits esprits exagèrent trop l'injustice que l'on fait aux Africains. Car, si elle était telle qu'ils le disent, ne serait-il pas venu dans la tête des princes d'Europe, qui font entre eux tant de conventions inutiles, d'en faire une générale en faveur de la miséricorde et de la pitié?" Montesquieu, *De l'esprit des lois* 15.5, 1: 265. Having numbered this list in the English translation, I have done likewise here in the French.

52. Ibid.

53. Ibid.

54. Voltaire, *Dictionnaire philosophique*, in *Oeuvres complètes*, 2: 604.

55. In fact, Montesquieu only cites Dampier's *Nouveau voyage* (1723) and Smith's *New Voyage to Guinea* (1744).

56. Montesquieu sometimes uses the term *barbare* to allude to black Africans, but within his overall breakdown of human types, they are classed as *sauvages*. See Montesquieu, *De l'esprit des lois* 18.11, 1: 308–9.

57. "Plusieurs choses gouvernent les hommes: le climat, la religion, les lois, les maximes du gouvernement, les exemples des choses passées, les moeurs, les manières; d'où il se forme un esprit général qui en résulte. . . . *La nature* et *le climat* dominent presque seuls sur les sauvages" (ibid. 19.4, 1: 329; emphasis added).

58. Arbuthnot, *Essay*, 153.

59. "L'air froid a resserré les extrémités des fibres extérieures de notre corps; cela aug-

mente leur ressort, et favorise le retour du sang des extrémités vers le coeur. Il diminue la longueur de ces mêmes fibres; il augmente donc encore par là leur force. L'air chaud, au contraire, relâche les extrémités des fibres, et les allonge; il diminue donc leur force et leur ressort. On a donc plus de vigueur dans les climats froids" (Montesquieu, *De l'esprit des lois* 14.2, 1: 245).

60. Montesquieu had already stated quite clearly in the *Lettres persanes* that human groups could degenerate due to a poor environment. Note the botanical metaphors: "Men are like plants, who never thrive unless they are well cultivated: among people who live in misery, the species decays and sometimes even degenerates [Les hommes sont comme les plantes, qui ne croissent jamais heureusement si elles ne sont bien cultivées: chez les peuples misérables, l'espèce perd et même quelquefois dégénère]" (*Lettres persanes*, 259).

61. See Montesquieu, *De l'esprit des lois* 14.2, 1: 246.

62. "On appelle mamelons pyramidaux les extrémités de tous les nerfs de la peau [The ends of all the nerves of the skin are called pyramidal nipples]" (*ENC*, 13: 594). A *houppe* is described as a "small *mamelon*" (ibid., 8: 326).

63. "imagination, le goût, la sensibilité, la vivacité" (Montesquieu, *De l'esprit des lois* 14.2, 1: 247).

64. "La chaleur du climat peut être si excessive que le corps y sera absolument sans force. Pour lors l'abattement passera à l'esprit même; aucune curiosité, aucune noble entreprise, aucun sentiment généreux; les inclinations y seront toutes passives; la paresse y fera le bonheur" (ibid., 1: 248).

65. "la plupart des châtiments [pour les habitants des pays chauds] y seront moins difficiles à soutenir que l'action de l'âme, et la servitude moins insupportable que la force d'esprit qui est nécessaire pour se conduire soi-même" (ibid., 1: 248).

66. Ibid., 15.7, 1: 267.

67. "Aristotle veut prouver qu'il y a des esclaves par nature, et ce qu'il dit ne le prouve guère. Je crois que, s'il y en a de tels, ce sont ceux dont je viens de parler. Mais, comme tous les hommes naissent égaux, il faut dire que l'esclavage est contre la nature, quoique dans certains pays il soit fondé sur une raison naturelle; et il faut bien distinguer ces pays d'avec ceux où les raisons naturelles mêmes le rejettent, comme les pays d'Europe où il a été si heureusement aboli" (ibid.).

68. Ibid.

69. "choque . . . moins la raison" (ibid.).

70. See Peabody, *"There Are No Slaves in France,"* 67. Ehrard, *Lumières et esclavage*, 159–60, also devotes a significant amount of time explaining the contradictions in Montesquieu's text, particularly defending the author's oscillations between contradictory theories.

71. Sala-Molins, *Misères des Lumières*, 67.

72. "C'est dans le cadre de [la] refléxion sur les fondements juridiques et idéologiques de l'esclavage et de la traite négrière en particulier qu'il convient de souligner l'étroite subordination de la théorie des climats dans *De l'esprit des lois* de Montesquieu à la légitimation de la servitude naturelle. Il faut également insister sur les glissements sémantiques qui font de l'ordre naturel pour les uns la nécessité d'être respecté comme des personnes car ils sont inscrits au préalable dans une visibilité anthropologique et politique, tandis qu'elle autorise pour les autres le claquement du fouet." Estéve, "Théorie des climats," in *Déraison, esclavage, et droit*, ed. Castro Henriques and Sala-Molins, 59.

73. "Il faut . . . borner la servitude naturelle à de certains pays particuliers de la terre." Montesquieu, *De l'esprit des lois* 15.8, 1: 267. In fact, when Montesquieu paused to consider what he was proposing, he perhaps consulted his *coeur,* as he put it, and asserted that there was perhaps no country on the planet where men could not be coerced to work without enslaving them. This qualification brings about one of the key moments in Montesquieu's discussion of slavery, a moment where he confronted the realities of the two operative epistemologies in his argument. After enumerating the progress that has been made in the area of forced labor—e.g., using free workers in Hungarian mines, using new machines to replace slave labor—Montesquieu further hesitated about the need for slaves, stating "there is maybe no climate on Earth where one cannot put free men to work. Because the laws were badly made, we found lazy men: because these men were lazy, we enslaved them" (il n'y a peut-être pas de climat sur la terre où l'on ne pût engager au travail des hommes libres. Parce que les lois étaient mal faites, on a trouvé des hommes paresseux: parce que ces hommes étaient paresseux, on les a mis dans l'esclavage). *De l'esprit des lois* 15.7, 1: 267. In this new view of slavery, which eschews the physiological and the ethnographic, civil society and its constraints are accused of being the source of indolence, and thus the explanation for slavery.

74. It might also be argued that Montesquieu acted, for all intents and purposes, as more of a reformer than an abolitionist. If, according to the author of *De l'esprit des lois,* slavery is the natural outgrowth of a particular (unfortunate) situation, then those people who oversee slave societies must improve or reform their behavior. This position culminates in the conceptually ambiguous passage in which Montesquieu establishes the *correct* conditions for taking a slave's life, *De l'esprit des lois* 15.7, 1: 275.

75. "A Achim tout le monde cherche à se vendre. Quelques-uns des principaux seigneurs n'ont pas moins de mille esclaves, qui sont des principaux marchands, qui ont aussi beaucoup d'esclaves sous eux, et ceux-ci beaucoup d'autres; on en hérite, et on les fait trafiquer. Dans ces états, les hommes libres, trop faibles contre le gouvernement, cherchent à devenir les esclaves de ceux qui tyrannisent le gouvernement" (ibid., 1: 266).

76. "L'auteur de *l'Esprit des lois* ajoute que, suivant le récit de Guillaume Dampier, 'tout le monde cherche à se vendre dans le royaume d'Achem.' Ce serait là un étrange commerce. Je n'ai rien vu dans le *Voyage de Dampier* qui approche d'une pareille idée. C'est dommage qu'un homme qui avait tant d'esprit ait hasardé tant de choses, et cité faux tant de fois." Voltaire, *Dictionnaire philosophique,* in *Oeuvres complètes,* 19: 604.

77. See the lengthy discussion of biblical monogenesis in chap. 2 above. Noah steps off the ark with his three sons, Ham, Shem, and Japheth, and "of them was the whole earth overspread" (Gen. 9:19 [KJV]).

78. Most famously, Voltaire derides the story of Noah and the Ark as a string of preposterous "miracles" in the *Dictionnaire philosophique* (*Oeuvres complètes,* 19: 475).

79. "Tous ces différents hommes, me dit-il, que vous voyez sont tous nés d'un même père; et de là il me conte une longue histoire" (Voltaire, *Traité de métaphysique,* in *Oeuvres complètes,* 14: 192).

80. "Je m'informe si un nègre et une négresse, à la laine noire et au nez épaté, font quelquefois des enfants blancs, portant cheveux blonds, et ayant un nez aquilin et des yeux bleus; si des nations sans barbe sont sorties des peuples barbus, et si les blancs et les blanches n'ont jamais produit des peuples jaunes" (ibid.).

81. "On me répond que non; que les nègres transplantés, par exemple en Allemagne, ne font que des nègres" (ibid., 192).

82. "Comme deux graines venues sur la même plante donnent un fruit dont les qualités sont différentes, quand ces graines sont semées en des terroirs différents, ou bien quand elles sont semées dans le même terroir en des années différentes: ainsi deux enfants qui seront nés avec leurs cerveaux composés précisément de la même manière, deviendront deux hommes différents pour l'esprit et pour les inclinations, si l'un de ces enfants est élevé en Suède et l'autre en Andalousie" (Dubos, *Réflexions*, 1: 570).

83. "perdrait enfin la couleur naturelle aux nègres" (ibid.).

84. "jamais homme un peu instruit n'a avancé que les espèces non mélangées dégénérassent, et qu'il n'y a guère que l'abbé Dubos qui ait dit cette sottise" (Voltaire, *Traité de métaphysique*, in *Oeuvres complètes*, 14: 192).

85. "Je vois des singes, des éléphants, des nègres, qui semblent tous avoir quelque lueur d'une raison imparfaite. Les uns et les autres ont un langage que je n'entends point, et toutes leurs actions paraissent se rapporter également à une certaine fin. Si je jugeais des choses par le premier effet qu'elles font sur moi, j'aurais du penchant à croire d'abord que de tous ces êtres c'est l'éléphant qui est l'animal raisonnable" (ibid., 191).

86. Ibid.

87. "L'homme est un animal noir qui a de la laine sur la tête, marchant sur deux pattes, presque aussi adroit qu'un singe, moins fort que les autres animaux de sa taille, ayant un peu plus d'idées qu'eux, et plus de facilité pour les exprimer; sujet d'ailleurs à toutes les mêmes nécessités; naissant, vivant, et mourant tout comme eux" (ibid., 191–92).

88. "Je suis donc forcé de changer ma définition et de ranger la nature humaine sous deux espèces la jaune avec des crins, et la noire avec de la laine" (ibid., 192).

89. "Mais à Batavia, Goa, et Surate, qui sont les rendez-vous de toutes les nations, je vois une grande multitude d'Européens, qui sont blancs et qui n'ont ni crins ni laine, mais des cheveux blonds fort déliés avec de la barbe au menton. On m'y montre aussi beaucoup d'Américains qui n'ont point de barbe: voilà ma définition et mes espèces d'hommes bien augmentées" (ibid.).

90. "Il me semble alors que je suis assez bien fondé à croire qu'il en est des hommes comme des arbres; que les poiriers, les sapins, les chênes et les abricotiers, ne viennent point d'un même arbre, et que les blancs barbus, les nègres portant laine, les jaunes portant crins, et les hommes sans barbe, ne viennent pas du même homme" (ibid.).

91. Roger, *Sciences de la vie*, 733, notes that Voltaire confuses the terms *race* and *espèce*.

92. "petit animal blanc comme du lait"; "un muffle taillé comme celui des Lapons, ayant, comme les nègres, de la laine frisée sur la tête" (Voltaire, "Relation," in *Oeuvres* [1830], 38: 521).

93. "Cet animal s'appelle un *homme*, parce qu'il a le don de la parole, de la mémoire, un peu de ce qu'on appelle raison, et une espèce de visage" (Voltaire, *Traité de métaphysique*, in *Oeuvres complètes*, 14: 189–90). The italics are in the original text.

94. See Terrall, *Man Who Flattened the Earth*, 302–9, for an excellent summary of this debate. On Voltaire's relationship to Maupertuis, Diderot, and Needham, among others, see Roger, *Sciences de la vie*, 732–48.

95. See Abanime, "Voltaire as an Anthropologist."

96. "les Blancs, les Nègres, les Albinos, les Hottentots, les Lapons, les Chinois, les

Américains, soient des races entièrement différentes" (Voltaire, *Essai sur sur les moeurs,* 1: 6).

97. "Les Albinos sont, à la vérité, une nation très petite et très rare: ils habitent au milieu de l'Afrique; leur faiblesse ne leur permet guère de s'écarter des cavernes où ils demeurent, cependant les Nègres en attrapent quelquefois, et nous les achetons d'eux par curiosité" (ibid.).

98. "Prétendre que ce sont des Nègres nains, dont une espèce de lèpre a blanchi la peau, c'est comme si l'on disait que les noirs eux-mêmes sont des blancs que la lèpre a noircis. Un Albinos ne ressemble pas plus à un Nègre de Guinée qu'à un Anglais ou à un Espagnol" (ibid.).

99. But denying the assertion that these white humans were simply degenerate blacks was only the first front in this battle; Voltaire had also to deny any *link* between these unnaturally pale creatures and the white race as well. See ibid., 7.

100. "La laine qui couvre [la] tête [des albinos] et qui forme leurs sourcils est comme un coton blanc et fin: ils sont au-dessous des nègres pour la force du corps et de l'entendement, et la nature les a peut-être placés après les nègres et les Hottentots, au-dessus des singes, comme un des degrés qui descendent de l'homme à l'animal" (ibid., 2: 319).

101. "Y a-t-il eu en effet des espèces de satyres, c'est-à-dire des filles ont-elles pu être enceintes de la façon des singes, et enfanter des animaux métis, comme les juments font des mulets et des jumars?" (Voltaire, "Des singularités de la nature," in *Oeuvres complètes,* 27: 186).

102. "les Albinos sont en si petit nombre, si faibles, et si maltraités par les Nègres, qu'il est à craindre que cette espèce ne subsiste pas encore longtemps" (Voltaire, *Essai sur les moeurs,* 1: 7).

103. Although he professed admiration for the Roman poet, Voltaire referred to Lucretius's atoms as "absurdities" in the article "Anti-Lucrèce" in his *Dictionnaire philosophique.*

104. While Voltaire believed that other "species," including the *Maure blanc* and the Hottentot, were significantly inferior to the *nègre,* he nonetheless cast this "race" as the most telling foil to white Europeans.

105. "les noirs sont une race de blancs noircie par le climat" (Voltaire, "Des singularités de la nature, in *Oeuvres complètes,*" 27: 184).

106. "la cause évidente de [la] noirceur inhérente et spécifique [des *nègres*]" (ibid.).

107. "il y a dans chaque espèce d'hommes, comme dans les plantes, un principe qui les différencie" (Voltaire, *Essai sur les moeurs,* 2: 335).

108. "membrane muqueuse, ce réseau que la nature a étendu entre les muscles et la peau"; "la race des épagneuls l'est des lévriers" (ibid., 305).

109. "Quiconque voudra faire disséquer un nègre (j'entends après sa mort) trouvera cette membrane muqueuse noire comme de l'encre de la tête aux pieds" (Voltaire, *La défense de mon oncle,* in *Oeuvres complètes,* 5: 403).

110. "Leurs yeux ronds, leur nez épaté, leurs lèvres toujours grosses, leurs oreilles différemment figurées, la laine de leur tête, la mesure même de leur intelligence, mettent entre eux et les autres espèces d'hommes des différences prodigieuses" (Voltaire, *Essai sur les moeurs,* 1: 6).

111. "La forme de leurs yeux n'est point la nôtre. Leur laine noire ne ressemble point à nos cheveux; et on peut dire que si leur intelligence n'est pas d'une autre espèce que

notre entendement, elle est fort inférieure. Ils ne sont pas capables d'une grande attention; ils combinent peu, et ne paraissent faits ni pour les avantages ni pour les abus de notre philosophie. Ils sont originaires de cette partie de l'Afrique, comme les éléphants et les singes; . . . ils se croient nés en Guinée pour être vendus aux blancs et pour les servir" (ibid., 2: 306).

112. "Pour qu'une nation soit rassemblée en corps de peuple, qu'elle soit puissante, aguerrie, savante, il est certain qu'il faut un temps prodigieux" (ibid., Introduction, §3, 1: 9).

113. Locke, *Two Treatises*, 319.

114. "La plupart des Nègres, tous les Cafres, sont plongés dans la . . . stupidité, et y croupiront longtemps" (Voltaire, *Essai sur les moeurs*, 1: 10).

115. Charles de Brosses's *Du culte des dieux fétiches ou Parallèle de l'ancienne religion de l'Egypte avec la religion actuelle de Nigritie* (1760) also evaluated belief systems in terms of a development from primitive to more advanced stages of religion, but Voltaire's tone was much more mocking.

116. "Tous les peuples furent . . . pendant des siècles ce que sont aujourd'hui les habitants de plusieurs côtes méridionales de l'Afrique, ceux de plusieurs îles, et la moitié des Américains. Ces peuples n'ont nulle idée d'un dieu unique, ayant tout fait, présent en tous lieux, existant par lui-même dans l'éternité. On ne doit pas pourtant les nommer athées dans le sens ordinaire, car ils ne nient point l'Etre suprême; ils ne le connaissent pas; ils n'en ont nulle idée. Les Cafres prennent pour protecteur un insecte, les Nègres un serpent" (Voltaire, *Essai sur les moeurs*, 1: 13).

117. Ibid., chap. 141, 2: 306.

118. "La nature a subordonné à ce principe ces différents degrés de génie et ces caractères des nations qu'on voit si rarement changer. C'est par là que les nègres sont les esclaves des autres hommes. On les achète sur les côtes d'Afrique comme des bêtes" (ibid., chap. 152, 2: 379–80).

119. Roger, *Les Sciences de la vie*, 457.

120. Lanson, *Voltaire*, trans. Wagoner, 73.

121. Savary des Brûlons and Savary, *Dictionnaire universel*. Similar short paragraphs were lifted from "Le Savary" for the articles "Asie," "Amériques," and "Europe."

122. "mangent leurs pères, mères, frères, et soeurs aussitôt qu'ils sont morts" (*ENC*, 1: 490). Ansico was the region in which the cannibalistic Anziques supposedly lived; see chap. 1, p. 152.

123. "plus il faudra de témoins pour les faire croire" (*ENC*, 1: 490).

124. "Si toutefois le pays pouvait suffire à une si horrible anthropophagie, et que le préjugé de la nation fût qu'il y a beaucoup d'honneur à être mangé par son souverain, nous rencontrerions dans l'histoire des faits appuyés sur le préjugé, et assez extraordinaire pour donner quelque vraisemblance à celui dont il s'agît ici" (ibid.).

125. Ladvocat, *Dictionnaire géographique portatif*. For a discussion of how Diderot uses ethnography to advance a particular message in the *Supplément*, see Curran, "Logics of the Human."

126. Dapper actually uses the notion of civilization to qualify the Beninians: "These *nègres* were much more civilized than those of the other coast." ([Ces *nègres* étaient] beaucoup plus civilisés que les autres de cette côte.) See Dapper, *Description de l'Afrique*, in *Objets interdits*, 230.

127. "ces peuples ne rendent aucun culte à Dieu; ils prétendent que cet être étant

parfaitement bon de sa nature, n'a pas besoin de prières ou de sacrifices" (*ENC*, 2: 204). Dapper also asserts that the *nègres* recognize the existence of an omnipotent Creator, but they do not believe he needs to be served; see *Description de l'Afrique*, in *Objets interdits*, 234.

128. For a discussion of geographical vices as well as their implications, see Miller, *Blank Darkness*, 8.

129. *Encyclopédie* contributors drew from a variety of sources, thereby exacerbating the problem of inconsistently spelled, overlapping, and often baffling designations for the continent. Indeed, some articles designate African peoples according to the earliest geographical breakdown of the continent, using Leo Africanus's sixteenth-century division of Africa into Barbary, Nigritia, Guinea and the *pays des nègres*. Other entries abandon this framework, preferring more specific geographical markers (lower Guinea, Ethiopia, Lower Ethiopia, Abyssinia, and Cafferia, etc.). Some of the confusion resulting from the alphabetical arrangement of such proto-ethnographical information was cleared up in the *Supplément à l'Encyclopédie*, where an ethnographic typology of sorts was supplied as an annex to the longer article "Afrique."

130. Seventeenth-century writers had also long used the term *caffre*, but without the racialized inflection found in the later eighteenth century. Olfert Dapper, an important source for *Encyclopédie* contributors, had, for example, mistakenly used *caffre* or *cafre* interchangeably with "Hottentot," writing: "Les Cafres, qui portent le nom de leur pays, n'ont ni religion ni connaissance de Dieu, et vivent presque comme des bêtes [The *Caffres*, who bear the name of their nation, have neither religion nor an understanding of God, and live almost like animals]" (Dapper, *Description de l'Afrique*, in *Objects Interdits*, 123). For a brief history of the significance of this term, see Lestringant and Carile, "Preface," in Chenu de Laujardière, *Relation d'un voyage à la côte des Cafres*.

131. See Buffon: "It is necessary to divide the blacks into different races, and it seems to me that we can reduce them to these two principal races: that of the *Nègres* and that of the *Cafres* [Il est . . . nécessaire de diviser les noirs en différentes races, et il me semble qu'on peut les réduire à deux principales, celle des *Nègres* et celle des Cafres]" (*HN*, 3: 453). Hottentots, within this scheme, were described as a variety of light-skinned Caffres. It should also be pointed out, however, that if the term *caffre* was accepted by many to identify lighter-skinned peoples living in southern Africa, the obviously slippery criteria for such a designation produced much confusion in the era's travelogues and natural history works. The *Encyclopédie* was no exception. Jaucourt consistently applied the term *Caffre* to those people living on the southern tip and eastern coast of Africa—according to him, "Hottentots" were "*Cafres*" and "Natal" was the land of the "*Cafres*"—but other contributors did not. The article "Cafrerie" contradicts him, saying that the "habitants de cette contrée sont *nègres* et idolâtres [inhabitants of this land are *nègres* and idolatrous]" (*ENC*, 2: 529; emphasis added).

132. "Depuis le tropique du cancer jusqu'à celui du capricorne l'Afrique n'a que des habitants noirs. Non seulement la couleur les distingue, mais ils diffèrent des autres hommes par tous les traits de leurs visages, des nez gros et plats, de grosses lèvres, et de la laine au lieu des cheveux" (*ENC*, 11: 76; Maupertuis, *Vénus physique*, 128).

133. Jaucourt also recapitulated some of the reigning theories on the skin pigmentation of the *nègre* in the article "Peau." Following (and restating) Buffon's monogenetic view of the human race, Jaucourt underlines the importance of environment for skin

pigmentation, refuting, along the way, the theory that African blood is blacker (Dr. Towns) or that it contains a darker bile (Barrère). *ENC,* 12: 215–20.

134. In contrast to epigenesis, preformationism is the belief in preformed seed germs.

135. "dans une carrière profonde, lorsque la veine de marbre blanc est épuisé, l'on ne trouve plus que des pierres de différentes couleurs" (*ENC,* 11: 77; Maupertuis, *Vénus physique,* 132). In *Vénus physique,* Maupertuis goes on to question this preformationist theory by providing a much more dynamic view of nature (and embryology), within which one could cross existing species and even create "espèces nouvelles."

136. To this somewhat essentialist ovism, Formey adds more pointed theories on African pigmentation by Barrère and Malpighi, the former of whom contended that the source of Africans' blackness came from the bile, while the latter maintained that it could be found in the skin itself, in the *corps muqueux*. Neither of these theories led Formey to advance a polygenesist explanation for different races. *ENC,* 11:78.

137. "d'un blanc livide comme les corps morts"; "au clair de la lune, comme des hibous" (ibid., 79).

138. Ibid.

139. "Ex Africa semper aliquid novi." See chap. 2, p. 88.

140. "peut-être que l'intérieur de l'Afrique, si peu connu des Européens, renferme des peuples nombreux d'une espèce entièrement ignorée de nous." *Espèce* could mean either type or species. *ENC,* 11: 79.

141. "on ne connait pas toutes les variétés et les bizarreries de la nature" (ibid.).

142. Diderot's article on the "Assiento" is a notable exception.

143. "trouvent le salut de leur âme dans la pertc de leur liberté"; "la culture des sucres, des tabacs, des indigos, etc." (*ENC,* 11: 80); Savary des Brûlons and Savary, *Dictionnaire universel,* 3: 552. See Peabody, *"There Are No Slaves in France,"* 60–61, for a discussion of the term *nègre* in eighteenth-century dictionaries.

144. According to the *Discours préliminaire,* Le Romain was an expert in the manufacture of sugar, which he communicated to his readers "as a philosophe and attentive observer" (*ENC,* 1: xliv).

145. This mid-century attitude to the slave trade has been characterized as one of *mauvaise conscience* by both Hoffmann, *Nègre romantique,* 50, and Roger Mercier, *Afrique noire,* who specifies the years 1735–69.

146. "ils sont nés vigoureux et accoutumés à une nourriture grossière"; "douceurs [aux Amériques] qui rendent la vie animale beaucoup meilleure que dans [l'Afrique]" (*ENC,* 11: 79).

147. "enclin[s] au libertinage, à la vengeance, au vol et au mensonge. Leur opiniatreté est telle qu'ils n'avouent jamais leurs fautes, quelque châtiment qu'on leur fasse subir" (ibid., 82).

148. "une espèce d'hommes extrêmement vicieuse, très rusée et d'un naturel paresseux [qui] pour s'exempter du travail, feignent des indispositions cachées, affectent des maux de tête, des coliques etc." (*ENC,* 15: 618).

149. Buffon's typology essentially restates Charlevoix's proto-ethnographic breakdowns by utility. See chap. 2, pp. 113–14.

150. "Les [esclaves] du cap Verd ou Sénégalais sont regardés comme les plus beaux de toute l'Afrique. . . . On les emploie dans les habitations au soin des chevaux et des

bestiaux, au jardinage et au service des habitations; Les Aradas, les Fonds, les Fouéda, et tous les nègres de la côte de Juda sont idolâtres . . . ces nègres sont estimés les meilleurs pour le travail des maisons. Les nègres Mines sont vigoureux et fort adroits pour apprendre des métiers. . . . La côte d'Angola, les royaumes de Loangue et de Congo fournissent abondamment de très beaux nègres. . . . Leurs inclinations pour les plaisirs les rendent peu propres aux occupations laborieuses, étant d'ailleurs paresseux, poltrons, et fort adonnés à la gourmandise. Les moins estimés de tous les nègres sont des Bambaras; leur mal propreté, ainsi que plusieurs grandes balaffres qu'ils se font transversalement sur les joues depuis le nez jusqu'aux oreilles, les rendent hideux. Ils sont paresseux, ivrognes, gourmands et grands voleurs" (*ENC*, 11: 81).

151. In the 1735 edition, these four varieties were not described in any detail, and the rigidity and extent of their difference was thus not outlined at all.

152. Slotkin, *Readings*, 178, emphasis his. Unlike other varieties, the description of the African is the only one that includes a description of the women, and is the only one with an additional description of the skin and lips.

153. Foucault, *Mots*, 143.

154. The short *Encyclopédie*'s own entry for "Race," for example, does not develop any taxonomical scheme; following past practice, race remained, for Jaucourt, a question of kinship or genealogy, "extraction, line, lineage; referring as much to ancestors as to descendants of the same family [extraction, lignée, lignage; ce qui se dit tant des ascendants que des descendants d'une même famille]" (*ENC*, 13: 740). This has been admirably summarized by Hudson, "From 'Nation' to 'Race.'"

155. "Nous ne voulons point ressembler à cette foule de Naturalistes [. . .] qui occupés sans cesse à diviser les productions de la Nature en genres et en espèces, ont consumé dans ce travail un temps qu'ils auraient beaucoup mieux employé à l'étude de ces productions même" (*ENC*, 1: xvj).

156. The article ends with a section dedicated to the body's gateway to the exterior world, the senses.

157. "Les Maures sont petits, maigres et de mauvaise mine, avec de l'esprit et de la finesse" (*ENC*, 8: 346; *HN*, 3: 456); "Les naturels du Mexique sont bien faits, dispos, bruns, et olivâtres" (*HN*, 3: 499; *ENC*, 8: 347). The exact wording in the *Histoire naturelle* is often different.

158. "il n'y a . . . eu originairement qu'une seule race d'hommes, qui s'étant multipliée et répandue sur la surface de la terre a donné à la longue toutes [s]es variétés" (*ENC*, 8: 348; *HN*, 3: 530).

159. These varieties are hardly inflexible categories; Buffon, for example, identifies the *Foules* as being an intermediary between *Maures* and *nègres*. See *HN*, 3: 456.

160. See Thompson, "Diderot." Proust, *Diderot*, 417, also emphasizes the optimistic elements of this discourse.

161. Hudson, "From 'Nation' to 'Race,'" 248.

162. "les plus blancs, les plus beaux, et les mieux proportionnés de la terre"; "le blanc paraît être la couleur primitive de la nature, que le climat, la nourriture et les moeurs altèrent, et font passer par le jaune et le brun, et conduisent au noir" (*ENC*, 8: 346–47).

163. "la dernière nuance des peuples basanés" (ibid., 8: 347).

164. *HN*, 3: 468; *ENC*, 8: 347.

165. This belief in a monogenesis was also reflected in the theological article "Adam,"

the "tige de tout le genre humain" (root of all mankind), by the orthodox Catholic abbé Edme-François Mallet; see *ENC,* 1: 125.

166. "n'ont pour ainsi dire que des idées d'un jour. Leurs lois n'ont d'autres principes que ceux d'une morale avortée, et d'autre consistance que celle que dans une habitude indolente et aveugle" (Didier Robert de Vaugondy, "Afrique," in *Encyclopédie méthodique par ordre de matières,* 1: 18). Significant portions of Vaugondy's text for the *Encyclopédie méthodique* had appeared earlier in 1776 in the *Suppléments* to Diderot's *Encyclopédie.* See Diderot et al., eds., *Supplément,* 5: 57.

167. The *Encyclopédie méthodique* had a much more considerable anti-slavery discourse than did the original *Encyclopédie.* Seeber, *Anti- Slavery Opinion in France,* 143, credits the *Encyclopédie méthodique* with "notable writers who gave particular prestige and impetus to the advancing cause of abolition."

168. "utile, instructif, [et] interessant" (Valmont de Bomare, *Dictionnaire raisonné universel* [1764], 1: xvii). Robbins, *Elephant Slaves,* 166.

169. "les nègres vendent aux Européens non seulement les esclaves nègres qu'ils ont pris en temps de guerre, mais encore leurs propres enfants. Souvent une mère négresse livre sa fille à un étranger pour une somme de *cauris*" (Valmont de Bomare, *Dictionnaire raisonné universel* [1764], 3: 576).

170. "plus vicieuse que celles des autres parties du Monde"; "[l]a perfidie, la cruauté, l'impudence, l'irréligion, la malpropreté et l'intempérence, semblent avoir étouffé chez eux tous les principes de la Loi naturelle et les remords de la conscience" (ibid. [1767–68], 4: 213).

171. "Leurs usages sont si extravagants et si déraisonnables, que leur conduite jointe à leur couleur, a fait douter pendant longtemps s'ils étaient véritablement des hommes issus du premier homme comme nous" (ibid., 214).

172. "L'on peut regarder les races des nègres comme des nations barbares et dégénérées ou avilies" (ibid., 214).

173. "Dans un Européen ou un *Blanc,* la lymphe est blanche, exceptée quand elle est mêlée de bile, car elle donne à la peau un teint jaune. Mais dans un *nègre,* dont la lymphe et la bile sont noires, la peau, selon quelques rapports, doit par une raison semblable être de la même couleur" (ibid. [1768], 6: 531).

174. "[L]es *nègres* varient entre eux par la nuance de leur teint, mais ils diffèrent encore des autres hommes par tous les traits de leur visage: des joues rondes, l'os de la pommette élevé, le front un peu bossu, le nez court, large, écrasé ou plat, de grosses lèvres, le lobe ou appendice de l'oreille petit, la laideur et l'irrégularité de la figure caractérisent leur extérieur. Les *Négresses* ont les reins écrasés et une croupe monstrueuse, ce qui donne à leur dos la forme d'une selle de cheval" (ibid. [1775], 6: 46).

175. "[La] couleur noire dans l'espèce humaine, est aussi accidentelle que le brun, le rouge, le jaune, l'olive et le basané. Nous devons regarder les Blancs comme la tige de tous les hommes" (ibid. [1791], 9: 223).

176. "L'on peut *jusqu'à un certain point* regarder les races des Nègres comme des nations barbares et dégénérées" (ibid., 214).

177. "une espèce de bétail" (ibid., 226).

178. See Tussac, *Cri des colons,* 160.

CHAPTER FOUR: The Natural History of Slavery, 1770–1802

1. "Histoire naturelle," *ENC,* 8: 226.

2. See Pluche's best-selling *Spectacle de la nature.*

3. David Brion Davis, *Problem of Slavery,* 455.

4. Jordan, *White Man's Burden,* 50, also maintains that economic justifications of slavery were bolstered by the comparative vulnerability of African society to "European aggressiveness and technology."

5. Some historians have attempted to temper such assertions by emphasizing the fact that perceived *differences* were perhaps less important than the relative "costs and availability of Black and White labor." See, e.g., Walvin *Questioning Slavery,* 16.

6. The number of books of natural history far exceeded the number of treatments of slavery during the first eighty years of the eighteenth century; while scholars have understandably given a great deal of attention to those works that focused on the question of human bondage—for example, John Wesley's *Thoughts on Slavery* (1774) or John Gabriel Stedman's *Narrative of a five years' expedition against the revolted Negroes of Surinam in Guiana on the wild coast of South America from the years 1772 to 1777* (1790)—such works are few and far between compared to the hundreds of natural history and proto-anthropology treatments of the black African that were published during the same era. Slavery was simply not a preoccupation in England and France until the 1770s and 1780s.

7. Long, *History of Jamaica,* 3: 375. In his "Of National Characters" (1748), David Hume had also affirmed that there was an original and *natural* distinction between whites and blacks (Hume, *Essays,* 208, n. 10).

8. "La perfectibilité n'est pas un don fait à l'homme en général, mais à la seule race blanche et barbue," Galiani wrote Madame d'Epinay in October 1776. Grimm and Diderot, *Correspondance littéraire,* 280.

9. "Tout ce qu'on dit des climats est une bêtise, un *no causâ,* l'erreur la plus commune de notre logique" (ibid.).

10. "Tout tient aux races" (ibid.). See also Gobineau, *Inequality of Human Races,* 50.

11. "une origine primitive" (Du Phanjas, *Théorie des êtres sensibles,* 1: 115).

12. "malheureuses victimes"; "de père en fils" (ibid.).

13. "couleur primitive"; "sensiblement" (ibid., 113).

14. See Eigen and Larrimore, eds., *German Invention of Race,* 51–52.

15. The various editions of Blumenbach's *De generis humani varietate nativa* can be found in his *Anthropological Treatises.*

16. Ibid., 98. Within the decontextualized race anthologies that generally "situate" Blumenbach, the naturalist often seems to grow organically out the polygenist thinkers he disdainfully rejected in his *De generis humani varietate nativa.* The implication is that the classificatory impulse associated with these "separate origins" thinkers overlaps with the classificatory impulse found in Blumenbach's monogenism. For the polygenists cited here, see La Peyrère, *Praeadamitae*; Hughes, *Natural History of Barbados*; Kames, *Sketches of the History of Man*; Tyssot de Patot, *Voyages et aventures de Jacques Massé*; Voltaire, *Essai sur les moeurs.*

17. Kidd, *Forging of the Races,* 85.

18. Blumenbach said he did this to refute thinkers who associated humankind with the orangutan. See *Generis humani varietate nativa,* pp. 88–89.

19. Ibid., 101.
20. Ibid., 105.
21. Ibid., 106.
22. Ibid., 310.
23. See Blumenbach's own qualification regarding the fact that this scheme would later grow to five races; the final category including the people of the "new southern world," "the Sunda, the Molucca, and the Philippine Islands" (Blumenbach, *Anthropological Treatises*, 100, n. 4).
24. See Bernasconi, "Kant and Blumenbach's Polyps," in *German Invention of Race*, ed. Eigen and Larrimore, 73–90.
25. None of these were polygenist schemes. All, very much like Kant, maintained that there was an original phylum, but also a significant hereditary transfer. In Kant's thought, this transfer involved "both 'germs' (*Keime*) and predispositions (*Anlagen*)." On this, see the editor's preface to Kant's *Of the Different Races of Human Beings*, in id., *Anthropology, History, and Education*, 83.
26. "Les différents degrés de chaleur dans les climats respectifs agissant constamment de génération en génération, communiquent enfin à ce principe huileux commun au sang, à la bile, au fiel, au sperme, au corps muqueux, une teinte proportionnée à la température locale. Pour mieux me faire entendre, je suppose qu'une homme blanc et une femme blanche, bien constitués, passent de notre zone tempérée dans une des contrées les plus chaudes de la zone torride; je suppose encore que ce couple se plie aux moeurs, aux usages des habitants indigènes, et ne prenne d'autre espèce de fièvre; leur épiderme se hâlera et se durcira; il se détachera par feuilles et par lambeaux. En effet, la chaleur excessive doit d'abord produire une plus grande agitation dans le sang et dans les autres humeurs, et l'éther animal du corps muqueux s'imbibera de plus de phlogistique" (Nauton, "Essai," in *Observations*, ed. Rozier and Mongez, 17: 178). Nauton numbered among those early-modern thinkers who believed that "phlogiston" existed in all combustible substances and was released during combustion.
27. "Les variétés et les nuances de la couleur du teint se font remarquer dans les principales liqueurs du corps humain" (ibid., 17: 179).
28. "une propriété, une affection de nature, qui se perpétuera de génération en génération, comme l'on voit les difformités ou les maladies des pères et des mères passer aux enfants" (ibid.). "On ne peut . . . disconvenir . . . que la couleur noire de la peau des nègres ne prenne son origine dans le cerveau [One cannot dispute the fact that the black color of *nègres*' skin originates in the brain]," Sigaud de la Fond substantiated in his *Dictionnaire de physique*, 3: 337.
29. "physicien-naturaliste"; "l'étude anatomique et physiologique de l'individu"; "l'espèce" (Nauton, "Essai," in *Observations*, ed. Rozier and Mongez, 17: 166).
30. "la différence du teint et de la couleur, celle de la forme et de la grandeur, [et] celle du tempérament, du naturel et du génie national" (ibid., 165).
31. "race gothico germanique"; "la race de l'Europe occidentale (Celtique)"; "la race nègre" (*Encyclopédie méthodique*, 5: 259).
32. Para du Phanjas, *Théorie des êtres sensibles*, 122.
33. See J. P. Berthout van Verchem, in *Mémoires de la Société des sciences physiques de Lausanne*, 1: 11–12: "Le climat et la nourriture sont les grandes et principales causes de ces divers changements; elles doivent influer sur plusieurs nations, et par conséquent former les différentes races d'hommes, ou *variétés générales*." (Climate and diet are the

primary causes of these diverse changes; they have an impact on various nations, and consequently, form the different races of man, or *general varieties.*)

34. "la couleur de la peau" (La Pérouse, *Voyage*, 1: 184).

35. "La liqueur spermatique des hommes plus ou moins basanés, la pulpe cérébrale et le sang, répondent . . . à la teinte de leur peau" (ibid.). See also Dunmore, *Where Fate Beckons*.

36. La Pérouse, *Voyage*, 1: 184.

37. It is hard to deduce exactly what the members of the *Société* may have thought of albinism. As had been the case since the 1730s, the albino was clearly a subject of fascination that many thinkers believed would provide clues about race, in this case, race's relationship to climate and heredity. By the time that the *Société* was providing guidelines for La Pérouse, many of these naturalists would have probably assumed that albinism was not specific to the newly designated *race nègre*, but a degenerative condition found in all races, albeit more frequently among indigenous peoples in tropical climates.

38. "le nègre n'est . . . pas seulement nègre à l'extérieur, mais encore dans toutes ses parties, et jusque dans celles qui sont les plus intérieures" (*Nouveau dictionnaire d'histoire naturelle*, 15: 433). The same sentence appears in the 1824 edition of Virey's *Histoire naturelle du genre humain*, 2: 38, but not in the first edition (1800).

39. The fundamental coherence of slavery and natural history is also plainly evident in how Labat chose to discuss the *nègre* in his *Nouveau voyage aux isles*. See chap. 1, p. 60.

40. "l'église et les rois" (Helvétius, *De l'esprit*, 252).

41. "maudit au nom de Dieu celui qui porte le trouble et la dissension dans les familles, [mais qui] bénit le négociant qui court la Côte d'Or ou le Sénégal?" (ibid.).

42. "La conversion des nègres entre pour très peu de choses dans les motifs qui portent les Européens de faire le commerce de la côte d'Afrique" (Boudet, *Journal oeconomique*, 135). This comment came in a review of an African travelogue that criticized the metaphysical justification of slavery.

43. "fait horreur à l'humanité" (De Pauw, *Recherches philosophiques*, 1: 43).

44. "nulles idées, nulles connaissances qui appartiennent à des hommes" (Rousselot de Surgy, *Mélanges intéressans*, 10: 165).

45. "fait[s] pour une condition supérieure à celle où ils sont réduits" (ibid., 164).

46. Despite the fact that the publication date of Mirabeau's *Ami des hommes* was 1756, it did not appear until 1757.

47. Mirabeau believed that "American colonies were blighted by a disease that threatened the liberty of Europe. . . . True progress required freedom of trade and a wider distribution of goods; yet colonization had led to a form of commercial exploitation that allowed the avarice and overconsumption of a small group to choke the sources of healthy growth. The principle of slavery had infected colonial commerce as well as colonial labor" (David Brion Davis, *Problem of Slavery*, 428).

48. The physiocrats regarded the right to liberty as one granted by nature, and thus independent of man-made laws. Social and civic institutions were meant only to honor and enforce the preeminent laws of nature, protecting citizens' liberty and property. Within this natural order, the physiocrats believed in agriculture as the foundation for the economic viability and growth of a nation. Slavery was thus seen as untenable for

two reasons: it denied the natural (economic) freedom of man and relegated that which should be the most honored tradition in society—agriculture—to a level that most men deem unworthy of their own labor. Some physiocrats used François Quesnay's 1758 *Tableau économique* to determine the high cost of slave labor in the Antilles, ultimately advocating fair compensation to free workers both in the colonies and in African nations with whom France could set up legitimate commercial relationships. See Røge, "La clef de commerce," 431–43.

49. "imbécile enfance" (Mirabeau, *Ami des hommes*, 3: 153).

50. "Nos esclaves de l'Amérique sont une race d'hommes à part, distincte et séparée de notre espèce par le trait le plus ineffaçable, je veux dire, la couleur" (ibid., 204).

51. "brutes ou doués d'un instinct qui nous est étranger" (ibid., 205).

52. "même les hommes les plus épais ont toujours assez de lumières pour sentir l'avantage de la liberté" (ibid.).

53. "les artisans d'Europe . . . prendront l'avance sur l'industrie des nègres, qui n'est jamais que d'exception parmi cette race d'hommes" (ibid., 206).

54. "consommation d'hommes" (Helvétius, *De l'esprit,* 3: 37).

55. "la cupidité et [le] pouvoir arbitraire [des] maîtres" (ibid.).

56. "il n'arrive point de barrique de sucre en Europe qui ne soit teint de sang humain" (ibid.).

57. "la nature humaine se trouve la même dans tous les hommes, il est clair que selon le droit naturel, chacun doit estimer et traiter les autres comme autant d'êtres qui lui sont naturellement égaux" (*ENC*, 5: 415).

58. Jaucourt asserts man's inborn freedom and equality, but quickly goes on to delineate the loss of that equality through man's desire for "les aisances de la vie . . . et des biens superflus [the luxuries of life . . . and superfluous goods]" (ibid., 934).

59. "L'esclavage fondé par la force, par la violence, et dans certains climats par excès de la servitude, ne peut se perpétuer dans l'univers que par les mêmes moyens [Slavery, established by force, violence and, in certain climates, by an excess of servitude, can only perpetuate itself in the universe by these same means]," Jaucourt writes (ibid., 398).

60. Wallace, *System.*

61. For an interesting side-by-side comparison of the two texts, including Wallace's landmark proclamation that "every one of those unfortunate men, who are pretended to be slaves, has a right to be declared to be free, for he never lost his liberty; he could not lose it; his prince had no power to dispose of him," see David Brion Davis, "New Sidelights," 586.

62. "viole la religion, la morale, les lois naturelles, et tous les droits de la nature humaine" (*ENC*, 16: 532).

63. Ibid.

64. "On dira peut-être [que ces colonies] seraient bientôt ruinées . . . si l'on y abolissait l'esclavage des nègres. Mais quand cela serait, faut-il conclure de-là que le genre humain doit être horriblement lésé, pour nous enrichir ou fournir à notre luxe? Il est vrai que les bourses des voleurs de grand chemin seraient vides, si le vol était absolument supprimé: mais les hommes ont-ils le droit de s'enrichir par des voies cruelles et criminelles? Quel droit a un brigand de dévaliser les passants? A qui est-il permis de devenir opulent, en rendant malheureux ses semblables? Peut-il être légitime de dépouiller

l'espèce humaine de ses droits les plus sacres, uniquement pour satisfaire son avarice, sa vanité, ou ses passions particulières? Non. . . . Que les colonies européennes soient donc plutôt détruites, que de faire tant de malheureux!" (ibid., 533).

65. "Que l'on mette les nègres en liberté, et dans peu de générations ce pays vaste et fertile comptera des habitants sans nombre. Les arts, les talents y fleuriront; et au lieu qu'il n'est presque peuplé que de sauvages et de bêtes féroces, il ne le sera bientôt que par des hommes industrieux" (ibid.). For the slight modifications made by Jaucourt, see David Brion Davis, "New Sidelights," 594.

66. "le premier à franchir le pas de l'antiesclavagisme à l'abolitionnisme" (Ehrard, *Lumières et Esclavage,* 180). See also Røge "Question of Slavery," in *L'economia come linguaggio,* ed. Albertone, 149–69.

67. *ENC,* 15: 618. Having reduced the African to a vicious "species" early in his article, Le Romain ultimately felt the need to argue for a more humane treatment of Africans, albeit for "practical" reasons. Implicitly castigating sugar plantation owners who tortured their slothful slaves unjustly, Le Romain was careful to inform his readership how to determine, without recourse to the whip, if naturally indolent Africans were feigning illness. See ibid., 619.

68. "ne connaissent ni pudeur ni retenue dans les plaisirs de l'amour" (*ENC,* 7: 1009). The inhabitants of Senegal live under a miserable and fraudulent monarch, Jaucourt writes, and "se volent réciproquement, et tâchent de se vendre les uns les autres aux Européens qui font commerce d'esclaves sur leurs côtes [steal one another and attempt to sell one another to the Europeans who run the slave trade on their coasts]" (ibid., 15: 13).

69. "beaucoup de sentiment" (*HN,* 15: 457; *ENC* 8: 347).

70. "Nous avons réduits [les nègres], je ne dis pas à la condition d'esclaves, mais à celle des bêtes de somme; et nous sommes raisonnables! Et nous sommes chrétiens!" (*ENC,* 8: 347).

71. Eighteenth-century literature that attempts to negotiate the complexities and moral unease of colonial empire and slavery includes portrayals of patrician African princes, miserable, suffering slaves, and violent *nègres* presented as justified in their revolt. See Hoffmann, *Nègre romantique,* and Cohen, *French Encounter.*

72. For a useful summary of the adaptations, see Little, "Oroonoko and Tamango."

73. "C'est à ce prix que vous mangez du sucre en Europe" (Voltaire, *Candide,* 72).

74. "singulier monument" (Mercier, *An 2440,* 127).

75. "Je sortais de cette place, lorsque vers la droite j'aperçus sur un magnifique piédestal un nègre, la tête nue, le bras tendu, l'oeil fier, l'attitude noble, imposante. Autour de lui étaient les débris de vingt septres. A ses pieds, on lisait ces mots: *Au vengeur du nouveau monde!"* (ibid.).

76. It should also be pointed out, however, that on the level of the representation of the black African himself, *L'An 2440* simply recasts older stereotypes or tropes. In particular, the leader of this revolt, much like the well-established Oroonoko figure, is portrayed as an anomaly within his comparatively inferior cohort; he is a superior being, the *nègre exceptionnel.* Furthermore, much like the African "type" in general, he is portrayed as closer to nature—in this case, to a violent, avenging nature.

77. "Mon séjour dans les Antilles et mes voyages en Afrique, m'ont confirmé dans une opinion que j'avais depuis longtemps. . . . Les négociants qui font la traite des

nègres, les colons qui les tiennent dans l'esclavage, ont de trop grands torts avec eux pour nous en parler vrai" (Saint-Lambert, "Ziméo," in id., *Contes*, 69).

78. "la première de nos injustices est de donner aux Africains un caractère général" (ibid.).

79. "Les gouvernements, les productions, les religions qui varient dans ces contrées immenses, ont nécessairement varié les caractères. Ici vous rencontrerez des Républicains qui ont la franchise, le courage, l'esprit de justice que donne la liberté. Là, vous verrez des nègres indépendants, qui vivent sans chefs et sans lois, aussi féroces et aussi sauvages que les Iroquois. Entrez dans l'intérieur des terres, ou même bornez-vous à parcourir les côtes, vous trouverez de grands Empires, le despotisme des princes et celui des prêtres, le gouvernement féodal, des monarchies réglées, etc. Vous verrez partout des lois, des opinions, des points d'honneurs différents; et par conséquent, vous trouverez des nègres humains, des nègres barbares; des peuples guerriers, des peuples pusillanimes; de belles moeurs, des moeurs détestables; l'homme de la nature, l'homme perverti, et nulle part l'homme perfectionné" (ibid.).

80. "[p]ortons-leur nos découvertes et nos lumières; dans quelques siècles ils y ajouteront peut-être, et le genre humain y aura gagné" (ibid., 73).

81. This split nonetheless foreshadows the more delineated breakdown of monogenist/anti-slavery versus polygenist/pro-slavery thinkers during the nineteenth century.

82. As Michèle Duchet notes, it was not until 1763 that France began to consider Africa as a potential commercial interest in its own right. See Duchet, *Anthropologie*, 46–48; Røge "La clef du commerce."

83. "les nègres sont des êtres *maltraités* de la nature, et non maudits de [la] justice [de Dieu]" (*HDI* [1770], 4: 119 ; emphasis added).

84. "des philosophes"; "des naturalistes célèbres"; "climat qu'ils habitent" (ibid.).

85. "Quelle que soit la cause primitive et radicale des variétés du coloris dans l'espèce humaine, on convient que la couleur du teint et de la peau, vient d'une substance gélatineuse qui se trouve entre l'épiderme et la peau. Cette substance est noirâtre dans les nègres, brune dans les peuples olivâtres ou basanées, blanche dans les Européens, parsemée de taches rougeâtres chez les peuples extrêmement blonds ou roux" (ibid., 120).

86. "L'anatomie a découvert dans les nègres, la substance du cerveau noirâtre, la glande pinéale comme toute noire, et le sang d'un rouge plus foncé que dans les blancs. Leur peau est toujours plus échauffée, et leur pouls plus vif. Aussi la crainte et l'amour sont-ils excessifs chez ce peuple; et c'est ce qui le rend plus efféminé, plus paresseux, plus faible, et malheureusement plus propre à l'esclavage. D'ailleurs ses facultés intellectuelles étant presque épuisées par les prodigalités de l'amour physique, il n'a ni mémoire ni intelligence, pour suppléer par la ruse à la force qui lui manque" (ibid., 121).

87. "Enfin, l'anatomie a trouvé l'origine de la noirceur des nègres dans les germes de la génération. Il n'en faut pas davantage, ce semble, pour prouver que les nègres sont une espèce particulière d'hommes car si quelque chose différencie les espèces, ou les classes dans chaque espèce, c'est assurément la différence des spermes" (ibid.).

88. This assertion also appears in the 1774 edition of *HDI*.

89. "climats qui ne sont propres qu'à certaines espèces"; "différence des climats"; "change la même espèce du blanc au noir"; "le soleil ne va point jusqu'à altérer et modifier les germes de la reproduction" (*HDI* [1770], 4: 121).

90. "sans fondement qu'on attribue au climat la couleur des nègres" (ibid.). To the degree that climate even factored into his overall assessment of the African species of man, Raynal stated only that to live in the torrid zone was to be less influenced by moral considerations.

91. Cohen, *French Encounter*, 85.

92. "Il meurt tous les ans en Amérique la septième partie des noirs qu'on y porte de Guinée. Quatorze cents mille malheureux qu'on voit aujourd'hui dans les colonies Européennes du nouveau monde sont les restes infortunés de neuf millions d'esclaves qu'elles ont reçu" (*HDI* [1770], 4: 160).

93. Raynal produced three discourses: "Une condamnation absolue de l'esclavage; de prudentes propositions de réformes en faveur des Noirs; des propositions de réformes en faveur de colons" (An absolute condemnation of slavery; prudent propositions of reforms in the favor of the blacks; propositions of reforms in favor of the colonizers), which "overlap," according to Ehrard, *Lumières et esclavage*, 56.

94. "beaucoup de douceur et d'humanité" (*HDI* [1770], 4: 161).

95. Michèle Duchet's analysis of these passages demonstrates how the source of this reformist discourse did not simply involve well-meaning philosophes taking aim against evil planters (although this is how this was often portrayed by the philosophes themselves), but was also the result of a new pragmatic reformist position that came from planters like Pierre-Victor Malouet. See Duchet's discussion of Bessner and Malouet in *Anthropologie*, 131–34.

96. "En accordant à ces malheureux la liberté, mais successivement, comme une récompense de leur économie, de leur conduite, de leur travail, ayez soin de les asservir à vos lois et à vos moeurs, de leur offrir vos superfluités. . . . Donnez-leur une patrie, des intérêts à combiner, des productions à faire naître, une consommation analogue à leurs goûts; et vos colonies ne manqueront pas de bras, qui soulagés de leurs chaînes, en seront plus actifs et plus robustes" (*HDI* [1770], 4: 173).

97. Curiously enough, this segment in the 1770, 4: 167–68, and 1774 editions of Raynal's *Histoire des deux Indes*, begins with an indictment of Montesquieu's ironic treatment of the justifications for slavery in *De l'esprit des lois*. In the more than two decades that had elapsed since Montesquieu provided his satirical rationalizations for the slave trade in 1748, the topic had become a matter of great solemnity, so much so that Raynal believed that even if Montesquieu had been on the right side of the question of slavery, he had sinned *esthetically* in treating the topic so lightly.

98. "Quiconque justifie un si odieux système, mérite du philosophe un silence plein de mépris, et du nègre un coup de poignard" (*HDI* [1770], 4: 167–68).

99. "Dira-t-on que celui qui veut me rendre esclave n'est point coupable, qu'il use de ses droits? Où sont-ils ses droits? qui leur a donné un caractère assez sacré pour faire taire les miens?" (ibid., 217); "Ne sentez-vous pas, malheureux apologistes de l'esclavage, que vous couvrez la terre d'assassins légitimes?" (ibid., 218). This idea came from Abbé Baudeau's refutation of Rousselot de Surgy in the *Ephémerides du citoyen,* October 10, 1766, 171.

100. "s'il existait une religion qui autorisât, qui tolérât, ne fut ce que par son silence, de pareilles horreurs; si d'ailleurs occupée de questions oiseuses ou séditieuses, elle ne tonnait pas sans cesse contre les auteurs ou les instruments de cette tyrannie; si elle faisait un crime à l'esclave de briser ses chaînes; si elle souffrait dans son sein le juge

inique qui condamne le fugitif à la mort; si cette religion existait, il faudrait en étouffer les ministres sous les débris de leurs autels" (*HDI* [1770], 4: 221).

101. "L'ouvrage de l'abbé Roubaud fut utilisé en 1780 pour effectuer un renversement d'optique de la discussion concernant les Africains, pour lutter contre le préjugé qui en faisait une race à part et pour réduire l'écart entre eux et les Européens" (Ann Thomson, "Diderot, Roubaud, l'esclavage," 76).

102. "Vers le huitième jour après leur naissance, les enfants [des Noirs] commencent à changer de couleur; leur peau brunit; enfin elle devient noire. Cependant la chair, les os, les viscères, toutes les parties intérieures, ont la même couleur chez les noirs et chez les blancs: la lymphe est également blanche et limpide; le lait des nourrices est partout le même" (*HDI* [1780], 6: 62).

103. This may also have to do with the historiography of the Enlightenment, which has sometimes tended to overreach when distinguishing between religious and antireligious spheres of thought.

104. Benezet's text was published in French as *Avertissement à la Grande Bretagne et à ses colonies, ou Tableau abrégé de l'état miserable des nègres esclaves dans les dominations anglaises* (n.p., 1767).

105. Benezet, *Caution and Warning,* 11.

106. On J.-B. Labat's adaptation of André de Brüe, see chap. 1, p. 62.

107. In his *Some Historical Account of Guinea,* Benezet also used the same basic strategy (and much of the same prose), citing first-hand accounts of Africans by region much like the geographical dictionaries of the day. Bypassing the numerous censorious comments contained in the works he was consulting, Benezet selectively cited a number of Africanist writers: William Smith in order to emphasize the good government of the "Jalofs"; Francis Moore as a means of praising the abstemious and morally upright "Fuli Blacks" (9); William Bosman to portray the industrious rice farmers and fishermen of Accra (19); and Michel Adanson, who suggested that the black African, although primitive, had the ability to become an excellent astronomer (14). The collective force of such ethnographic snapshots allowed Benezet to affirm that Africans were "a humane sociable people, whose faculties are as capable of improvement as those of other men (2). When coupled with legal refutation of slavery taken from Montesquieu, exposés on the cruelty of the slave plantations in Jamaica and Barbados, and the "shocking inhumanity" of the way the trade was conducted in Africa, Benezet's apologia functioned as an essential element within his larger negation of pro-slavery arguments.

108. For a discussion of the iconography of abolition, see Peggy Davis, "La réification de l'esclave noir," in *L'Afrique du siècle des Lumières,* ed. Gallouët et al., 237–53.

109. "Quoique je ne sois pas de la même couleur que vous, je vous ai toujours regardé comme mes frères. La nature vous a formés pour avoir le même esprit, la même raison, les mêmes vertus que les blancs. Je ne parle ici que ceux d'Europe, car pour les blancs des Colonies, je ne vous fais pas l'injure de les comparer avec vous" (Condorcet, *Réflexions,* iii–iv).

110. "Ce n'est ni au climat, ni au terrain, ni à la constitution physique, ni à l'esprit national qu'il faut attribuer la paresse de certains peuples; c'est aux mauvaises lois qui les gouvernent" (ibid., 22).

111. "peuple doux, industrieux, [et] sensible"; "paresseux, stupides et corrompus"; "le sort de tous les esclaves" (ibid., 28).

112. "une grande stupidité"; "ce n'est pas à eux que nous en faisons le reproche, c'est à leurs maîtres" (ibid, 35).

113. Wheeler, *Complexion of Race*, 256.

114. Clarkson, *Essay*, 133. His emphasis.

115. "leur manière de vivre dans leur patrie est si misérable, qu'ils sont très heureux qu'on les en tire" (Frossard, *Cause des esclaves nègres*, 1: 141).

116. Dorigny and Gainot, *Société des amis des noirs*, 157, quoting minutes of the Société's meeting on April 8, 1788.

117. "Je . . . suis sûr que, des larmes de joie couleront des . . . yeux des Nègres [après leur libération]. Non, non, ce ne sont pas des vengeances qu'ils méditeront quand ils verront tomber leurs fers. Donnons-leur du pain, qu'ils goûtent enfin le repos, qu'ils aient enfin la liberté d'embrasser leurs enfants, de jouir avec eux et leurs épouses des douceurs de la vie domestique et ils seront loin de s'occuper des vengeances. Au contraire ils nous regarderont, nous aimeront comme leurs libérateurs. Le Nègre est tendre mari, bon père, la vengeance n'habite pas dans une âme qu'animent ces sentiments, à qui on rend le droit d'en jouir" (ibid.).

118. In addition to Equiano, African writers who chronicled their experiences included Ukawsaw Gronniosaw, John Marrant, and Quobna Ottobah Cugoano. Substantial portions of these narratives are published with good introductions in Potkay and Burr, eds., *Black Atlantic Writers of the Eighteenth Century*.

119. Equiano, *Interesting Narrative*, xix, n.5.

120. Gates, *Figures in Black*, xxiv.

121. The number of members who came to meetings was far lower, however. See Dorigny and Gainot, *Société des amis des noirs*, 43–44.

122. Blackburn, *Overthrow of Colonial Slavery*, 176.

123. This vote in favor of rights for *des hommes de couleur libres* took place in April 1792.

124. *Abolitions de l'esclavage*, ed. Dorigny, 7–8. Dorigny also cautions against the (postcolonial) ahistoricism that equates this decision solely with economic interests, noting that this vote abolished slavery in all the French colonies, not simply in Saint-Domingue. See also Benot, *Révolution française*, 1–10, for a more pessimistic reading of this decision, and, more recently, Dobie, *Trading Places*, 14–15.

125. "le XVIIIe siècle n'a pas libéré les esclaves, mais à leur sujet il a libéré la pensée" (Ehrard, *Lumières et esclavage*, 214).

126. Sonthonax, who was a member of the Amis des noirs, was sent to Saint-Domingue in 1791. Initially charged with bringing order to the island, including inducing the slaves to return to their plantations, Sonthonax is best known for freeing the slaves of northern Saint-Domingue as a strategic move within his larger fight against the English, who were threatening to enter into an alliance with white colonialists in the fall of 1793.

127. As has been noted, the Amis des noirs were not the first to indict the colonial system for impairing the moral character and intellect of the black African, but this charge was gaining traction among the French public in the early 1790s, when legislation on the colonial project was a real possibility. Benjamin-Sigismond Frossard, a member of the Amis des noirs, advocated such legislation in his *La cause des esclaves nègres* (1789); moreover, a series of anti-slavery novels appeared during the same era, including Joseph Lavallée's *Le Nègre comme il y a peu de blancs* and Lecointe-Marsillac's,

Le More-Lack. In addition to advocating for an end to slavery, many of these novels put forward a positive understanding of a perfectible African.

128. Moreover, as Marguerite-Elie Guadet reported to the Assemblée nationale in February, 1792, "The accounts received of the disturbances in Saint-Domingue undoubtedly leave us in much uncertainty; . . . but the insurrection no sooner broke out than it was attributed to the *Amis des Noirs*" (Garran de Coulon and Guadet, *Inquiry*, 23).

129. Finkelman and Miller, eds., *Macmillan Encyclopedia of World Slavery*, 2: 535.

130. Engaging with this flight of the imagination directly, in a clear rewriting of the *singulier monument* of Mercier's *L'an 2440*, the author of *Le danger* proposed to erect a statue to any *négrophile* able to free the *nègres* without them seeking revenge on their white masters (7).

131. Looking back at the demise of French empire—Saint-Domingue in chaos, Martinique in the hands of the British (March 1794), Guadeloupe's freed slaves revolting against their former masters (December 1794)—François Barbé-Marbois lamented that the *négrophile* mentality was unable to grasp the "leçons des événements" (Barbé-Marbois, *Réflexions*, 1: 243).

132. This comes from the baron de Wimpffen. See Noël, *Etre noir en France au XVIIIe siècle*, 174–75.

133. "Les nègres se portent indifféremment à tous les actes de perfidie et de scélératesse. . . . Ces peuples, trop méchants pour goûter un établissement national, portent bien justement la peine de l'avoir négligé: ils ont perdu les sentiments de la nature; ils ont rétrogradé, parce qu'ils n'ont pas su avancer dans la civilisation. Leurs écarts et leurs bassesses les ont rendus le jouet des nations étrangères; et l'âme fausse, méchante et perfide de ces peuples, fuyant toute instruction salutaire, s'est jetée dans la fange de la crédulité la plus superstitieuse, en caressant des fétiches, des devins, et en s'environnant de sortilèges" (Mercier, *Fragments*, 1: 204).

134. Raimond, *Extrait d'une lettre*, 19. Like many documents of this era, Raimond's text is as much linked to self-presentation and contemporary politics as it was to absolute views of race. In this case, it was clear that Raimond was positioning himself as one of the free and powerful *hommes de couleur* of Saint-Domingue, the only group that had any hope of reestablishing order on the island. Raimond was later one of the collaborators in the drafting of a new constitution for Saint-Domingue in 1801.

135. See the assessment of this difficult period in Dorigny and Gainot, *Société des amis des noirs*, 301–27.

136. See Dubois, *Avengers of the New World*, 32.

137. "Comment a-t-on pu accorder la liberté à des Africains, à des hommes qui n'avaient aucune civilisation, qui ne savaient seulement pas ce que c'était que la France?" (Saint-Hilaire, *Napoléon*, 1: 112).

138. During the Terror, Moreau de Saint-Méry, a member of the club Massiac, was forced to flee to Philadelphia, where he wrote his *Description topographique . . . de l'Isle de Saint-Domingue*. This precious resource for scholars offers a casta painting–like typology of 128 racial mixtures, from the blackest black to the purist white, based on blood types. Couched in the neutral language of natural history, Saint-Méry's breakdown reveals, not only a belief in measurable categories of *négritude*, but a hidden fantasy of racial whitening, improvement, and, ultimately, the eradication of black populations though the introduction of white sperm. On Saint-Méry, see Garraway, *Libertine Colony*, 262, 268, and Burbank and Cooper, *Empires in World History*, 228.

139. See Dubois, *Avengers of the New World,* 254. In an exchange of correspondence between Toussaint and Napoleon in 1801, "Louverture had suggested to the consul that he [Napoleon] send emissaries back to the colony to discuss the terms of the constitution. Bonaparte, however, did not send a 'negotiator.' Instead, he 'sent an army'" (ibid.). Ultimately, some 77,000 soldiers had been sent to Saint-Domingue by 1803. See Régent, *La France et ses esclaves,* 265–66, for the military statistics related to this portion of the war.

140. Dubois, *Avengers of the New World,* 275.

141. Ibid., 293.

142. "Ceux qui veulent la destruction de nos colonies sentent bien qu'il n'est plus possible de plaider avec succès la cause de cette race barbare" (*Mercure de France,* no. 22 [April 1805]: 441).

143. See Jacques, "From Savages and Barbarians to Primitives"; "d'une nature dégradée". . . "vouée tantôt à une stupidité, tantôt au délire de l'imagination la plus extravagante" (Perreau, *Etudes,* 1: 157).

144. "anéantissement" (Gérando, *Des signes,* 2: 466).

145. "ne voit dans les productions de la nature que ce qu'elles peuvent avoir de relatif aux besoins de ses sens" (ibid., 15).

146. Lavater, *Essai sur la physiognomie.* The French translation of the German is considered a primary text because Lavater had a hand in the translation and because it contains enlarged sections.

147. In 1817, Georges Cuvier, Buffon's intellectual heir, painted a damning view of the *race nègre* that combined geography, esthetics, anatomy, and comparative anatomy, as well as a pessimistic assessment of the *nègre*'s future: "La race nègre est confinée au midi de l'Atlas; son teint est noir, ses cheveux crépus, son crane comprimé, et son nez écrasé; son museau saillant et ses grosses lèvres, la rapprochent manifestement des singes: les peuplades qui la composent sont toujours restées barbares [The *race nègre* borders on the southern Atlas [Mountains]; its complexion is black; its hair, nappy; its skull, compressed; its nose, flat; [and] its jutting mug and thick lips are manifestly like the apes': the peoples that constitute this race have always remained uncivilized]" (Cuvier, *Règne animal,* 1: 95). While the ideas contained in this ostensibly monogenist understanding of the *nègre* do not represent any real break with the past, the tone definitely does. Unlike earlier natural history texts, Cuvier did not separate his portrait of the *nègre* from the anti-black vitriol that had been confined in an earlier era to properly pro-slavery texts.

148. "le nègre diffère spécifiquement de toutes les autres races humaines." See Virey, *Histoire naturelle du genre humain,* 1: 202.

149. "le ciel doux et fertile de l'Asie paraît avoir été jadis le berceau du genre humain, comme celui des religions; mais l'espèce nègre et les races américaines ont sans doute pris naissance dans des contrées différentes" (ibid., 1: 228–29); see also "Nègre" in Deterville's *Nouveau dictionnaire d'histoire naturelle,* 15: 433.

150. "fonctions cérébrales" (Virey, "Nègre," in Deterville, *Nouveau dictionnaire d'histoire naturelle,* 15: 456).

151. "radicalement différent de l'espèce blanche" (ibid., 448).

152. "[se sont] répandues sur tout la surface de la terre" (La Métherie, *De la perfectibilité et de la dégénérescence,* 3: 359). This is an important text that synthesizes many of the seeds of polygenesis from the earlier century.

153. Kidd, *Forging of the Races*, 27.

154. "surpassent [les Européens] en qualités morales" (Cullion, *Examen*, 2: 177). See Bernardin de Saint-Pierre, *Etudes*, 1: 637.

155. "[Les négrophiles] nous accordent les avantages de l'esprit, pour s'autoriser à nous calomnier sur les qualités du coeur bien plus estimables. [Bernardin de Saint Pierre] fait plus, c'est aux nègres qu'il attribue la supériorité sur nous par les qualités du coeur. Par le raisonnement, il est absurde d'avancer que ce qu'on appelle abusivement bonté dans un être stupide soit véritablement de la bonté, ce n'est qu'une négation de méchanceté" (Cullion, *Examen*, 2: 177–78).

156. "il n'y a rien à attendre d'une race d'hommes aussi inférieure" (ibid., 295); "ne sont bons en effet que pour la servitude"; "l'esclavage est donc [leur] état naturel" (ibid., 296).

157. "Ce peuple est la lie, l'écume de la race humaine; qu'il soit donc au dernier rang; qu'il serve. La nature a donc prononcé sur son sort" (ibid., 290).

158. Grégoire, *De la littérature des Nègres*. Grégoire had thrown down the gauntlet by attacking the nexus of ideas being reused to justify the African's servitude in the Caribbean. In addition to refuting the increasingly dominant polygenist explanation of the different human races, Grégoire praised the African as having great potential and cited a number of prominent African writers to prove it. Although this book was allowed to be published, later abolitionist efforts by Grégoire were censured. See Sepinwall, *Abbé Grégoire and the French Revolution*.

159. To add to his argument, Grégoire also enumerated the "race's" high moral standards, heroism, and refinement in the arts.

160. In 1810, two years after Grégoire's *De la littérature des nègres* appeared, Richard Tussac published his *Cri des colons*, the most comprehensive anti-*négrophile* text produced during this era.

Coda

1. "Ce n'est pas assez pour nous de vivre avec nos contemporains, et de les dominer. Animés par la curiosité et par l'amour-propre, et cherchant par une avidité naturelle à embrasser à la fois le passé, le présent et l'avenir, nous désirons en même temps de vivre avec ceux qui nous suivront, et d'avoir vécu avec ceux qui nous ont précédé. De-là l'origine et l'étude de l'Histoire, qui nous unissant aux siècles passés par le spectacle de leurs vices et de leurs vertus, de leurs connaissances et de leurs erreurs, transmet les nôtres aux siècles futurs" (*ENC*, 1: ij).

2. See the interesting discussion of the link between past and future in Brewer, *Enlightenment Past*, 72–82.

3. See Curtin, *Atlantic Slave Trade*. Among French historians, see Daget, *Répertoire*.

4. See Daget and Renault, *Traites négrières*; Lovejoy, *Transformations*; Hair, *Atlantic Slave Trade* and *Africa Encountered*; Thornton, *Africa and Africans*; Law, *Ouidah* and *Slave Coast*.

5. See Rediker, *Slave Ship*; Harms, *The Diligent*; Smallwood, *Saltwater Slavery*.

6. See Hall, *Slavery and African Ethnicities*; Gilroy, *Black Atlantic*.

7. See Peabody and Grinberg, *Slavery, Freedom, and the Law*; Peabody, *"There Are No Slaves in France."*

8. In 1963, Fanon's *Les damnés de la terre* was published in an English translation by Constance Farrington under the title *The Wretched of the Earth*.

9. Benot, *Démence coloniale, Révolution francaise,* and *Lumières;* Duchet, *Anthropologie;* Roger Mercier, *Afrique noire;* Hoffmann, *Nègre romantique.* See also Cohen, *French Encounter.*

10. Kidd, *Forging of Race,* 79, summarizes this critique: "[T]here has been a generalized non-specific charge that the Enlightenment, the principal prop of modern Western intellectual life, was the achievement of several generations of periwigged white males who complacently assured the superiority of white European culture to the values of extra-European civilisations and gave at the very least implicit, sometimes very explicit, support to campaigns for overseas empire and colonialism." Kidd goes on to say that "[m]uch of this is, to a point, fair comment" (ibid.), but not all critics have been as open to an undermining of Enlightenment philosophy. Ehrard, *Lumières et esclavage,* 16, seeks to refute the accusation that the Enlightenment put up with slavery, an institution that was "si manifestment contraires à ses principes [so clearly contrary to its principles]."

11. This is the general thesis of Sala-Molins, *Misères des Lumières.*

12. "The individual is wholly devaluated in relation to the economic powers, which at the same time press the control of society over nature to hitherto unsuspected heights" (Adorno and Horkheimer, *Dialectic of Enlightenment,* xiv).

13. *Race, Writing, and Difference,* ed. Gates, 8.

14. Goldberg, *Racist Culture,* 29.

15. "Le pays n'est pas indifférent à la culture des hommes; ils ne sont tout ce qu'ils peuvent être que dans les climats tempérés" (Rousseau, *Emile,* in id., *Oeuvres complètes,* 4: 266–67).

16. "Il paraît encore que l'organisation du cerveau est moins parfaite aux deux extrêmes. Les nègres ni les lapons n'ont pas le sens des européens" (ibid.).

17. This monolithic view surfaced often in the era's thought; such was the case in the very popular *Universal History* (which was translated almost simultaneously into French). "[W]ith respect [to black Africans], one might reasonably expect to find, in such a vast extensive tract of land, and so great a variety of climates, nations, and governments, a proportionable diversity of inhabitants, in regard to their qualifications both of body and mind. . . . Our readers, therefore, will doubtless be much surprised to find, on the contrary, a general uniformity run through all those various regions and people; so that, if any difference be found between any of them, it is only in the degrees of the same qualities, and what is more strange still, those of the worst kind; it being a common known proverb, that all people of the globe have some good as well as ill qualities, except the *Africans*" (*Modern Part of an Universal History,* 14: 17; emphasis in original).

18. This is quite different from the properly stadial-minded theorists who put forward what Lévi-Strauss called the century's "false evolutionism," a theory according to which the African and other "savage" peoples were designated as "pre-historic," "static," "children," and "uncivilized." See Lévi-Strauss, *Race et histoire,* 23–24.

19. Sala-Molins, *Code noir,* 49.

20. Ibid., 54.

21. See Rapoport, *Origins of Violence,* which argues that "to be objective, one does not need to be morally neutral" (572).

Works Cited

Primary Sources

Adanson, Michel. *Histoire naturelle du Sénégal: Avec la relation abrégée d'un voyage fait en ce pays*. Paris: Claude-Jean-Baptiste Bauche, 1757.

Africanus, Joannes Leo. *The History and Description of Africa and the Notable Things Therein Contained, written by al-Hassan ibn Mohammed al-Wezaz, al-Fasi, a Moor, baptized as Giovanni Leone, but better known as Leo Africanus*. Translated by John Pory. 1600. Edited by Robert Brown. London: Hakluyt Society, 1896.

Annales typographiques ou notice du progrès des connoissances humaines. Paris: Vincent, 1761.

Arbuthnot, John. *An Essay Concerning the Effects of Air on Human Bodies*. London: J. Tonson, 1733.

Aristotle. *Generation of Animals*. London: William Heinemann, 1943.

———. *The Complete Works of Aristotle*. Edited by Jonathan Barnes. Princeton: Princeton University Press, 1994.

Augustine, Saint, bishop of Hippo. *City of God*. Translated by Henry Bettenson. New York: Penguin Classics, 2003.

Avity, Pierre d'. *Les Estats, empires, royaumes, et principautez du monde*. Paris: Pierre Chevalier, 1625.

Azurara, Gomes Eannes de. *The Chronicle of the Discovery and Conquest of Guinea*. Translated by Charles Raymond Beazley and Edgar Prestage. New York: Burt Franklin, 1963. See also under Zurara below.

Barbé-Marbois, François. *Réflexions sur la colonie de Saint-Domingue ou Examen approfondi des causes de sa ruine*. Paris: Garney, 1796.

Barbot, Jean. "A Description of the Coasts of North and South-Guinea, and of Ethiopia inferior, vulgarly Angola . . . And a new Relation of the Province of Guiana." In *A Collection of Voyages and Travels*. London: n.p., 1732.

Barrère, Pierre. *Dissertation sur la cause physique de la couleur des nègres, de la qualité de leurs cheveux, et de la dégénération de l'un et de l'autre*. Paris: P.-G. Simon, 1741.

———. *Observations anatomiques, tireés des ouvertures d'un grand nombre de cadavres*. Perpignan: J. B. Reynier, 1753.

Barros, João de, and Diogo de Couto. *Da Asia de João de Barros e de Diogo de Couto*. Lisbon: Livraria Sam Carlos, 1973.

Battel, Andrew. *The Strange Adventures of Andrew Battell of Leigh, in Angola and the Adjoining Regions*. Edited by E. G. Ravenstein. London: Hakluyt Society, 1901.

Baudry des Lozières, Louis-Narcisse, *Les égarements du nigrophilisme*, N.p., 1802.

Benezet, Anthony. *A Caution and Warning to Great Britain and her Colonies, in a Short*

Representation of the Calamitous State of the Enslaved Negroes in the British Dominions. Philadelphia: Henry Miller, 1766.

———. *Avertissement à la Grande Bretagne et à ses colonies, ou Tableau abrégé de l'état miserable des nègres esclaves dans les dominations anglaises.* N.p., 1767.

———. *Some Historical Account of Guinea, Its Situation, Produce, and the General Disposition of its Inhabitants, with an Inquiry into the Rise and Progress of the Slave Trade.* London: J. Phillips, 1788.

Behn, Aphra. *Oroonoko, or, The Royal Slave.* 1668. Edited by Catherine Gallagher. Boston: Bedford/St. Martin's, 2000.

Bernardin de Saint-Pierre, Henri. *Etudes de la nature.* 1784. 2 vols. Paris: Lefèvre, 1836.

Bernier, François. "Nouvelle division de la terre, par les différentes espèces ou races d'hommes qui l'habitent." *Journal des sçavans,* April 1684. Paris: Jean Cusson.

Berthelin, Pierre-Charles. *Abrégé du Dictionnaire universel françois et latin, vulgairement appelé Dictionnaire de Trévoux.* Paris: Libraires associés, 1762.

The Bible Designed to Be Read as Living Literature: The Old and New Testaments in the King James Version. Edited by Ernest Sutherland Bates. 1936. New York: Simon & Schuster, 1993.

Bibliothèque universelle des dames: Neuvième classe: Physique de l'homme. Paris: n.p., 1787.

Blumenbach, Johann Friedrich. *De generis humani varietate nativa.* Göttingen: Vandenhoek & Ruprecht, 1775. 3rd ed., 1795.

———. *The Anthropological Treatises of Johann Friedrich Blumenbach.* London: Longman, Green, Longman, Roberts, & Green, 1865.

Bosman, William [Willem]. *A New and Accurate Description of the Coast of Guinea.* London: J. Knapton, 1721.

Boudet, Antoine. *Journal oeconomique, ou, Mémoires, notes et avis sur l'Agriculture, les Arts, le Commerce, et tout ce qui peut y avoir rapport à la Santé, ainsi qu'à la conservation et à l'augmentation des biens des familles, etc.* Paris: Antoine Boudet, 1767.

Brice, Andrew. *A universal geographical dictionary, or, Grand gazetter: of general, special, ancient and modern geography . . . more especially of the British dominions and settlements throughout the world.* London: J. Robinson and W. Johnston, 1759.

Brissot de Warville, Jacques-Pierre. *Examen critique des voyages dans l'Amérique septentrionale, de M. le marquis de Chatellux . . . dans laquelle on réfute principalement ses opinions sur les quakers, sur les nègres, sur le peuple et sur l'homme.* London: n.p., 1786.

Brosses, Charles, de. *Du culte des dieux fétiches ou Parallèle de l'ancienne religion de l'Egypte avec la religion actuelle de Nigritie.* Paris: n.p, 1760.

Buffon, Georges-Louis Leclerc, comte de. *Histoire naturelle, générale et particulière.* Paris: Imprimerie royale, 1749–88.

Ca' da Mosto, Alvise da. *The Voyages of Cadamosto and Other Documents on Western Africa in the Second Half of the Fifteenth Century.* Translated and edited by G. R. Crone. London: Hakluyt Society, 1937.

Caillié, René. *Journal d'un voyage à Temboctou et à Jenné, dans l'Afrique centrale.* Paris: Imprimerie royale, 1830.

Cavazzi, Giovanni Antonio. *Istorica descrizione de' tre' regni, Congo, Matamba, et Angola.* Edited by Fortunato Alamandini. Bologna: Giacomo Monti, 1687.

———. *Relation historique de l'Ethiopie occidentale: Contenant la description des royaumes de Congo, Angolle, et Matamba.* Translated by Jean-Baptiste Labat. Paris: C.-J.-B. Delespine, 1732.

Chambers, Ephraim. *Cyclopaedia or, An universal dictionary of arts and sciences.* London: J. and J. Knapton, 1728.

Charlevoix, Pierre-François-Xavier de, and Jean-Baptiste Le Pers. *Histoire de l'isle Espagnole ou de Saint-Domingue, écrite particulièrement sur des mémoires manuscrits du P. Jean-Baptiste le Pers, Jésuite, missionaire à Saint-Domingue, et sur les pièces originales qui se conservent au Dépôt de la Marine, par le P. Pierre-François-Xavier de Charlevoix.* 2 vols. Paris: H.-L. Guérin, 1730–31.

Chateaubriand, François-René de. *Du génie du christianisme.* 2 vols. Paris: Flammarion, 1963.

Clarkson, Thomas. *An Essay on the Slavery and Commerce of the Human Species, Particularly the African.* London: Joseph Crushank, 1786.

Code Noir ou Recueil d'édits, déclarations et arrêts concernant les esclaves nègres de l'Amérique. Paris: Libraires associés, 1685.

Condorcet, Jean-Antoine-Nicolas de Caritat, marquis de. *Réflexions sur l'esclavage des nègres par M. Schwartz.* Neufchâtel: Société typographique, 1781.

Cortés, Hernán. *Letters from Mexico.* Edited by Anthony Pagden. New Haven: Yale University Press, 1986.

Cullion, Valentin de. *Examen de l'esclavage en général, et particulièrement de l'esclavage des nègres dans les colonies françaises de l'Amérique.* Paris: Desenne, 1802.

Cuvier, Georges. *Le règne animal distribué d'après son organisation.* Paris: Deterville, 1817.

Dampier, William. *Nouveau voyage autour du monde où l'on décrit en particulier l'isthme* [sic] *de l'Amerique, plusieurs côtes et isles des Indes occidentales, les isles du cap Verde . . . etc.* Rouen: Jean-Baptiste Machuel, 1723.

Le danger de la liberté des nègres. Paris: Imprimerie patriotique, 1791.

Dapper, Olfert. *Naukeurige Beschrijvingen der Afrikaensche gewesten* (Exact Descriptions of the African Lands). Amsterdam: van Meurs, 1668. Translated as *Description de l'Afrique* in *Objets interdits,* 89–357. Paris: Fondation Dapper, 1989. See also under Ogilby below.

Dazille, Jean-Barthélemy. *Observations sur les maladies des nègres, leurs causes, leurs traitements et les moyens de les prévenir.* Paris: Didot, 1776.

Delisle de Sales, Jean-Baptiste-Claude. *Essai philosophique sur le corps humain: Pour servir de suite à la "Philosophie de la nature."* 3 vols. Amsterdam: Arkstée & Merkus, 1773–74.

———. *De la philosophie de la nature ou Traité de morale de l'espèce humaine.* London: n.p., 1777.

Demanet, Abbé. *Nouvelle histoire de l'Afrique françoise, enrichie de cartes et d'observations astronomiques et géographiques.* Paris: Duchesne, 1767.

Dictionnaire de l'Académie françoise. Lyon: Benoît Duplain père et Joseph Duplain fils, 1772.

Dictionnaire universel françois et latin . . . vulgairement appelé "Dictionnaire de Trévoux." 1704. Paris: Rollin, 1732.

Diderot, Denis. *Oeuvres complètes de Denis Diderot.* Edited by Herbert Dieckmann, Jacques Proust, Jean Varloot, et al. Paris: Hermann, 1975–.

———. "Observations sur Hemsterhuis." In *Diderot: Oeuvres,* ed. Laurent Versini. Paris: R. Laffont, 1994.

Diderot, Denis, and Jean Le Rond d'Alembert, eds. *Encyclopédie, ou, dictionnaire raisonné des sciences, des arts, et des métiers.* Paris: Panckouche, 1751–72.

Diderot, Denis, Jean Le Rond d'Alembert, and Pierre Mouchon, eds. *Supplément à l'Encyclopédie, ou Dictionaire raisonné des sciences, des arts, et des métiers*. Amsterdam: M. Ray, 1776–77.

Du Tertre, Jean-Baptiste. *Histoire générale des Antilles habitées par les François*. Paris: Thomas Jolly, 1647–71.

Dubois-Fontanelle, Joseph-Gaspard. *Anecdotes africaines depuis l'origine ou la découverte des différents royaumes qui composent l'Afrique, jusqu'à nos jours*. Paris: Vincent, 1774.

Dubos, Jean-Baptiste. *Réflexions critiques sur la poésie et sur la peinture*. Paris: P.-J. Mariette, 1733.

Duchesne, Henri-Gabriel et Pierre-Joseph Macquer. *Manuel du naturaliste, ouvrage dédié à Monsieur de Buffon*. Paris: G. Desprez, 1771.

Encyclopédie méthodique par ordre de matières. Géographie moderne. 3 vols. Paris: Panckoucke, 1782–88.

Equiano, Olaudah. *The Interesting Narrative of the Life of Olaudah Equiano, or Gustavus Vassa, the African, written by himself*. New York: Norton, 2001.

Erxleben, Johann Christian Polykarp. *Systema regni animalis per classes, ordines, genera, species, varietates cum synonymia et historia animalium*. Leipzig: Weygandianis, 1777.

Fenning, Daniel, and Joseph Collyer. *A New System of Geography, or, A general Description of the World containing a Particular and Circumstantial Account of all the Countries, Kingdoms and States of Europe, Asia, Africa, and America*, London: S. Crowder, 1764.

Féraud, Jean-François, Abbé. *Dictionnaire critique de la langue française*. 3 vols. Marseille: J. Mossy père et fils, 1787–88.

Fontenelle, Bernard de. "Observations de physique générale." In *Histoire de l'Académie royale des sciences*. Paris: Imprimerie royale, 1734 [1736].

Formey, Johann Heinrich Samuel. *Nouvelle bibliothèque germanique, ou Histoire littéraire de l'Allemagne, de la Suisse, et des pays du nord*. 26 vols. Amsterdam: J. Schreuder & Pierre Mortier le jeune, 1746–60.

Forster, John Reinhold. *Observations Made During a Voyage Round the World*. London: G. Robinson, 1778.

Frossard, Benjamin-Sigismond. *La cause des esclaves nègres et des habitans de la Guinée*. Lyon: Aimé de la Roche, 1789.

Furetière, Antoine. *Dictionnaire universel contenant generalement tous les mots françois tant vieux que modernes*. 3 vols. Rotterdam: Arnout & Reinier Leers, 1690.

Garran de Coulon, Jean-Philippe, and Marguerite-Elie Guadet, *An Inquiry into the Causes of the Insurrection of the Negroes in the Island of St. Domingo: To which are added, Observations of M. Garran de Coulon on the same subject, read in his Absence by M. Guadet*. London: J. Johnson, 1792.

Gérando, J.-M. de. *Des signes et de l'art de penser considérés dans leurs rapports mutuels*. Paris: Goujon, 1800.

Gobineau, Arthur, comte de. *Essai sur l'inégalité des races humaines*. Paris: Pierre Belfond, 1853. Translated as *The Inequality of Human Races* (New York: Putnam, 1915).

Green, John. *A New General Collection of Voyages and Travels consisting of the most Esteemed Relations, which have been hitherto published in any language: comprehending everything remarkable in its kind, in Europe, Asia, Africa, and America*. 4 vols. London: Thomas Astley, 1745–47.

Grégoire, Henri. *De la littérature des Nègres*. Paris: Maradan, 1808. Facsimile reprint, Paris: Perrin, 1990.

Grimm, Friedrich Melchoir, and Denis Diderot. *Correspondance littéraire, philosophique et critique, adressée à un souverain d'Allemagne, depuis 1770 jusqu'en 1782*. Paris: F. Buisson, 1812.

Helvétius, Claude-Adrien. *De l'homme: De ses facultés intellectuelles et de son éducation.* Paris: Société typographique, 1775.

———. *De l'esprit.* Paris: Fayard, 1988.

Herodotus. *The Histories*. Translated by Robin Waterfield. Oxford: Oxford University Press, 2008.

Hippocrates. *Hippocrates on airs, waters, and places*. London: Wyman & Sons, 1881.

Histoire de l'Académie royale des sciences de Paris. Paris: Imprimerie royale et al., 1699–1790. Paris: Martin & Guérin, [1702] 1703.

Holbach, Paul-Henri Dietrich, baron d'. *Système de la nature.* Paris: Etienne Ledoux, 1770.

Hughes, Griffith. *The Natural History of Barbados, in Ten Books*. London: Griffith Hughes, 1750.

Hume, David. "Of National Characters" (1748). In *Essays: Moral, Political, and Literary.* Indianapolis: Liberty Classics, 1987.

Kames, Henry Home, Lord. *Sketches of the History of Man.* Edinburgh: Bell & Bradfute, 1813.

Kant, Immanuel. *Of the Different Races of Human Beings*. 1775. Translated by Holly Wilson and Günter Zöller. In Kant, *Anthropology, History, and Education*, ed. Robert B. Louden and Günter Zöller. The Cambridge Edition of the Works of Immanuel Kant. Cambridge: Cambridge University Press, 2007.

Labat, Jean-Baptiste. *Nouvelle relation de l'Afrique occidentale, contenant une description exacte du Sénégal et des pays situés entre le Cap-Blanc et la rivière de Serrelionne*. Paris: G. Cavelier, 1728.

———. *Nouveau voyage aux isles de l'Amérique*. Paris: P. F. Giffart, 1722.

———. *Voyage du Chevalier Des Marchais en Guinée, isles voisines, et à Cayenne, fait en 1725, 1726 et 1727: Contenant une description très exacte et très étenduë de ces pays, et du commerce qui s'y fait. Enrichi d'un grand nombre de cartes et de figures en tailles douces.* Paris: Saugrain l'ainé, 1730.

———. *Voyage aux isles: Chronique aventureuse des Caraïbes, 1673–1705*. Edited by Michel Le Bris. Paris: Phébus, 1993.

La Courbe, Michel Jajolet de. *Premier voyage du sieur de la Courbe fait à la côte de l'Afrique en 1685*. Edited by Prosper Cultru. Paris: E. Champion, 1913.

Ladvocat, Jean-Baptiste. *Dictionnaire géographique portatif . . . traduit de l'anglois sur la 13e édition de Laurent Echard.* Paris: Didot, 1749.

La Fosse, Eustache de. *Voyage d'Eustache Délafosse sur la côte de Guinée, au Portugal et en Espagne (1479–1481)*. Edited by Denis Escudier. Paris: Chandeigne, 1992.

La Métherie, Jean-Claude de. *De la perfectibilité et de la dégénérescence des êtres organisés.* Paris: Courcier, 1806.

Lanni, Dominique. *Fureur et barbarie: Récits de voyage chez les Cafres et les Hottentots, 1665–1721*. Paris: Cosmopolo, 2003.

La Pérouse, Jean-François de Galaup, comte de. *Voyage de La Pérouse autour du monde*. Edited by L.-A. Milet-Mureau. 4 vols. Paris: Imprimerie de la République, an V [1797].

La Peyrère, Isaac de. *Praeadamitae*. Amsterdam: Louis and Daniel Elzevier, 1655.

Lavallée, Joseph. *Le Nègre comme il y a peu de blancs*. Paris: Buisson, 1790.

Lavater, Johann Caspar. *Essai sur la physiognomie, destiné à faire connaître l'homme et à le faire aimer.* The Hague: Jacques van Karnebeek and J. van Cleef, 1781–1803.

Le Cat, Claude-Nicolas. *Traité des sens*. Paris: G. Cavelier, 1742.

———. *Traité de la couleur de la peau humaine en général, de celle des nègres en particulier, et de la métamorphose d'une de ces couleurs en l'autre, soit de naissance, soit accidentellement; ouvrage divisé en trois parties*. Amsterdam: n.p., 1765.

Lecointe-Marsillac. *Le More-Lack ou Essai sur les moyens les plus doux et les plus équitables d'abolir la traite et l'esclavage des nègres d'Afrique*. Paris: Prault, 1789.

Le François, Laurent, Abbé, and A. D. Fer. *Methode abregée et facile pour apprendre la géographie*. The Hague: P. Husson, 1706.

Le François, Laurent, Abbé. *Méthode abrégée et facile pour apprendre la géographie, où l'on décrit la forme du gouvernement de chaque pays, ses qualités, les moeurs de ses habitans.* Paris: Brocas [De l'Imprimerie de P.-M. Delaguette], 1781.

Leeuwenhoek, Antoni van. *The Collected Letters of Antoni van Leeuwenhoek, 1673–1676.* Amsterdam: Swets and Zeitlinger, 1939.

Lelarge de Lignac, Joseph Adrien. *Lettres à un Américain sur l'Histoire générale et particulière de M. de Buffon*. Hamburg: n.p., 1751.

Le Maire, Jacques-Joseph [Jacob]. *Les Voyages du sieur Le Maire aux Isles Canaries, Cap-Verd, Sénégal et Gambie*. Paris: Jacques Collombut, 1695. Translated as *A voyage of the Sieur Le Maire to the Canary Islands, Cape-Verd, Senegal and Gamby, under Monsieur Dancourt, Director-General of the Royal African Company* (London: F. Mills and W. Turner, at the Rose and C[r]own without Temple-Bar, 1696).

Lenglet-Dufresnoy, Nicolas. *Géographie des enfans, ou Méthode abrégée de la géographie.* Amsterdam: La Compagnie, 1744.

Lestringant, Frank, and Paolo Carile. "Preface." In Guillaume Chenu de Laujardière, *Relation d'un voyage à la côte des Cafres (1686–1689)*. Paris: Editions de Paris, 1996.

Le Vaillant, François. *Voyage dans l'intérieur de l'Afrique par le Cap de Bonne-Espérance, dans les années 1780, 81, 82, 83, 84 & 85*. Paris: Leroy, 1790.

———. *Second voyage dans l'intérieur de l'Afrique par le Cap de Bonne-Espérance, dans les années 1783, 84, 85*. Amsterdam: Libraires associés, 1797.

Lévi-Strauss, Claude. *Race et histoire*. Paris: Denoël, 1987.

Liceti, Fortunio. *De monstrorum caussis, natura, et differentiis libri duo*. Padua: Paulum Frambottum, 1634

Linnaeus, Carl. *Systema naturae*. 1735. Facsimile of the first edition. Nieuwkoop, The Netherlands: B. de Graaf, 1964.

Lobo, Jerónimo. *A Voyage to Abyssinia*. Translated from the French by Samuel Johnson. Edited by Joel J. Gold. New Haven: Yale University Press, 1985.

Locke, John. *Two Treatises of Government*. 1689. Edited by Peter Laslett. Cambridge: Cambridge University Press, 1967.

———. *An Essay Concerning Human Understanding*. 1690. In *The Clarendon Edition of the Works of John Locke*, ed. Peter H. Nidditch, John W. Yolton, et al. Oxford: Clarendon Press, 1975–.

Long, Edward. *The History of Jamaica or General Survey of the Antient* [sic] *and Modern State of That Island*. London: T. Lowndes, 1774.

Lopes, Duarte. *Relatione del reame di Congo e della circonvicine contrade*. Translated by Filippo Pigafetta. Rome: Bartolomeo Grassi, 1591.

———. *Regnum Congo, hoc est, Vera descriptio regni Africani, quod tam ab incolis quam Lusitanis Congus appellatur.* Frankfurt a/M: Johann Theodor and Johann Israel de Bry, 1598.

———. *A Report of the Kingdom of Congo and the Surrounding Countries; drawn out of the writings and discourses of the Portuguese, Duarte Lopez, by Filippo Pigafetta, in Rome, 1591.* Translated and edited by Margarite Hutchinson. 1881; New York: Negro Universities Press, 1969.

———. *Le royaume de Congo et les contrées environnantes (1591). Translated* by Willy Bal. Paris: Chandeigne, 2002.

Lucretius. *On the Nature of Things.* Translated by Martin Ferguson Smith. Indianapolis: Hackett, 2001.

Maupertuis, Pierre-Louis Moreau de. *Les Oeuvres de Mr de Maupertuis.* Dresden: G. C. Walther, 1752.

———. *Vénus physique suivi de la Lettre sur le progès des sciences.* Edited by Patrick Tort. Paris: Aubier Montaigne, 1980.

Meckel, Johann Friedrich. "Recherches anatomiques, sur la nature de l'épiderme, et du réseau, qu'on appelle Malpighien"; "Sur la diversité de couleur dans la substance médullaire du cerveau des négres"; "Description d'une maladie particulière du péritoine." In *Mémoires de l'Académie royale des sciences et des belles lettres de Berlin,* 79–113. Berlin: Ambroise Haude, 1755.

Mémoires de la Société des sciences physiques de Lausanne. Lausanne: Mourer, 1783.

Mémoires pour l'Histoire des sciences et des beaux arts. Paris: Chauert, 1738.

Mercator, Gerhard, Jodocus Hondius, and Henry Hexham. *Atlas or A Geographical Description of the Regions, Countries and Kingdoms of the World.* Amsterdam: Henry Hondius and John Johnson, 1636.

Mercier, Louis-Sébastien. *Fragments de politique et d'histoire.* Paris: Buisson, 1792.

———. *L'an 2440: Rêve s'il en fut jamais.* 1771. Paris: France Adel, 1977.

Merolla, Girolamo. *A Voyage to Congo, and several other Countries chiefly in Southern Africa.* London: H. Lintot and J. Osborne, 1682.

Millin, Aubin Louis, ed. *Magasin encyclopédique, ou Journal des sciences, des lettres et des arts.* Paris: Fuchs, 1797.

Mills, Charles W. *The Racial Contract.* Ithaca, NY: Cornell University Press, 1997.

Mirabeau, François Victor Riquetti, marquis de. *L'ami des hommes,* ou Traité de la population. Paris, 1757.

Mocquet, Jean. *Voyage à Mozambique et Goa: La relation de Jean Mocquet (1607–1610).* Paris: Chandeigne, 1996.

The Modern Part of an Universal History from the Earliest Account of Time. 44 vols. London: S. Richardson et al., 1759–66.

Monsieur de J***. "Explication physique de la noirceur des nègres." In *Mémoires pour l'Histoire des sciences et des beaux arts,* 1153–1205. Paris: Chaubert, 1738.

Montesquieu, Charles-Louis de Secondat, baron de. *De l'esprit des lois.* 1748. Paris: Garnier frères, 1973.

———. *Lettres persanes.* Paris: Garnier, 1975.

Moreau de Saint-Méry, Médéric Louis-Elie. *Description topographique, physique, civile, politique, et historique de la partie française de l'Isle de Saint-Domingue.* Philadelphia: l'auteur, et al., 1797–98.

Morenas, François. *Dictionnaire portatif comprenant la géographie et l'histoire universelle, la chronologie, la mythologie, l'astronomie, la physique, l'histoire naturelle et toutes ses parties, la chimie, l'anatomie, l'hydrographie, et la marine*. Avignon: L. Chambeau, 1760.

Moréri, Louis, ed. *Le grand dictionnaire historique, ou le mélange curieux de l'histoire sacrée et profane, qui contient en abrégé l'histoire fabuleuse des dieux et des héros de l'Antiquité païnne*. Paris: Libraires associés, 1759.

Nauton, L'Abbé. "Essai sur la cause physique de la couleur des différents habitants de la Terre." In *Observations sur la physique et sur l'histoire naturelle et sur les arts*, ed. Abbé François Rozier and J.-A. Mongez. Paris: Au bureau du Journal de physique, 1781.

Nouveau dictionnaire d'histoire naturelle appliquée aux arts à l'agriculture . . . à la médecine. Par une société de naturalistes et d'agriculteurs. Edited by Jean-François-Pierre Deterville. 24 vols. Paris: Deterville, 1803–4.

Ogilby, John. *Africa being an Accurate Description of the Regions of Aegypt, Barbary, Lybia, and Billedulgerid, the Land of Negroes, Guinee, Aethiopia and the Abyssines*. London: Printed by Tho. Johnson, 1670.

Para Du Phanjas, François. *Théorie des êtres sensibles, ou, Cours complet de physique*. 4 vols. Paris: Charles-Antoine Jombert, 1772.

Paré, Ambroise. *Oeuvres complètes d'Ambroise Paré*. Paris: J.-B. Baillière, 1840.

Park, Mungo. *Travels in the Interior Districts of Africa*. Durham, NC: Duke University Press, 2000.

Pauw, Cornelius de. *Recherches philosophiques sur les Américains, ou Mémoires intéressants pour servir à l'histoire de l'espèce humaine*. Berlin: G. J. Decker, 1768.

Perreau, Jean-André. *Etudes de l'homme physique et moral considéré dans ses différents âges*. Paris: Imprimerie des Annales d'Agriculture, 1797.

Phillips, Thomas. "A Journal of a Voyage made in the *Hannibal* of London, Ann. 1693–1694." In *Documents Illustrative of the History of the Slave Trade to America*, ed. Elizabeth Donnan. Washington, DC: Carnegie Institution of Washington, 1930.

[Anon.] "Philosophical Enquiries concerning the Americans by M. de Pauw; with a Dissertation upon *America* and its Inhabitants, by Dom Pernetty, and the Defense of the Author of the Enquiry against the Dissertation." *Critical Review*, ed. Tobias Smollet, no. 19 (1771): 220–22.

Picart, Bernard. *Cérémonies et coutumes religieuses de tous les peuples du monde representées par des figures dessinées de la main de Bernard Picard: avec une explication historique, et quelques dissertations curieuses*. Amsterdam: J. F. Bernard, 1723–37.

Pluche, Antoine. *Le Spectacle de la nature, ou Entretiens sur les particularités de l'Histoire naturelle qui ont paru les plus propres à rendre les jeunes gens curieux et à leur former l'esprit*. 9 vols. Paris: Vve Estienne, 1732–50.

Poivre, Pierre. *Voyage d'un philosophe ou Observations sur les moeurs et les arts des peuples de l'Afrique et de l'Asie et de l'Amérique*. Maestricht: Dufour & Roux, 1779.

Le Pour et contre: Ouvrage périodique d'un goût nouveau, dans lequel on s'explique librement sur tout ce qui peut intéresser la curiosité du public, en matière de sciences, d'arts, de livres, d'auteurs, &c. sans prendre aucun parti, et sans offenser personne. Edited by Antoine François Prévost, Pierre-François Guyot Desfontaines, and Charles-Hugues Le Febvre de Saint-Marc. 20 vols. Paris: Didot, 1733–40.

Prévost, Antoine François. *Voyages du capitaine Robert Lade en différentes parties de l'Afrique, de l'Asie et de l'Amérique*. Paris: Didot, 1744.

———. *Histoire générale des voyages, ou nouvelle collection de toutes les Relations de voyages*

par mer et par terre qui ont été publiées jusqu'à présent dans les différentes langues de toutes les nations connues. 19 vols. Paris: Didot, 1746–70.
Proust, Jacques. *Diderot et l'Encyclopédie*. Paris: Albin Michel, 1995.
Purchas, Samuel. *Purchas his pilgrimage. Or Relations of the World and the Religions observed in all ages and places discovered*. London: William Stansby, 1613.
Raimond, Julien. *Extrait d'une lettre sur les malheurs de Saint-Domingue*. Paris: Au jardin égalité, 1793.
Ramsay, James. *An Essay on the Treatment and Conversion of African Slaves in the British Sugar Colonies*. London: James Phillips, 1784.
Ramusio, Giovanni Battista. *Delle navigationi e viaggi* . . . 3 vols. Venice: Giunti, 1563.
Raynal, Guillaume-Thomas. *Histoire philosophique et politique des établissements et du commerce des Européens dans les deux Indes*. 4 vols. Amsterdam: n.p., 1770; 8 vols. The Hague: Gosse fils, 1774. 10 vols. Geneva: Jean-Leonard Pellet, 1780.
Reisch, Gregor. *Margarita Philosophica*. Freiburg im Breisgau: Joannis Schotti, 1503.
Riolan, Jean, the Younger. *Encheiridium anatomicum et pathologicum, in quo ex naturali constitutione partium recessus à naturali statu demonstratur, ad usum theatri anatomici adornatum: Figuris elegantissimis indiceque accuratissime exornatum*. Leiden: Adrian Wyngaerden, 1649.
———. *Manuel anatomique et pathologique, ou abrégé de toute l'anatomie, et des usages que l'on peut tirer pour la connaissance, et pour la guérison des maladies*. Paris: Gaspar Meturas, 1661.
Robinet, J. B. *Considérations philosophiques de la gradation naturelle des formes de l'être ou les Essais de la Nature qui apprend à faire l'homme*. Paris: Charles Saillant, 1768.
Roger, Jacques. *Les sciences de la vie dans la pensée française du XVIIIe siècle*. Paris: Albin Michel, 1963.
———. *The Life Sciences in Eighteenth-Century French Thought*. Translated by Robert Ellrich. Stanford: Stanford University Press, 1997.
Roubaud, Pierre-Joseph-André. *Histoire générale de l'Asie, de l'Afrique et de l'Amérique*. 15 vols. Paris: Desventes de la Doué, 1770–75.
Rousseau, Jean-Jacques. *Emile, ou, De l'éducation*. 1762. In *Oeuvres complètes*, vol. 4. Paris: Gallimard, 1969.
———. *Discours sur l'origine et les fondements de l'inégalité parmi les hommes*. 1755. In *Oeuvres complètes. Oeuvres philosophiques et politiques*, vol. 3. Paris: Seuil, 1971.
Rousselot de Surgy, Jacques-Philibert. *Mélanges intéressans et curieux ou Abregé d'histoire naturelle, morale, civile, et politique de l'Asie, l'Afrique, l'Amérique et des Terres Polaires*. 10 vols. Yverdon: Durand, 1763–65.
Rozier, François, Abbé, and Jean-André Mongez, ed. *Observations sur la physique, sur l'histoire naturelle et sur les arts, avec des planches en taille-douce*. Paris: Au bureau du Journal de physique, 1781.
Saint-Hilaire, Emile Marco de. *Napoléon au conseil d'Etat*. Brussels: A. Lebèque & Sacré fils, 1843.
Saint-Lambert, Jean-François. *Contes de Saint-Lambert*. Paris: Librarie des bibliophiles, 1883.
Santos, João dos. *Ethiopia oriental*. Evora, Portugal: Manoel de Lyra, 1609.
Savary des Brûlons, Jacques, and Philémon-Louis Savary. *Dictionnaire universel de commerce, contenant tout ce qui concerne le commerce qui se fait dans les quatres parties du monde*. Paris: J. Estienne, 1723–30.

Savary, Jacques. *Le Parfait Negociant, ou Instruction générale pour ce qui regarde le commerce de marchandises de France et des pays étrangers.* Paris: L Billaine, 1675.

Schreber, Johann [Jean] Christian Daniel von. *Histoire naturelle des quadrupèdes.* Erlang[en], Germany: Wolfgang Walther, 1775–80.

Sigaud de la Fond, Joseph-Aignan. *Dictionnaire de physique.* Paris: Rue et Hôtel Serpente, 1781.

Smith, William. *A new voyage to Guinea: describing the customs, manners, soil, climate, habits, buildings, education, manual arts, agriculture, trade, employments, languages, ranks of distinction, habitations, diversions, marriages, and whatever else is memorable among the inhabitants, likewise an account of their animals, minerals, etc.* London: John Nourse, 1744.

———. *Nouveau voyage de Guinée.* Paris: Durant & Pissot, 1751.

Strabo, *The Geography of Strabo.* Translated by Horace Leonard Jones and J. R. Sitlington Sterrett. London: W. Heinemann, 1985.

Swift, Jonathan. *The Poems of Jonathan Swift.* Edited by Harold Williams. Oxford: Oxford University Press, 1958.

Tauvry, Daniel. *Nouvelle anatomie raisonnée, ou les usages de la structure du corps de l'homme, et de quelques autres animaux, suivant les lois des mécaniques.* 1690. Paris: Laurent d'Houry, 1720.

Tussac, Richard. *Cri des colons contre un ouvrage de M. l'évêque et sénateur Grégoire.* Paris: Delaunay, 1810.

Tyson, Edward. *Orang-outang, sive Homo Sylvestris or, The Anatomy of a Pygmie compared with that of a Monkey, and Ape, and a Man. To which is added a Philosophial Essay concerning the Pygmies, the Cynocephali, the Satyrs, and Sphinges of the Ancients wherein it will appear that they are all either Apes or Monkeys, and not Men, as formerly pretended.* London: T. Bennett and D. Brown, 1699. Reproduced in *A Philological Essay Concerning the Pygmies of the Ancients* (Austin, TX: Book Lab, 1991).

Tyssot de Patot, Simon. *Voyages et aventures de Jacques Massé.* Bordeaux: J. L'Aveugle, 1710.

Vaissette, Joseph, Gilles Robert de Vaugondy, and Didier Robert de Vaugondy. *Géographie historique, ecclésiastique et civile, ou, Description de toutes les parties du globe terrestre.* Paris: Desaing et al., 1755.

Valmont de Bomare, Jacques-Christophe. *Dictionnaire raisonné universel de l'histoire naturelle: Contenant l'histoire des animaux, des végétaux, et des minéraux, et celle des corps célestes, des météores, et des autres principaux phénomenes de la nature.* 5 vols. Paris: Didot-le-jeune, 1764. 6 vols. Paris: Lacombe, 1767–68. 9 vols. Lyon: Jean-Marie Bruysset, 1775. 15 vols. Lyon: Bruysset frères, 1791.

Virey, Joseph-Julien. *Histoire naturelle du genre humain, ou Recherches sur ses principaux fondemens physiques et moraux; précédées d'un Discours sur la nature des êtres organiques et sur l'ensemble de leur physiologie. On y a joint une Dissertation sur le sauvage de l'Aveyron. . . .* Paris: F. Dufart, an IX [1800]. Paris: Brochard, 1824.

Voltaire, François Marie Arouet de. "Relation touchant un Maure blanc amené d'Afrique à Paris en 1744." 1745. In *Oeuvres de Voltaire, avec préfaces, avertissements, notes.* Paris: Lefèvere, 1830.

———. *Oeuvres complètes.* Edited by Jean Michel Moreau et al. Paris: Garnier frères, 1877–85.

———. *Essai sur les moeurs et l'esprit des nations et sur les principaux faits de l'histoire depuis Charlemagne jusqu'à Louis XIII.* 1769. 2 vols. Paris: Garnier frères, 1963.

———. *Candide*. Paris: Didier, 1972.

———. *Romans et contes*. Paris Gallimard, 1972.

Wallace, George. *A System of the Principles of the Laws of Scotland*. Edinburgh: G. Hamilton and J. Balfour, 1760.

Walrond, Humphrey. "An Act for the Better Ordering and Governing of Negroes, *Barbados 1661*." In *Slavery*, ed. Stanley Engerman, Seymour Drescher, and Robert Paquette, 105–13. New York: Oxford University Press, 2001.

Zurara, Gomes Eanes de. *Chronique de Guinée (1453) de Gomes Eanes de Zurara*. Edited by Jacques Paviot. Paris: Chandeigne, 1994.

———. *Chronicle of the Discovery and Conquest of Guinea*. Cambridge: Cambridge University Press, 2010.

Secondary Sources

Abanime, Emeka P. "Voltaire as an Anthropologist: The Case of the Albino." *Studies on Voltaire and the Eighteenth Century* 143, 85–104. Oxford: Voltaire Foundation, 1976.

Les abolitions de l'esclavage: De L.F. Sonthonax à V. Schoelcher: 1793–1794–1848. Actes du colloque international tenu à l'université de Paris VIII les 3, 4 et 5 février 1994. Edited by Marcel Dorigny. Saint-Denis: Presses universitaires de Vincennes; Paris: UNESCO, 1995.

Achebe, Chinua. *Things Fall Apart*. 1958. New York: Anchor Books, 1994.

Adorno, Theodor W., and Max Horkheimer. *Dialectic of Enlightenment*. Translated by John Cumming. London: Verson, 1979.

L'Afrique du siècle des Lumières: Savoirs et representations. Edited by Catherine Gallouët, David Diop, Michèle Bocquillon, and Gérard Lahouati, 167–96. Oxford: Voltaire Foundation, 2009.

Aravamudan, Srinivas. *Tropicopolitans: Colonialism and Agency, 1688–1804*. Durham, NC: Duke University Press, 1999.

Armstrong, Nancy, and Leonard Tennenhouse, ed. *The Violence of Representation: Literature and the History of Violence*. London: Routledge, 1989.

Augstein, Hannah Franziska. "Introduction." In *Race: The Origins of an Idea, 1760–1850*, ed. id. Bristol, UK: Thoemmes Press, 1996.

Barnard, Alan. *Hunters and Herders of Southern Africa: A Comparative Ethnography of the Khoisan Peoples*. Cambridge: Cambridge University Press, 1992.

Benot, Yves. *La démence coloniale sous Napoléon*. Paris: La Découverte, 1992. Paris: La Découverte, 1987.

———. *La Révolution française et la fin des colonies*. Paris: La Découverte, 1987.

———. *Les Lumières, l'esclavage, la colonisation*. Paris: La Découverte, 2005.

Bernasconi, Robert. "Kant and Blumenbach's Polyps: A Neglected Chapter in the History of the Concept of Race." In *The German Invention of Race*, ed. Sara Eigen and Mark Larrimore, 73–90. Albany: State University of New York Press Press, 2006.

Bernasconi, Robert and Tommy L. Lott, ed. *The Idea of Race*. Indianapolis: Hackett, 2000.

Bilé, Serge. *La Légende du sexe surdimensionné des noirs*. Monaco: Le Serpent à plumes, 2005.

Blackburn, Robin. *The Overthrow of Colonial Slavery, 1776–1848*. London: Verso, 1988.

Boogaart, Ernest van den. "Books on Black Africa: The Dutch Publications and Their

Owners in the Seventeenth and Eighteenth Centuries." In *European Sources for Sub-Saharan Africa Before 1900, Use and Abuse*, ed. Beatrix Heintze and Adam Jones. *Paideuma* 33. Stuttgart: Franz Steiner Verlag Wiesbaden, 1987.

Boulle, Pierre Henri. *Race et esclavage dans la France de l'Ancien Régime*. Paris: Perrin, 2007.

Braude, Benjamin. "The Sons of Noah and the Construction of Ethnic and Geographical Identities in the Medieval and Early Modern Periods." *William and Mary Quarterly* 54, no. 1 (January 1997): 103–42.

Brewer, Daniel. *The Enlightenment Past: Reconstructing Eighteenth-Century French Thought*. New York: Cambridge University Press, 2008.

Broc, Numa. *La géographie de la Renaissance, 1420–1620*. Paris: Editions du C.T.H.S., 1986.

Burbank, Jane, and Frederick Cooper. *Empires in World History: Power and the Politics of Difference*. Princeton: Princeton University Press, 2010.

Butel, Paul. *Histoire des Antilles françaises: XVIIIe–XXe siècle*. Paris: Perrin, 2002.

Campbell, Gordon Lindsay. *Lucretius on Creation and Evolution: A Commentary on De rerum natura* 5. Oxford classical monographs. Oxford: Oxford University Press. 2003.

Chowdhry, Gheeta, and Mark Beeman. "Situating Colonialism, Race, and Punishment." In *Race, Gender, and Punishment: From Colonialism to the War on Terror*, ed. Mary Bosworth and Jeanne Flavin, 13–31. New Brunswick, NJ: Rutgers University Press, 2007.

Cohen, William B. *The French Encounter with Africans: White Response to Blacks, 1530–1880*. Bloomington: Indiana University Press, 1980.

Curran, Andrew. *Sublime Disorder: Physical Monstrosity in Diderot's Universe*. Oxford: Voltaire Foundation, 2001.

———. "Pourquoi étudier la représentation de l'Afrique dans la pensée du dix-huitième siècle." In *L'Afrique du siècle des Lumières: Savoirs et representations*, ed. Catherine Gallouët, David Diop, Michèle Bocquillon, and Gérard Lahouati, xix–xxix. Oxford: Voltaire Foundation, 2009.

———. "Logics of the Human in Diderot's *Supplément au Voyage de Bougainville*." In *New Essays on Diderot*, ed. James Fowler. Cambridge: Cambridge University Press, 2011.

———. "Imaginer l'Afrique au siècle des Lumières." In *Le Problème de l'altérité dans la culture européenne: Anthropologie, politique, et religion aux XVIIIe et XIXe siècles*, ed. Guido Abbattista and Rolando Minuti. Naples: Bibliopolis, 2006.

Curtin, Philip D. "Forward to Part I." In *Africa Remembered: Narratives by West Africans from the Era of the Slave Trade*, ed. Philip D. Curtin. Madison: University of Wisconsin Press, 1967.

———. *The Atlantic Slave Trade: A Census*. Madison: University of Wisconsin Press, 1969.

Daget, Serge. *Répertoire des expéditions négrières françaises à la traite illégale (1814–1850)*. Nantes: Centre de recherche sur l'histoire du monde atlantique, Université de Nantes, 1988.

Daget, Serge, and François Renault, *Les traités négrières en Afrique*. Paris: Karthala, 1985.

Davis, David Brion. "New Sidelights on Early Antislavery Radicalism." *William and Mary Quarterly* 28, no. 4 (October 1971): 585–94.

———. *Slavery and Human Progress*. New York: Oxford University Press, 1984.

———. *The Problem of Slavery in Western Culture*. New York: Oxford University Press, 1988.

———. *Inhuman Bondage: the Rise and Fall of Slavery in the New World.* Oxford: Oxford University Press, 2006.

Davis, Peggy. "La réification de l'esclave noir dans l'estampe sous l'Ancien Régime et la Révolution" In *L'Afrique du siècle des Lumières: Savoirs et representations*, ed. Catherine Gallouët, David Diop, Michèle Bocquillon, and Gérard Lahouati, 237–54. Oxford: Voltaire Foundation, 2009.

Dictionary of Canadian Biography, ed. Marc La Terreur, Ramsay Cook, Francess G. Halpenny, and Jean Hamelin. Toronto: University of Toronto Press, 1974.

Diop, David. "La mise à l'épreuve d'un régime de véridiction sur 'la paresse et la négligence des nègres' dans le *Voyage au Sénégal* (1757) d'Adanson." In *L'Afrique du siècle des Lumières: Savoirs et representations*, ed. Catherine Gallouët, David Diop, Michèle Bocquillon, and Gérard Lahouati, 000-00, 15–29. Oxford: Voltaire Foundation, 2009.

Dobie, Madeleine. *Trading Places: Colonization and Slavery in Eighteenth-Century French Culture.* Ithaca, NY: Cornell University Press, 2010.

Dorigny, Marcel, and Bernard Gainot. *La Société des amis des noirs, 1788–1799: Contribution à l'histoire de l'abolition de l'esclavage.* Paris: UNESCO, 1998.

Douthwaite, Julia. *The Wild Girl, Natural Man, and the Monster.* Chicago: University of Chicago Press, 2002.

Droixhe, Daniel, and Pol-P. Gossiaux, eds. *L'homme des lumières et la découverte de l'autre.* Etudes sur le XVIIIième siècle 3. Brussels: Editions de l'Université de Bruxelles, 1985.

Dubois, Laurent. *Avengers of the New World: The Story of the Haitian Revolution.* Cambridge, MA: Harvard University Press, 2004.

———. *A Colony of Citizens: Revolution and Slave Emancipation in the French Caribbean, 1787–1804.* Chapel Hill: University of North Carolina Press, 2004.

Duchet, Michèle. *Anthropologie et histoire au siècle des lumières: Buffon, Voltaire, Rousseau, Helvétius, Diderot.* Paris: Flammarion, 1977.

Dunmore, John. *Where Fate Beckons: The Life of Jean-François de la Pérouse.* Fairbanks: University of Alaska Press, 2007.

Eche, Antoine. "L'image ethnographique africaine de l'*Histoire générale des voyages.*" In *L'Afrique du siècle des Lumières: Savoirs et representations*, ed. Catherine Gallouët, David Diop, Michèle Bocquillon, and Gérard Lahouati, 207–22. Oxford: Voltaire Foundation, 2009.

Ehrard, Jean. *Lumières et esclavage: L'esclavage colonial et l'opinion publique en France au XVIIIe siècle.* Paris: André Versaille, 2008.

Eigen, Sara, and Mark J. Larrimore, ed. *The German Invention of Race.* Albany: State University of New York Press, 2006.

Elphick, Richard. *Khoikhoi and the Founding of White South Africa.* New Haven: Yale University Press, 1977.

Eltis, David. "The Volume of the Transatlantic Slave Trade: A Reassessment." *William and Mary Quarterly* 58, no. 1 (January 2001): 17–46.

Eltis, David, and David Richardson, eds. *Extending the Frontiers: Essays on the New Transatlantic Slave Trade Database.* New Haven: Yale University Press, 2008.

Emmer, Pieter C. "The Dutch and the Slave Americas." In *Slavery in the Development of the Americas*, ed. David Eltis, Frank D. Lewis, and Kenneth L. Sokoloff, 70–88. Cambridge: Cambridge University Press, 2004.

Enders, Armelle. *Histoire de l'Afrique lusophone*. Paris: Chandeigne, 1994.

Erick, Noël. *Etre noir en France au XVIIIe siècle*. Paris: Tallandier, 2006.

Erickson, Paul A., with Liam D. Murphy. *A History of Anthropological Theory*. Ontario: Broadview Press, 1998.

Estève, Laurent. "La théorie des climats ou l'encodage d'une servitude naturelle." In *Déraison, esclavage, et droit: Les fondements idéologiques et juridiques et la traite negrière et de l'esclavage*, ed. Isabel de Castro Henriques and Louis Sala-Molins, 59–68. Paris: UNESCO, 2002.

Evans, William McKee. "From the Land of Canaan to the Land of Guinea: The Strange Odyssey of the Sons of Ham." *American Historical Review* 85, no. 1 (February 1980): 15–43.

Eze, Emmanuel Chukwudi, ed. *Race and the Enlightenment: A Reader.* Cambridge, MA: Blackwell, 1997.

Fanon, Frantz. *The Wretched of the Earth*. Translated by Constance Farrington. 1963. New York: Grove Press, 2007. Originally published in French as *Les damnés de la terre* (1961).

Fauvelle-Aymar, François-Xavier. *L'invention du Hottentot: Histoire du regard occidental sur les Khoisan (XVe–XIXe siècle)*. Paris: Sorbonne, 2002.

Festa, Lynn. *Sentimental Figures of Empire in Eighteenth-Century Britain and France*, Baltimore: Johns Hopkins University Press, 2009.

Finkelman, Paul, and Joseph C. Miller, eds. *Macmillan Encyclopedia of World Slavery*. New York: Macmillan, 1998.

Foucault, Michel. *Les mots et les choses: Une archéologie des sciences humaines*. Paris: Gallimard, 1966.

———. *L'archéologie du savoir*. Paris: Gallimard, 1969.

Fox, Christopher, Roy Porter, and Robert Wokler, ed. *Inventing Human Science: Eighteenth Century Domains*. Berkeley: University of California Press, 1995.

Garraway, Dorris. *The Libertine Colony: Creolization in the Early French Caribbean*. Durham, NC: Duke University Press, 2005.

Garrett, Aaron. "Human Nature." In *The Cambridge History of Eighteenth-Century Philosophy*, ed. Knud Haakonssen. Cambridge: Cambridge University Press, 2006.

Gates, Henry Louis, Jr. *Figures in Black*. New York: Oxford University Press, 1987.

———, ed. *Race, Writing, and Difference*. Chicago: University of Chicago Press, 1986.

Geggus, David. "Sex Ratio, Age and Ethnicity in the Atlantic Slave Trade: Data from French Shipping and Plantation Records." *Journal of African History* 30 (1989): 23–44.

Gilroy, Paul. *The Black Atlantic: Modernity and Double Consciousness*. Cambridge: Harvard University Press, 1993.

Goldberg, David Theo. *Racist Culture: Philosophy and the Politics of Meaning*. Oxford: Blackwell, 2002.

Gordon, Daniel. "Introduction: Postmodernism and the French Enlightenment." In *Postmodernism and the Enlightenment*, ed. id. New York: Routledge, 2001.

Gossiaux, Pol-Pierre. "Anthropologie des Lumières: Culture 'naturelle' et racisme rituel." In *L'homme des lumières et la découverte de l'autre*, ed. Daniel Droixhe and Pol-Pierre Gossiaux, 49–69. Etudes sur le XVIIIème siècle 3, Brussels: Editions de l'Université de Bruxelles, 1985.

Graille, Patrick. "Portrait scientifique et littéraire de l'hybride au siècle des Lumières." *Eighteenth-Century Life* 21, no. 2 (1997): 70–88.

———. "L'Afrique noire illustrée dans les récits des voyageurs, traducteurs, et compilateurs français du XVIIIe siècle" In *L'Afrique du siècle des Lumières: Savoirs et representations*, ed. Catherine Gallouët, David Diop, Michèle Bocquillon, and Gérard Lahouati, 167–96. Oxford: Voltaire Foundation, 2009.

Green, Toby. *Inquisition: The Reign of Fear.* New York: Thomas Books, 2009.

Hacking, Ian. *Historical Ontology*. Cambridge, MA: Harvard University Press, 2002.

Hair, P. E. H. *The Atlantic Slave Trade and Black Africa*. London: Historical Association, 1978.

———. *Africa Encountered: European Contacts and Evidence, 1450–1700*. Aldershot, UK: Variorum, 1997.

Hall, Gwendolyn Midlo. *Slavery and African Ethnicities in the Americas: Restoring the Links*. Chapel Hill: University of North Carolina Press, 2005.

Hannaford, Ivan. *Race: The History of an Idea in the West*. Baltimore: Johns Hopkins University Press, 1996.

Harms, Robert. *The Diligent: A Voyage through the Worlds of the Slave Trade*. New York: Basic Books, 2002.

Hoffmann, Léon-François. *Le nègre romantique: Personnage littéraire et obsession collective*. Paris: Payot, 1973.

Holmes, Rachael. *African Queen: The Real Life of the Hottentot Venus*. New York: Random House, 2007.

Hoquet, Thierry. "Une animalité en pièces. Spécimens empaillés et bocaux d'organes dans les planches de l'*Histoire naturelle* de Buffon (1749–1767)." *Revue d'esthétique* 40 (2001): 15–23.

———. *Buffon: Histoire naturelle et philosophie*. Paris: Honoré Champion, 2005.

———. *Buffon/Linné: Eternels rivaux de la biologie?* Paris: Dunod, 2007.

Horta, Jose da Silva. "A representação do Africano no literatura de viagens do Senegal à Serra Leoa (1453–1508)." *Mare Liberum* no. 2 (1991): 209–339.

Hudson, Nicholas. "From 'Nation' to 'Race': The Origin of Racial Classification in Eighteenth-Century Thought." *Eighteenth-Century Studies* 29 (1996): 247–64.

———. "'Hottentots' and the Evolution of European Racism." *Journal of European Studies* 34 (December 2004): 308–32.

———. "The 'Hottentot Venus,' Sexuality, and the Changing Aesthetics of Race, 1650–1850." *Mosaic* 41 (March 2008): 19–41.

Ibrahim, Annie. "La notion de moule intérieur dans les théories de la génération au XVIIIe siècle." In *Archives de philosophie* 50, no. 4 (1987): 555–80.

Jacques, T. Carlos. "From Savages and Barbarians to Primitives: Africa, Social Typologies, and History in Eighteenth-Century French Philosophy." *History and Theory* 36, no. 2 (1997): 190–215.

Jahoda, Gustav. *Images of Savages: Ancient Roots of Modern Prejudice in Western Culture*. London: Routledge, 1999.

Jones, Adam. "Olfert Dapper et sa description de l'Afrique." In *Objets interdits*, 73–81. Paris: Fondation Dapper, 1989.

Jordan, Winthrop D. *The White Man's Burden: Historical Origins of Racism in the United States*. New York: Oxford University Press, 1974.

Jordanova, Ludmilla. "Sex and Gender." In *Inventing Human Science: Eighteenth Century Domains*, ed. Christopher Fox, Roy Porter, and Robert Wokler. Berkeley: University of California Press, 1995.

Kadish, Doris Y., ed. *Slavery in the Caribbean Francophone World: Distant Voices, Forgotten Acts, Forged Identities*. Athens: University of Georgia Press, 2000.

Katzew, Ilona. *Casta Painting: Images of Race in Eighteenth-Century Mexico*. New Haven: Yale University Press, 2005.

Kennedy, Randall. *Nigger: The Strange Career of a Troublesome Word*. New York: Vintage, 2002.

Kidd, Colin. *The Forging of the Races*. Cambridge: Cambridge University Press, 2006.

Kieckhefer, Richard. *Magic in the Middle Ages*. Cambridge: Cambridge University Press, 2000.

Klaus, Sidney N. "A History of the Science of Pigmentation." In *The Pigmentary System: Physiology and Pathophysiology*, ed. James J. Nordlund. New York: Oxford University Press, 1998.

Kom, Ambroise, and Lucienne Ngoué, ed. *Le code noir et l'Afrique*. Ivry-sur-Seine: Nouvelles du Sud, 1991.

Lanson, Gustave. *Voltaire*. Translated by Robert A. Wagoner. New York: Wiley, 1966.

Laqueur, Thomas W. *Making Sex: Body and Gender from the Greeks to Freud*. Cambridge, MA: Harvard University Press, 1990.

Law, Robin. *The Slave Coast of West Africa, 1550–1750: The Impact of the Atlantic Slave Trade on an African Society*. New York: Oxford University Press, 1991.

———. *Ouidah: The Social History of a West African Slaving "Port", 1727–1892*. Athens: Ohio University Press, 2004.

Little, Roger. "Oroonoko and Tamango: A Parallel Episode." *French Studies* no. 46 (1992): 26–32.

———. *Nègres blancs: Représentations de l'autre autre: essai*. Paris: L'Harmattan, 1995.

Lively, Adam. *Masks: Blackness, Race, and the Imagination*. New York: Oxford University Press, 1998.

Lott, Tommy L. *The Invention of Race: Black Culture and the Politics of Representation*. Oxford: Blackwell, 1999.

Lovejoy, Paul E. *Transformations in Slavery: A History of Slavery in Africa*. Cambridge: Cambridge University Press, 1983.

Macksey, Richard. "The History of Ideas." In *The Johns Hopkins Guide to Literary Theory and Criticism*, ed. Michael Groden and Martin Kreiswirth, 388–92. Baltimore: Johns Hopkins University Press, 1994.

Magner, Lois N. *A History of the Life Sciences*. New York: Marcel Dekker, 2002.

Mark, Peter A. *"Portuguese Style" and Luso-African Identity: Precolonial Senegambia, Sixteenth-Nineteenth Centuries*. Bloomington: Indiana University Press, 2002.

M'Bokolo, Elikia. *Noirs et blancs en Afrique equatoriale: Les sociétés côtières et la pénétration française, vers 1820–1874*. Paris: Mouton, 1981.

McClellan, James E., III, *Colonialism and Science: Saint Domingue in the Old Regime*. Baltimore: Johns Hopkins University Press, 1992.

Medeiros, François de. *L'Occident et l'Afrique (XIIe–XVe siècle)*. Paris: Karthala, 1985.

Meijer, Miriam Claude. *Race and Aesthetics in the Anthropology of Petrus Camper (1722–1789)*. Amsterdam: Rodopi, 1999.

Mercier, Roger. *L'Afrique noire dans la littérature française, les premières images, XVII–XVIIIème siècles*. Dakar: Université de Dakar, 1962.
Merians, Linda. *Envisioning the Worst: Representations of "Hottentots" in Early-Modern England*. Newark: University of Delaware Press, 2001.
Meyer, Jean. *Les Européens et les autres de Cortès à Washington*. Paris: A. Colin, 1975.
———. *Histoire du sucre*. Paris: Desjonquères, 1989.
Miller, Christopher L. *Blank Darkness: Africanist Discourse in French*. Chicago: University of Chicago Press, 1985.
———. *Theories of Africans: Francophone Literature and Anthropology in Africa*. Chicago: University of Chicago Press, 1990.
———. *The French Atlantic Triangle: Literature and Culture of the Slave Trade*. Durham, NC: Duke University Press, 2008.
Moitt, Bernard. *Women and Slavery in the French Antilles, 1635–1848*. Bloomington: Indiana University Press, 2001.
Morgan, Kenneth. *Slavery, Atlantic Trade and the British Economy, 1660–1800*. Cambridge: Cambridge University Press: 2000.
Mostert, W. P. *Demography: Textbook for the South African Student*. Pretoria: Human Sciences Research Council, 1998.
Mudimbe, Valentin Y. *The Invention of Africa: Gnosis, Philosophy, and the Order of Knowledge*. Bloomington: Indiana University Press, 1988.
———. *The Idea of Africa*. Bloomington: Indiana University Press, 1994.
———. "Romanus Pontifex (1454) and the Expansion of Europe." In *Race, Discourse, and the Origin of the Americas: A New World View*, ed. Vera Lawrence Hyatt and Rex Nettleford. Washington, DC: Smithsonian Institution Press, 1995.
Munford, Clarence J. *The Black Ordeal of Slavery and Slave Trading in the French West Indies, 1625–1715*. Lewiston, NY: Mellen, 1991.
Noël, Erick. *Etre noir en France*. Paris: Tallandier, 2006.
O'Neal, John C. *Changing Minds: The Shifting Perception of Culture in Eighteenth-Century France*. Newark: University of Delaware Press, 2002.
Ogot, B. A., ed. *Histoire générale de l'Afrique*. Vol. 5. *L'Afrique du XVe au XVIII siècle*. Paris: UNESCO, 1999.
Park, Mungo. *Travels in the Interior Districts of Africa*. Edited by Kate Ferguson Marsters. Durham, NC: Duke University Press, 2000.
Parry, John Horace. *The Age of Reconnaissance: Discovery, Exploration, and Settlement, 1450–1650*. 1963. Berkeley: University of California Press, 1982.
Peabody, Sue. *"There are No Slaves in France": The Political Culture of Race and Slavery in the Ancien Régime*. Oxford: Oxford University Press, 1996.
Peabody, Sue, and Keila Grinberg. *Slavery, Freedom, and the Law in the Atlantic World: A Brief History with Documents*. New York: Palgrave Macmillian, 2007.
Porter, Roy, ed. *The Cambridge History of Science*. Vol. 4. *Eighteenth-Century Science*. Cambridge: Cambridge University Press, 2003.
Potkay, Adam, and Sandra Burr, eds. *Black Atlantic Writers of the Eighteenth Century: Living the New Exodus in England and the Americas*. New York: St. Martin's Press, 1995.
Quenum, Alphonse. *Les églises chrétiennes et la traite atlantique du XVe au XIXe siècle*. Paris: Karthala, 1993.
Quinlan, Sean. "Colonial Bodies, Hygiene, and Abolitionist Politics in Eighteenth-

Century France." In *Bodies in Contact: Rethinking Colonial Encounters in World History*, ed. Tony and Antoinette M. Burton Ballantyne. Durham, NC: Duke University Press, 2005.

Rapoport, Anatol. *The Origins of Violence: Approaches to the Study of Conflict*. 1989. Piscataway, NJ: Transaction Publishers, 1995.

Rawson, Claude. *God, Gulliver, and Genocide: Barbarism and the European Imagination, 1492–1945*. Oxford: Oxford University Press, 2001.

Rediker, Marcus. *The Slave Ship: A Human History*. New York: Penguin, 2007.

Reeser, Todd W. *Moderating Masculinity in Early-Modern Culture*. Chapel Hill: University of North Carolina Press, 2006.

Régent, Frédéric. *La France et ses esclaves: De la colonisation aux abolitions, 1620–1848*. Paris: Grasset, 2007.

Renault, François, and Serge Daget. *Les traites négrières en Afrique*. Paris: Karthala, 1985.

Roach, Joseph. *Cities of the Dead*. New York: Columbia University Press, 1996.

Robbins, Louise E. *Elephant Slaves and Pampered Parrots: Exotic Animals in Eighteenth-Century Paris*. Baltimore: Johns Hopkins University Press, 2002.

Røge, Pernille. "'La clef de commerce'—The Changing Role of Africa in France's Atlantic Empire, ca. 1760–1797." *History of European Ideas* 34 (2008): 431–43.

———. "The Question of Slavery in Physiocratic Political Economy." In *L'economia come linguaggio della politica nell'Europa del Settecento*, ed. Manuela Albertone. Milan: Feltrinelli, 2009.

Rogoziński, Jan. *A Brief History of the Caribbean: From the Arawak and the Carib to the Present*. New York: Facts on File, 1992.

Russell-Wood, A. J. R. *The Portuguese Empire, 1415–1808: A World on the Move*. Baltimore: Johns Hopkins University Press, 1998.

Saïd, Edward. *Orientalism*. New York: Random House, 1979.

Sala-Molins, Louis. *Le Code Noir ou le calvaire de Canaan*. Paris: Presses universitaires de France, 1987.

———. *Les misères des Lumières: Sous la raison, l'outrage*. Paris: Robert Laffont, 1992.

Schiebinger, Londa. *Nature's Body: Gender in the Making of Modern Science*. New Brunswick, NJ: Rutgers University Press, 2004.

Seeber, Edward Derbyshire. *Anti-Slavery Opinion in France during the Second Half of the Eighteenth Century*. New York: Burt Franklin, 1971.

Segal, Ronald. *Islam's Black Slaves: The Other Black Diaspora*. New York: Farrar, Straus & Giroux, 2001.

Sepinwall, Alyssa Rachel Goldstein. *The Abbé Grégoire and the French Revolution: The Making of Modern Universalism*. Berkeley: University of California Press, 2005.

Sharpley-Whiting, T. Denean. *Black Venus: Sexualized Savages, Primal Fears, and Primitive Narratives in French*. Durham, NC: Duke University Press, 1999.

Sloan, Philip. "The Gaze of Natural History." In *Inventing Human Science: Eighteenth-Century Domains*, ed. Christopher Fox, Roy Porter, and Robert Wokler. Berkeley: University of California Press, 1995.

Slotkin, James Sydney. *Readings in Early Anthropology*. New York: Routledge, 2004.

Smallwood, Stephanie E. *Saltwater Slavery: A Middle Passage from Africa to American Diaspora*. Cambridge, MA: Harvard University Press, 2007.

Snowden, Frank. *Blacks in Antiquity: Ethiopians in the Greco Roman Experience*. Cambridge: Belknap Press, Harvard University Press, 1970.

———. *Before Color Prejudice: The Ancient View of Blacks*. Cambridge, MA: Harvard University Press, 1983.

van Stekelenburg, A. V. "Ex Africa semper aliquid novi: A Proverb's Pedigree." *Akroterion: Kwartaalblad vir die Klassieke in Suid-Afrika* 33 (1988): 114–20.

Terrall, Mary. *The Man Who Flattened the Earth: Maupertuis and the Sciences in the Enlightenment*. Chicago: University of Chicago Press, 2002.

Thomas, Hugh. *The Slave Trade: The Story of the Atlantic Slave Trade*. New York: Simon & Schuster, 1997.

Thomson, Ann. "Diderot, le matérialisme et la division de l'espèce humaine." *Recherches sur Diderot et l'Encyclopédie* 26 (1999): 197–211.

———. "Diderot, Roubaud et l'esclavage." *Recherches sur Diderot et sur l'Encyclopédie* 35 (2003): 69–93.

Thomson, John Arthur. *The Outline of Science: A Plain Story Simply Told*. New York: Putnam, 1922.

Thornton, John. *Africa and Africans in the Making of the Atlantic World, 1400–1800*. Cambridge: Cambridge University Press, 1998.

Tort, Patrick. *L'ordre et les monstres: Le débat sur l'origine des déviations anatomiques au XVIIIe siècle*. Paris: Syllepse, 1998.

Walvin, James. *Questioning Slavery*. London: Routledge, 1996.

———. *Making the Black Atlantic: Britain and the African Diaspora*. London: Cassell, 2000.

Wheeler, Roxann. *The Complexion of Race: Categories of Difference in Eighteenth-Century British Culture*. Philadelphia: University of Pennsylvania Press, 2000.

White, Hayden. *The Content of the Form: Narrative Discourse and Historical Representation*. Baltimore: Johns Hopkins University Press, 1990.

Williams, Eric. *From Columbus to Castro: The History of the Caribbean*, 1492–1969. New York: Vintage Books, 1984.

Wokler, Robert. "Anthropology and Conjectural History in the Enlightenment." In *Inventing Human Science: Eighteenth Century Domains*, ed. Roy Porter, Christopher Fox, and Robert Wokler. Berkeley: University of California Press, 1995.

Wynter, Sylvia. "1492: A New World View." In *Race, Discourse, and the Origin of the Americas*, ed. Vera Lawrence Hyatt and Rex Nettleford. Washington, DC: Smithsonian Institution Press, 1995.

Zhiri, Oumelbanine. "Leo Africanus, Translated and Betrayed." In *The Politics of Translation in the Middle Ages and Renaissance*, ed. Renate Blumenfeld-Kosinski, Luise von Flotow, and Daniel Russell. Ottawa: University of Ottawa Press, 2001.

Index

Page numbers in italics refer to illustrations and tables.